AF341977

SKELETAL MUSCLE PHYSIOLOGY, CLASSIFICATION AND DISEASE

MUSCULAR SYSTEM - ANATOMY, FUNCTIONS AND INJURIES

Additional books in this series can be found on Nova's website
under the Series tab.

Additional e-books in this series can be found on Nova's website
under the e-book tab.

HUMAN ANATOMY AND PHYSIOLOGY

Additional books in this series can be found on Nova's website
under the Series tab.

Additional e-books in this series can be found on Nova's website
under the e-book tab.

SKELETAL MUSCLE

PHYSIOLOGY, CLASSIFICATION AND DISEASE

MARK WILLEMS
EDITOR

New York

LIBRARY OF CONGRESS CATALOGING-IN-PUBLICATION DATA

ISBN: 978-1-62417-271-7

Library of Congress Control Number: 2012951577

Published by Nova Science Publishers, Inc. † *New York*

CONTENTS

Preface — vii

Chapter 1 — Rotation of Myosin Lever Arms during Isometric Contraction of Skeletal Myofibrils — 1
K. Midde, R. Rich, V. Hohreiter, S. Raut, R. Luchowski, C. Hinze, R. Fudala, I. Gryczynski, Z. Gryczynski, and J. Borejdo

Chapter 2 — Mechanisms and Consequences of Skeletal Muscle Aging — 25
Ju Li and Irina M. Conboy

Chapter 3 — Mitochondrial Respiratory Chain Uncoupling, Oxidative Stress and Skeletal Muscle Energetics — 41
T.N. Tran, E. Noll, O. Collange, J. Bouitbir, A. L. Charles, M. Hengen, M. Kindo, J. Zoll, P. Diemunsch and B. Geny

Chapter 4 — The Emerging Role of Mitochondria in Inflammatory Myopathies — 55
Alain Meyer, Joffrey Zoll, Anne Laure Charles, François Piquard, Béatrice Lannes, Jacques-Eric Gottenberg, Jean Sibilia and Bernard Geny

Chapter 5 — Involvement of Inflammation on Skeletal Muscle Ischemia-Reperfusion Deleterious Effects — 71
M. Guillot, J. Pottecher, J. Boisramé-Helms, A. Meyer, A. L. Charles, Z. Mansour, P. Diemunsch, J. Zoll and B. Geny

Chapter 6 — Exercise-Induced Mitochondrial Biogenesis in Skeletal Muscle: Mechanisms of Activation through Peroxisome Proliferator-Activated Receptor Gamma Co-Activator 1 alpha (PGC-1α) — 91
Jonathan P. Little and Brendon Gurd

Chapter 7 Limb-girdle Muscular Dystrophies:
Different Types and Diagnosis **105**
Marija Meznaric and Karin Writzl

Chapter 8 Stretch-activated Channels in Muscular Dystrophy **129**
Ella W. Yeung

Chapter 9 Cellular and Molecular Mechanisms Regulating
the Hypertrophy and Atrophy of Skeletal Muscle **141**
Kunihiro Sakuma and Akihiko Yamaguchi

Chapter 10 Skeletal Muscle Loss and the Role of Branched-Chain
Amino Acids **195**
Stefan M. Pasiakos and James P. McClung

Chapter 11 Roles of Myostatin Propeptide in Promoting
Skeletal Muscle Growth and Metabolism **207**
Jinzeng Yang

Index **219**

PREFACE

It is essential for our quality of life to have healthy muscles. Tragically, the loss of a single protein can have dramatic effects on muscle functioning and quality of life. This book is about skeletal muscles, its physiological complexity and molecular functioning in health and disease. Although the normal functioning of skeletal muscle is well understood, less is known about diseased muscles. Our ability to move involves activity of skeletal muscles. We take the production of skeletal muscle force and power by the energy requiring process of cross-bridge formation for granted. The complexity of the physiology of skeletal muscles is highlighted by the experimental study in Chapter 1 by Borejdo and co-workers with a focus on the rotational change of lever arms of the cross-bridges and the probability distributions using polarized fluorescence. Although we may take the mechanical properties of skeletal muscles for granted, it is everybody's experience that aging changes our moving ability. During aging, skeletal muscles show a decline in their mass and function. We are getting weaker. Chapter 2 by Li and Conboy provides an excellent review of the complex changes at the physiological, cellular and molecular level that are associated with the aging process of skeletal muscles. A proposed mechanism for aging, i.e. the production of reactive oxygen species may cause damage to our cells and age individuals over time and may also play a role in disease of skeletal muscles. Although reactive oxygen species are generated by the *normal* energy producing process of oxidative phosphorylation in the mitochondria, mitochondrial uncoupling, i.e. the process of proton return to the matrix without the production of ATP, may be associated with many muscle pathologies. This link between mitochondrial uncoupling and mitochondrial physiopathology is reviewed in Chapter 3, Chapter 4 (inflammatory myopathies), and Chapter 5 (ischemia-reperfusion injury) by Geny and co-workers. Early studies in the late sixties and seventies provided evidence for a change in mitochondrial content exercise training. However, the molecular mechanisms for the change in mitochondrial content are still unresolved. Chapter 6 by Little and Gurd reviews the role of peroxisome proliferator-activated receptor γ co-activator 1α (PGC-1α) in mitochondrial biogenesis with implications for aging (Chapter 2) and muscle diseases with mitochondrial impairments (Chapter 4). A review by Meznaric and Writzl in Chapter 7 is on the classification of limb-girdle muscular dystrophies and potential limitations of genetic diagnosis. Muscle weakness is common in people with limb-girdle muscular dystrophy and no treatments for the disease are as yet available. A role for stretch-activated ion channels may contribute to the progressive muscle weakness in muscular dystrophy and reviewed by Yeung in Chapter 8. Many disease conditions, but also normal aging, are characterized by a loss of skeletal muscle mass (e.g. Chapters 2 and 8), a consequence of a disbalance between protein synthesis and protein breakdown. Any disbalance of these processes can result in hypertrophy or atrophy of skeletal muscles. The final three chapters deal with many aspects of the regulation of skeletal muscle mass. The molecular mechanisms for hypertrophy and atropy are extensively reviewed in Chapter 9 by Sakuma and Yamaguchi. This is followed by a review by Pasiakos and McClung in Chapter 10 on the regulation of skeletal muscle mass and the effect of supplementation with branch-chain amino acids on protein synthesis and breakdown. Finally, the specific role of myostatin, an inhibitor of muscle growth, with implications for chronic disease conditions, is presented in Chapter 11 by Yang.

This text provides a comprehensive overview of the complexity of the changes that can occur in skeletal muscles during health and disease. It is these changes in skeletal muscles that have an enormous impact on the quality of human life. The range of topics is of interest to researchers in skeletal muscle, health professionals dealing with musculoskeletal diseases and graduate students in Medicine, Exercise and Health related programmes.

In: Skeletal Muscle
Editor: Mark Willems

ISBN: 978-1-62417-271-7
© 2013 Nova Science Publishers, Inc.

Chapter 1

ROTATION OF MYOSIN LEVER ARMS DURING ISOMETRIC CONTRACTION OF SKELETAL MYOFIBRILS

K. Midde[1], R. Rich[1], V. Hohreiter[3], S. Raut[1], R. Luchowski[4],C. Hinze[1], R. Fudala[1], I. Gryczynski[1], Z. Gryczynski[2],and J. Borejdo[*]

[1]Department of Molecular Biology & Immunology and Center for Commercialization of Fluorescence Technologies, University of North Texas, Health Science Center, Fort Worth, Texas, US
[2]Department of Physics and Astronomy, Texas Christian University, Fort Worth, TX, US
[3]OriginLab Corporation, One Roundhouse Way, Northampton, MA, US
[4]Department of Physics, Maria Curie-Sklodowska University, Lublin, Poland

ABSTRACT

Muscle contraction is brought about by the coupling of chemical energy of ATP hydrolysis to conformational changes in myosin. The chemical changes in myosin are well understood, but the corresponding conformational changes occurring *ex-vivo* are not. In order to obtain information about conformation it is essential to observe only a few cross-bridges. Observing a large number averages out individual contributions making it impossible to extract kinetic information under steady-state conditions. To minimize the number of observed cross-bridges, only one in 60,000 lever arms in myofibrils were labeled with fluorescent myosin essential light chain 1 (LC1). When such sparsely labeled myofibril is observed through a small aperture of a confocal microscope, ~8 fluorescent lever arms are observed in a volume smaller than a single half-sarcomere. Conformation was measured by recording normalized differences between parallel and perpendicular components of the fluorescence of LC1 (polarized fluorescence, PF). We measured in a single half-sarcomere during isometric contraction *ex-vivo*, the rate of rotational change of the lever arms and the probability distributions of their orientations.

[*] Email address: Julian.Borejdo@unthsc.edu
University of North Texas, Health Science Center, 3500 Camp Bowie Blvd, Fort Worth, TX

Isometric contraction involved extrusion of massive amount of solvent from the myofilament space, necessitating normalization of PF data. The results indicated that during isometric contraction lever arms rotated at the rate of ~10 s^{-1} and that the probability distributions of PF were narrower during contraction than relaxation suggesting that the lever arms in contracting muscle were not completely disorganized.

INTRODUCTION

Muscle contraction is brought about by the coupling of chemical energy of ATP hydrolysis to mechanical changes in myosin. The cycle of chemical changes in myosin during contraction of skeletal muscle is well understood [1-3], but to understand corresponding mechanical changes, it is necessary to observe muscle on mesoscopic scale (i.e. on a scale small enough for measurements to be affected by fluctuations around the average [4]). In the case of muscle, this submicron scale can be made to contain only a few myosin cross-bridges. If the number of observed molecules is too large (e.g. >100) the signal (i.e. fluorescence) is averaged out, making it impossible to extract the contribution of an individual cross-bridge. We want to observe the un-averaged behavior of cross-bridges during isometric contraction *ex-vivo*. In particular, we want to know whether mechanical changes occur at all in a lever arm of myosin (and if yes at what rate) and the probability distributions of their orientations. Until now, mechanical changes in myosin *ex-vivo* have only been observed from 10^{11}-10^{9} cross-bridges averaged over many sarcomeres. This is because techniques that are commonly used to measure conformation in muscle, such as fluorescent microscopy, Electron Paramagnetic Resonance, or small angle X-ray diffraction report only the mean conformation of a large number of cross-bridges.

The averaging problem has been overcome *in vitro* by observing individual motor molecules. Thus, by using the fluorescence of rhodamine incorporated into smooth myosin, it was possible to determine its conformational states on a coverslip [5]. The creative application of total internal reflection fluorescence (TIRF) made it possible to measure the orientation of the myosin light chain [6]. Using quantum dots, it has been possible to observe diffusive movement of processive kinesin-1 on microtubules [7]. The motion of a single molecule of kinesin labeled with Cy3 and myosin VI labeled with GFP or Cy3 was determined to occur in hand-over-hand fashion [8]. The motion of myosin V, because of its large tail domain, has been investigated extensively. Thus, the three-dimensional structural dynamics of myosin V were measured by single-molecule fluorescence polarization [9], and processive motion of the lever arm and of head movements of myosin V have been observed simultaneously [10]. Stepping and structural dynamics of unconventional myosin X have been determined [11].

Observation of individual molecules *EX-VIVO*, on the other hand, presents a severe challenge because the number of molecules observed by even the most sensitive of the techniques, fluorescence, is too large. The number of molecules is equal to the concentration of molecules in a cell multiplied by the detection volume (DV) of the fluorescence microscope. There have been numerous attempts to minimize both. The concentration problem was solved by randomly activating individual fluorophores with light pulses and exploiting the fact that the center of a Point Spread Function (PSF) of the microscope objective can be determined with much higher precision than its Full Width at Half Maximum

(FWHM). These pointillism methods can isolate single molecule and generate outstanding superresolution images (for review see [12]), but they cannot follow fast events such as those occurring in contracting muscle. The volume problem was solved by Stimulated Emission Detection (STED) [13, 14] where DV is as small as 10 attoliters. STED microscopes produce spectacular images of the insides of a cell, but like pointillism microscopy, are too slow to follow rapid conformational changes.

To achieve required time resolution in mesoscopic experiments, we have adopted an approach that minimizes concentration and volume at the same time. The number of observed myosin molecules is minimized by labeling muscle sparsely, and the volume is minimized by observing muscle through a small confocal aperture. We measure conformation by recording polarization of fluorescence from a few molecules located in a single half-sarcomere by recording parallel ($I_\parallel$) and perpendicular ($I_\perp$) components of fluorescent light emitted by a fluorophore bound to myosin LC1. The normalized difference between these components, Polarized Fluorescence (PF), is a sensitive indicator of the orientation of the transition dipole of the fluorophore [15-22]. Isometric contraction involved extrusion of massive amount of solvent from the myofilament space, necessitating normalization of PF. The technique allows observation of 8 myosin molecules in contracting muscle with μs time resolution. Using this technique, we show that the lever arms of isometrically contracting muscle rotate on a millisecond time scale and assume relatively narrow range of orientations.

MATERIALS AND METHODS

Chemicals and Solutions

SeTau-647-mono-maleimide was from SETA BioMedicals (Urbana, IL). Alexa Fluor 488, hydrazine (cat# A10436) and non-targeted Quantum Dots (QD) (21 nm size, cat # Q21031MP) were from Molecular Probes (Eugene, OR) and were used as cell tracers. All other chemicals were from Sigma-Aldrich (St Louis, MO). The composition of the solutions were: EDTA-rigor solution: 50 mM KCl, 5 mM EDTA, 10 mM TRIS-HCl pH=7.5; Mg-rigor solution: 50 mM KCl, 2 mM $MgCl_2$, 10 mM TRIS-HCl pH=7.5); Ca-rigor solution: 50 mM KCl, 0.1 mM $CaCl_2$, 10 mM TRIS-HCl pH=7.5. Contracting solution: the same as Ca-rigor plus 5 mM Mg and 5 mM ATP; relaxing solution: the same as Ca-rigor except that 0.1 mM Ca was replaced with 2 mM EGTA. The contracting solution contained a regenerative system (Ca-rigor, 100 μM ATP, 20 mM creatine phosphate and 10 units/mL of 1mg/mL creatine kinase.

LC1

The procedure for making LC1 was the same as previously described [23]: briefly, a pQE60 vector containing recombinant LC1 with a single cysteine residue (Cys 178) was donated by Dr. Susan Lowey (University of Vermont). The plasmid DNA was transformed into *E.coli* M15 competent cells and recombinant clones were selected by ampicillin resistance. The sequencing of LC-cDNA insert of the clones was confirmed by the sequencing of both strands (Iowa State University of Science and Technology). The LC1 protein was over expressed in strain *E.coli* M15 and cultured in Luria broth containing 100 µg/mL of ampicillin, by the induction with IPTG. His-tagged LC1 protein was affinity purified on Ni-NTA column following the manufacturer's protocol. The imidazole eluted fractions were run on SDS-PAGE followed by a Western analysis with Anti-LCN1 antibodies (Abcam, CA). Fractions containing LC1 were pooled together and dialyzed against a buffer (50 mM KCl and 10 mM phosphate buffer (pH 7.0)). Dialyzed protein showed a single ~25-kDa band on SDS-PAGE after Coomassie staining. Protein concentration was determined using the Bradford assay. In some experiments, commercial skeletal human LC1 was used (Prospec, Ness Ziona, Israel).

LC1 Labeling

SeTau was used immediately after dissolving. Purified LC1 was dialyzed against the 50mM KCL and 10mM phosphate buffer (pH 7.0) and fluorescently labeled by incubating with a 5 molar excess of dye overnight on ice. Unbound dye was eliminated by passing labeled LC1 through a Sephadex G50 column.

Degree of Labeling of LC1

The degree of labeling was determined by comparing the concentrations of protein and dye in SeTau-LC1. Protein concentration was determined by the Bradford assay and SeTau concentration was determined from the peak absorbance (ε=230,000 $M^{-1}cm^{-1}$ at 635 nm measured on a Varian Eclipse spectrometer; Palo Alto, CA). The dye and protein were complexed in a 0.5:1 ratio.

Preparation of Myofibrils

Rabbit *psoas* muscle bundles were first washed with an ice-cold EDTA-rigor solution for 0.5 hr followed by an extensive wash with Mg-rigor solution followed by Ca-rigor solution. The fiber bundle was then homogenized using a Heidolph Silent Crusher S homogenizer for 20 s (with a break to cool after 10 s) in Mg-rigor solution.

LC1 Exchange Into Myofibrils

The low exchange ratio, necessary for mesoscopic experiments, was achieved with very low SeTau-LC1 concentration (5-10 nM). The exchange procedure was the same as in [24], except that exchange solution contained 15mM KCl, 5 mM EDTA, 5 mM DTT, 10 mM KH_2PO_4, 5 mM ATP, 1 mM TFP, and 10 mM imidazole, pH 7 and the exchange was carried out by incubating myofibrils at 30°C for 20 min (not usual 37°C for 1/2 hr. 1 mg/mL of freshly prepared myofibrils were used.

Cross-Linking

In the presence of Ca^{2+} and ATP, the myofibrils shorten, making it impossible to record polarized intensities during contraction. The myofibrils were prevented from shortening by cross-linking with a water-soluble cross-linker 1-ethyl-3-[3-(dimethylamino) propyl] carbodiimide (EDC) [25, 26]. 1 mg/mL of myofibrils in Ca-rigor solution containing 20 mM EDC were incubated for 20 min at room temperature [27]. The reaction was stopped by adding 20 mM DTT. The pH of the solution (7.4) remained unchanged throughout the 20 min reaction. The cross-linking had no effect on the probability distribution or ATPase [27]. The absence of shortening was verified by imaging myofibrils with differential contrast microscopy, and with fluorescence microscopy after labeling the myofibrils with a 10 nM rhodamine-phalloidin [27].

Solvent Extrusion Experiments

1 mg/mL myofibrils in appropriate solution were mixed in a Vortex with 10 nM hydrazine or 20 nM QD's in Ca^{2+}-rigor solution. Extrusion was observed in TIRF microscope equipped with differential interference contrast (DIC) optics. The diffusion was observed in PicoQuant microscope. Fluorescence was collected by Olympus 100x NA=1.3 objective (UPlanFL N), passed through 30 μm confocal aperture and projected into photosensitive area of Avalanche PhotoDiodes (APD, Perkin Elmer, Fremont, CA).

Imaging and Data Collection

The method of observation of a few molecules of myosin in an *ex-vivo* muscle was as follows: The myosin Light Chain 1 (LC1) is labeled with a fluorescent dye and exchanged with the native LC1 of a myofibril. The exchange is deliberately inefficient so that only one in ~60,000 myosin molecules carries the fluorophore. A small volume within the labeled myofibril (~1 femtoL, single half-sarcomere), is observed by confocal microscopy. For a myofibril that is inefficiently labeled, this volume contains only 6-9 fluorescent cross-bridges. This point is made schematically in Figure 1A: The exciting light beam (green hourglass) is focused to a diffraction limited volume positioned on an A-band of a myofibril. The entrance aperture of the objective was slightly overfilled resulting in a better image at the expense of narrowness of FWHM [28]. The Elliptical Confocal Volume (ECV) of the microscope (outlined by the dashed line) was estimated by measuring FWHM of 200 nm fluorescent beads imaged in X-Y and X-Z planes (Figure 1B). The FWHM's of the resulting Gaussians were 400 nm and 700 nm, respectively, and defined the waist (w_o) and height (z_o) of the elliptical $ECV = (\pi/2)^{3/2} \, w_o^2 z_o = 0.6 \, \mu m^3$. Since $ECV = (1/2)^{3/2} \cdot DV$ [29] this gives $DV = 1.7 \, \mu m^3$. ECV agrees well with the theoretical calculated value: $ECV = (\pi/2)^{3/2} \, \omega_o^2 z_o = 0.5 \, \mu m^3$ where a waist ω_o is 0.5 μm [i.e. equal to the diameter of the confocal pinhole divided by the magnification of the objective (60x)] and $z_o = 1$ μm is equal to the thickness of a typical myofibril. The DV is less than the volume of a typical half-sarcomere.

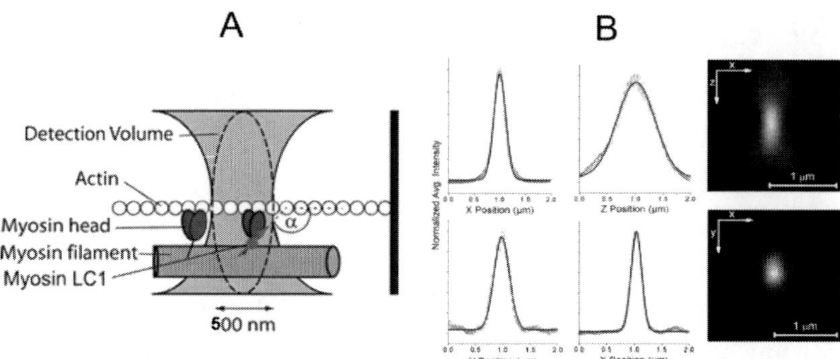

Figure 1. A: The principle of observing few molecules in muscle. A small fraction of myosin carries fluorescently labeled essential Light Chain 1 (LC1, red) characterized by a single transition dipole (red arrow). The dynamics of the fluorophore is determined by periodically measuring polarization of fluorescence and calculating its autocorrelation function. **The dispersion, Δα, of the lever arm orientations is calculated from parallel ($\parallel$) and perpendicular ($\perp$) components** the emitted fluorescent light. **B**: The beam profiles along X-Z and X-Y axes.

The details of data collection were the same as previously described [30]. Briefly, fluorescence was acquired on a PicoQuant MT 200 confocal system (PicoQuant, Berlin, Germany) coupled to an Olympus IX 71 microscope. Excitation was provided by a 635nm pulsed laser. It is the time-resolved fluorescence microscope system capable of Fluorescence Lifetime Imaging with Single Molecule Detection (SMD) sensitivity. The data is collected by Time-Correlated Single Photon Counting electronics in Time-Tagged Time-Resolved mode where each photon is recorded individually. Even though the exciting laser light was linearly polarized, a polarizer was inserted before the entrance to the microscope to avoid slight depolarization by the single mode optical fiber. The direction of polarization was parallel to the myofibrillar axis. Excitation light passed through the dichroic cube and was focused on a sample by the Olympus 60x, 1.2 NA water immersion objective. The fluorescent light passed through a 30 µm pinhole before being split by a 50-50 prism, and the now separated light beams were detected by separate Avalanche Photodiodes (APDs) with respective $\parallel$ and $\perp$ oriented analyzers. The APD's were carefully calibrated before each polarization experiment to give identical intensities when exposed to an isotropic solution of the 50 nM rhodamine 700 dye (for this control, a dye with long fluorescence lifetime has to be used. The laser polarization was at a magic angle with respect to long axis of a myofibril. The data was binned together to smooth it and give it the time resolution of 10 ms. Let $_{\parallel}I_{\parallel}$ be the polarized intensity obtained with the exciting and detected light polarized parallel to the myofibril axis, and $_{\parallel}I_{\perp}$ be the polarized intensity obtained with exciting and detected light polarized parallel and perpendicular to the myofibril axis, respectively [31]. The polarization was PF= $P_{\parallel}$ = $(_{\parallel}I_{\parallel}-_{\parallel}I_{\perp})/(_{\parallel}I_{\parallel}+_{\parallel}I_{\perp})$. Channels 2 and 1 were used to detect $_{\parallel}I_{\perp}$ and $_{\parallel}I_{\parallel}$, respectively.

Time Resolved Anisotropy and Lifetime Measurements

The fluorescent lifetime and fluorescence anisotropy of SeTau were measured by the time-domain technique using a FluoTime 200 fluorometer (PicoQuant, Inc.) at room temperature. The excitation was provided by a 635-nm pulsed diode laser, and the observation was conducted through a 670 nm monochromator with a supporting 660-nm long pass filter. The FWHM of the pulse response function was 68 ps and the time resolution was better than 10 ps. The intensity decays were analyzed by a multi-exponential model using FluoFit software (PicoQuant, Inc.). Supplementary Figures 1S-A and -B shows anisotropy and fluorescent lifetime of this dye.

SeTau-LC1 is Immobilized by a Myosin Lever Arm

To test whether SeTau was rigidly immobilized on the surface of LC1 to ensure that the orientation of the transition dipole of the fluorophore reflects the orientation of the lever arm, we measured the decay of anisotropy of SeTau-LC1 exchanged into myofibrils. Anisotropy is defined as $r = (I_\parallel - I_\perp)/(I_\parallel + 2I_\perp)$. The decay of anisotropy of free SeTau, SeTau bound to LC1 and SeTau-LC1 after the exchange to myofibrils is shown in supplementary Figures 1S-A, C & D. The corresponding fluorescence lifetimes are shown in Figures 1S-B, D & F. The decay of free SeTau and SeTau-LC1 were best fit by a single exponential curve $r(t) = R_{INF} + R_1 \cdot \exp(-t/\theta)$, where R_{INF} (anisotropy at infinite time), R_1 (initial anisotropy) and Θ (rotational correlation time) were similar (details Figure 1S). The anisotropy decay comprised entirely of the fast decay suggesting that the dye was not immobilized by LC1. The decay of SeTau-LC1 bound to myofibrillar myosin (Figure 1S-E) was quite different: it was best fit by the exponential curves $r(t) = R_{INF} + R_1 \cdot \exp(-t/\theta_1) + R_2 \cdot \exp(-t/\theta_2)$ with slow correlation time of θ_1 = 223 ns and fast time of, θ_2 = 1.268 ns. The fast decay comprised 47% of the positive anisotropy components, indicating that 53% of the dye was immobilized by myosin. The mobile fraction decayed with the correlation time characteristic of the dye alone.

Statistical Analysis

Non-linear curve fitting was performed in Origin v. 8.5, which uses the Levenberg-Marquardt algorithm to perform chi-square minimization. Chi-square minimization optimizes a parameterized fitting function with respect to a particular set of data by iteratively adjusting the fitting function's parameter values in order to minimize residuals. Residuals are the point-wise deviations between the fitting function (i.e., the theoretical curve) and the experimental data. Gaussian (normal) fitting function was used. We saw no difference in the results when the Voigt or Lorentzian model was used.

RESULTS

A typical image of a rigor myofibril is shown in Figure 2. As expected, only the A-bands were fluorescently labeled. Figure 2A is the fluorescent lifetime image. Figure 2B is the distribution of lifetimes showing that the average lifetime of SeTau bound to myosin in myofibrils is 3.4 ns. This is long in comparison with other red-excitable dyes and makes SeTau particularly suitable for polarization measurements. The lifetime increased when SeTau was bound to myofibrillar myosin (it increased from 2.4 ns for in solution, supplementary Figure 1S, to 3.4 ns in muscle). The red circle in Figure 2A is a 2D projection of the confocal aperture on the image plane and indicates the waist of the DV from which the signal is detected. Its diameter is ~0.5 μm (confocal aperture/magnification of objective, 30 μm/60x). Figures 2C & D show intensity images of the myofibril. Emission polarization is parallel (Figure 2C) & perpendicular (Figure 2D) to the direction of polarization of exciting light. Intensity images indicate that the fluorescence is highly polarized as expected from the anisotropic sample.

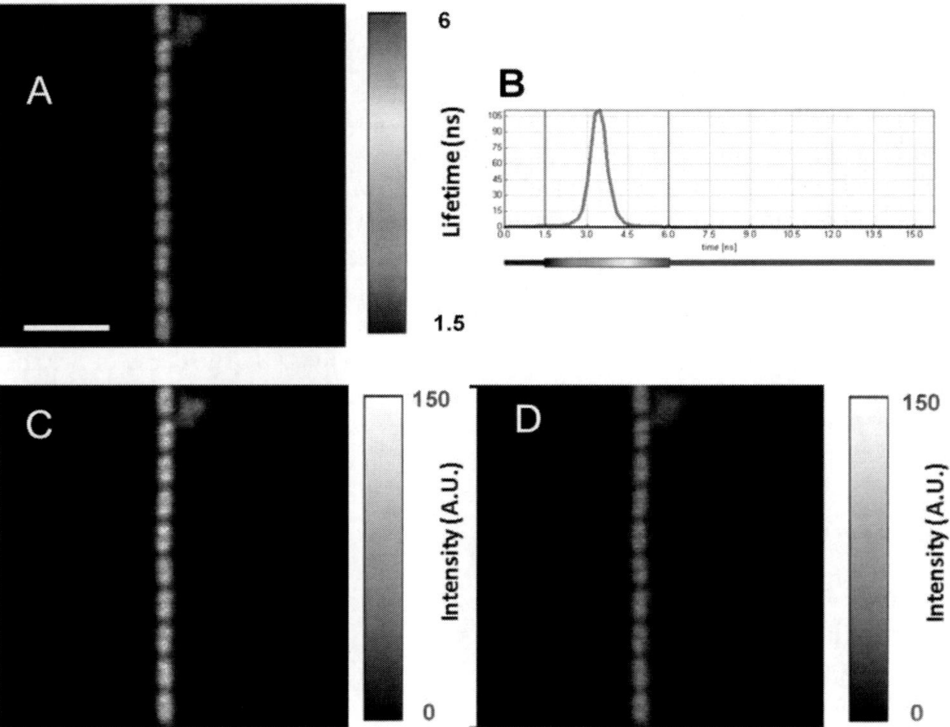

Figure 2. Images of a rigor myofibril. A: lifetime image, with the scale indicated by a color bar to the right of the image. The red circle is a projection of the confocal aperture on the sample plane. It indicates the waist of the detection volume from which the signal is collected. Bar 5 μm. B: the distribution of fluorescent lifetimes. C & D: intensity images. Emission polarization is parallel (C) & perpendicular (D) to the vertical direction of polarization of exciting light. The intensity scale is indicated in red at right of images. Absolute white and black are 255 and 0, respectively. 1 mg/mL myofibrils were exchanged with 5 nM SeTau647-LC1 for 20 min at 30°C. Sarcomere length = 2.0 μm. The images were acquired on a PicoQuant Micro Time 200 confocal lifetime microscope. The sample was excited with a 635 nm pulse of light and the emission collected on an APD through a LP700 filter.

Number of Observed Molecules

To determine how many molecules contributed to the observed fluorescence, we measured the fluorescence fluctuations for selected fluorophore concentrations. From the autocorrelation function (ACF) of these fluctuations, it is possible to determine the number of molecules contributing to fluctuations: the value of the autocorrelation function at delay time

0 [G(0)] is equal to the inverse of the number of molecules N contributing to the signal, N=1/G(0) [32-34]. The ACF's were obtained from solutions of SeTau in the range 5-20 nM. The autocorrelation functions obtained at concentrations of 5, 10, 15 and 20 nM are shown in Figure 3, panels A-D. The 1/G(0) is plotted vs. counts in Figure 3E (top scale, red symbols). Figure 3E also shows the number of molecules calculated from knowing the volume and concentration. The two agree fairly well. The parallel (ch 1, ●) and perpendicular (ch 2, ○) intensities in Figure 3E are nearly the same showing that polarization of fluorescence of free dye is nearly 0, consistent with the fact that fluorescence lifetime of SeTau is fairly long.

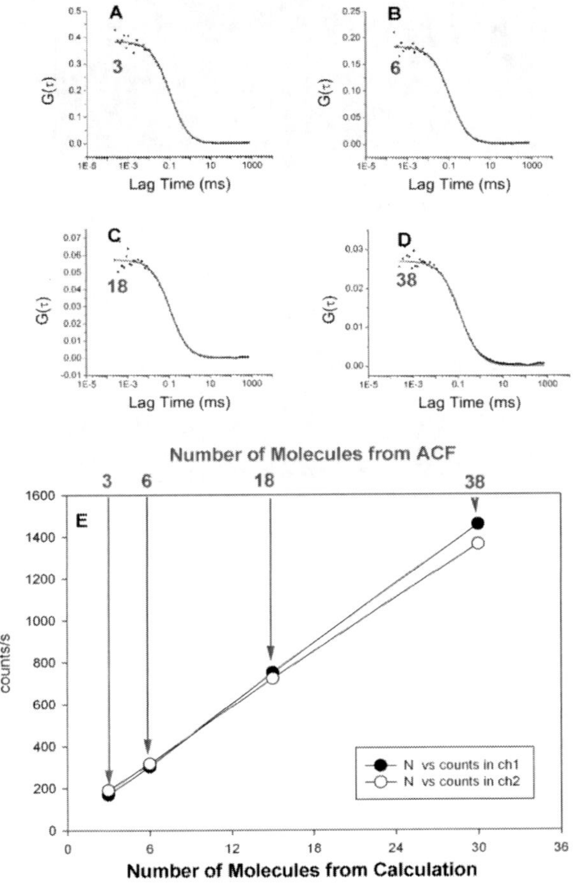

Figure 3. Estimating the number of molecules in the detection volume. A-D: autocorrelation functions of 5, 10, 15 & 20 nM SeTau in rigor solution. The numbers of molecules in the DV were obtained from the inverse of the autocorrelation value extrapolated to a lag time 0 are shown in red. Average diffusion coefficient of SeTau and its correlation time were 250 μm^2/s and 10.5 ms, respectively. E: (top scale, red) - the number of molecules in the DV calculated from ACF. (bottom scale,black) - the number of molecules in the DV calculated by multiplying DV by molar concentration of the dye.

The actual number of SeTau-LC1 molecules in the DV centered on the A-band of rigor myofibril can now be estimated from the photon rate collected in a typical experiment. Figure 4 shows a typical time course of fluorescence collected from an experiment in which myofibril was exchanged with 10 nM SeTau-LC1. The actual intensities were 2.7 and 1.4

counts/ms in parallel and perpendicular channels, respectively. The total average count in DV was 270+2*140=550 counts/s.

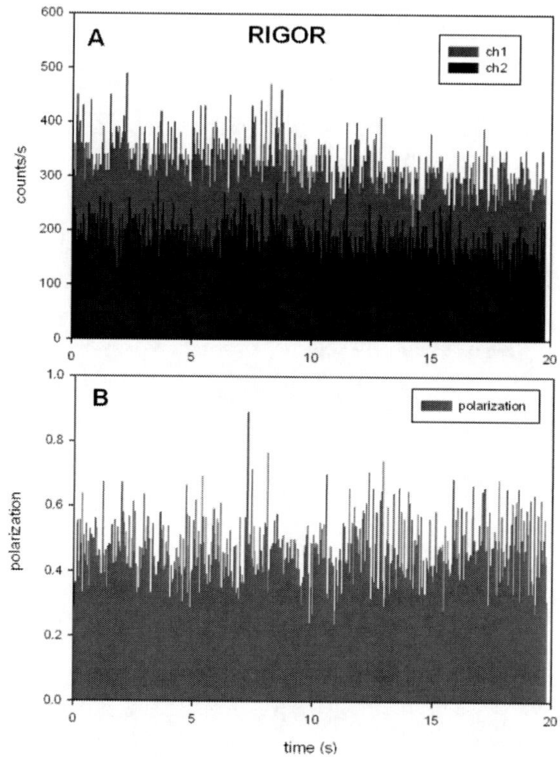

Figure 4. A: Typical time course of fluorescence of rigor muscle. The average number of counts/s was 270 and 140 in parallel (red) and perpendicular (blue) channels, respectively. The total average count rate was 270+2*140=550 counts/s, showing that the number of fluorescent cross-bridges in the DV was ~8. **B:** Polarization of fluorescence (**the normalized difference between the number of** ∥ and ⊥ polarized photons) is constant throughout the experiments, even though individual polarized intensity components decline somewhat because of bleaching. PF remains constant because both orthogonal components of fluorescence bleach at the same rate.

Comparison of experimental value of 550 counts/s with calibration curve of Figure 3E suggests that we observed ~8 fluorescent cross-bridges. It should be emphasized, however, that as long as the number of cross-bridges is mesoscopic, the exact number does not matter, i.e. 8 molecules should give the same result as 50 molecules etc.

Cross-Bridge Rotations

To examine the nature of PF fluctuations such as seen in Figure 4A, we measured the autocorrelation function of polarized fluorescence. The autocorrelation function of PF's is the time average of PF's multiplied by the value of PF's a delay time later. Its decay characterizes rapidity of rotational motions of the lever arm. During rigor, lever arms are stationary and the autocorrelation of fluctuations is expected to be flat. During contraction, lever arms undergo power-stroke cycles and PF changes cyclically. As mentioned before, it is not at all obvious that lever arms rotate at all during contraction. Imposition of isometric conditions by cross-linking may prevent cross-bridges from reaching neighboring actin target zones while detached. Moreover, since autocorrelation function of a periodic signal is periodic [35] analysis of autocorrelation function can reveal whether the changes are synchronous. Figure 5 compares autocorrelation functions in the three states. A flat correlation function, such as that observed in rigor (red), arises when there are no correlations between fluorescence intensities at any time within the time of the measurement. We conclude that the transition dipoles of the

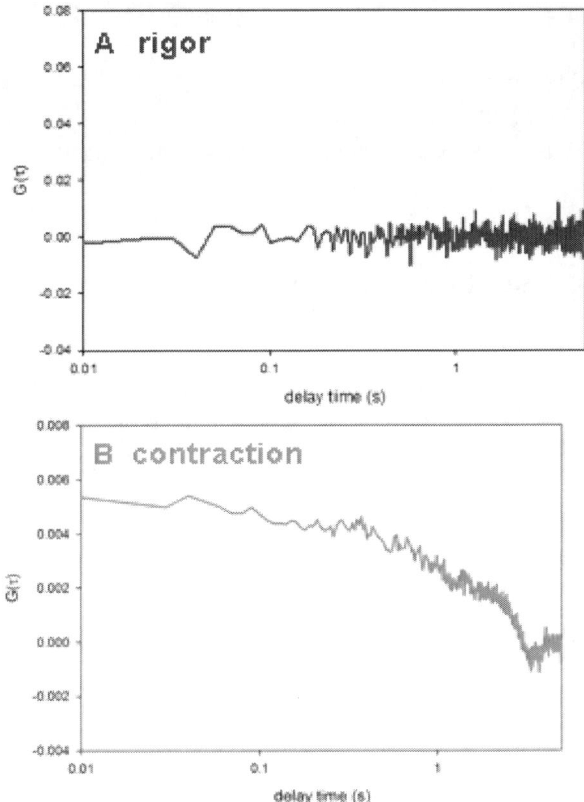

Figure 5. Normalized autocorrelation functions of fluctuations of polarized fluorescence of lever arms in rigor (red) and contracting (green) myofibrils. Note that the scale in **A** is 10x larger than in **B**. This is because fluorescent signal of rigor myofibril (before solvent extrusion, see below) is small, hence fluctuations are large.

dye do not change orientation at all [32, 36]. This suggests that LC1 moieties are close enough to myosin heads to be immobilized by binding of heads to thin filaments. The large fluctuations of polarization (note that vertical scale in Fig 5A is 10x bigger then in Figure 5B) are entirely due to stochastic noise. It is large because mean signal in rigor is small. In contraction (green), the lever arms undergo power-stroke cycles resulting in fluctuations in orientations of LC1. Non-flat correlation function arises because the transition dipoles change orientations. The other possible source of fluctuations - fluctuations in the number of fluorophores in the DV [32, 36] are not significant because PF is only sensitive to rotations and myofibrils were unable to twist.

The kinetic information was extracted from the correlation function by constructing a model relating the rate of changes of polarization of fluorescence of the lever arm to the various rates of a decay of an autocorrelation function [37]. In a simple two-state model, such as shown in supplementary Figure 2S, the correlation function can be shown to be [37]:

$$R_2(t) = \frac{\left(a_1 k_2 + a_2 k_1\right)^2}{\left(k_1 + k_2\right)^2} + k_1 k_2 \frac{\left(a_1 - a_2\right)^2}{\left(k_1 + k_2\right)^2} \exp\left[-\left(k_1 + k_2\right)t\right]$$

where k_1, a_1 and a_2, k_2 are forward and reverse rate constants and fluorescence intensities associated with conformational change. Substituting appropriate 0 for a_1 and 1 for a_2 yields three parameter exponential fit shown in supplementary Figure 2S that yields $k_1 = 10.82$ s^{-1} and $k_2 = 0.21$ s^{-1}.

Probability Distributions of Cross-Bridges

In order to make a statistically valid comparison between hundreds of half-sarcomeres examined in the present study in this chapter, it is important to recognize that contracting and rigor myofibrils each give rise to fluorescent signal of different strengths. Since the signal fluctuations are random, the width of a probability distribution depends on the square root of the signal strength [35, 38]. The relative (with respect to the mean) value of FWHM of probability distribution is small for strong signals and large for weak signals. Therefore, to make meaningful comparisons between the signals from contracting and relaxed myofibrils, the signals have to be normalized with respect to the total fluorescence intensity.,

The reason the signals from rigor, relaxed and contracting myofibrils are different is that contraction involves massive extrusion of solvent from the myofilament space. This extrusion increases effective concentration of fluorophore in the DV. We originally thought that solvent extrusion would be inhibited by cross-linking and that normalization would not be necessary. But this is not the case - massive extrusion makes a big difference to observed intensities. To estimate extent of solvent extrusion in cross-linked myofibrils, we added quantum dots (QD) to myofibrils in rigor solution and observed them in a Total Internal Reflection Fluorescence (TIRF) microscope. We used 21 nm diameter non-targeted Quantum Dots which were coated with polyethylene glycol (PG) minimizing any non-specific interactions with sarcomeric proteins (PG does not contain reactive functional groups). Figure 6A shows a DIC image of unlabeled myofibrils in rigor solution. The quality of the DIC image is degraded by the light scattering and refraction by QD's, but the main features of myofibril can be recognized. The fluorescence image of the same field (B) shows uniform distribution of QD (QD's are bright dots). Myofibrils are not fluorescently labeled and are invisible. The fluorescence image obtained a few seconds after adding contracting solution is shown in Figure 6C. The fluorescent field becomes not uniform. It has dark, non-fluorescent areas that which are confined to spaces around myofibrils. Dark regions are the areas that QD are absent from. They have been removed by flow of solvent extruded from myofilament space. Images Figures 6A & C are superimposed in Figure 6D proving that the dark regions with no fluorescence form only around myofibrils. The solvent extrusion was so rapid that it was impossible to capture the time course of the process. There was no solvent extrusion at all when rigor solution was replaced by relaxing solution (not shown).

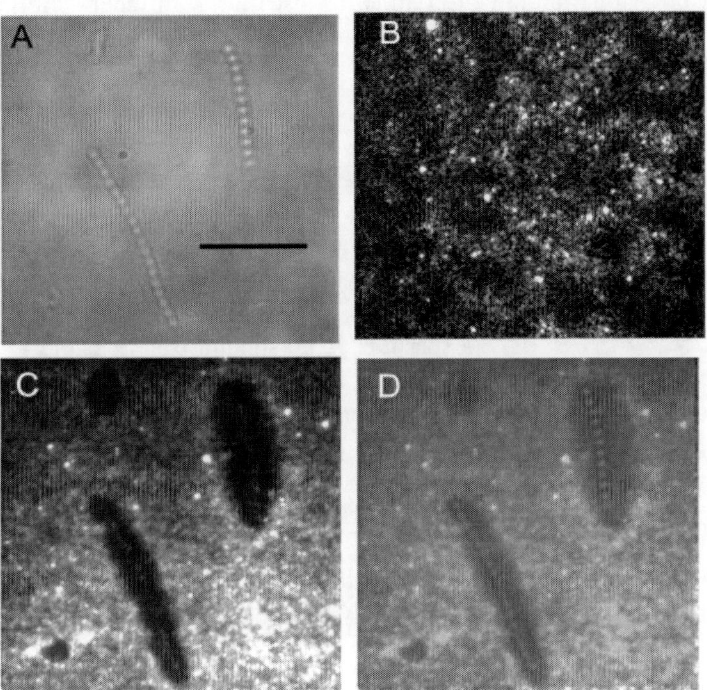

Figure 6. Solvent extrusion from cross-linked myofibrils. A: Nomarski image of cross-linked myofibrils in rigor solution in the presence of 20 nM QD (10,000 dilution of Molecular Probes stock) **B**: Fluorescent image of myofibrils in rigor solution, **C**: Fluorescent image of myofibrils in after replacing rigor with contracting solution. A halo formation around the myofibril indicates extrusion of QD from the myofilament space. **D**: superposition of A and C. λ_{ex}=532 nm, observed through 100x, NA=1.49 oil immersion objective. Scale bar: 15 μm.

When myofibrils are labeled with a fluorophore, the extrusion causes condensation of the fluorophore and results in the increase of intensity of fluorescent light in the labeled region. We estimated order-of-magnitude of condensation by measuring the number of fluorescent tracer molecules in half-sarcomeres of rigor and contracted myofibrils. To this end, we added hydrazine tracer to myofibrils and measured number of molecules of tracer inside myofilament space during rigor and contraction. The half-sarcomeres of rigor myofibrils contained on average 1.6 molecules of 10 nM hydrazine (supplementary Figure 3S). The half-sarcomeres of contracted myofibrils contained on average 17 molecules of 10 nM hydrazine tracer (supplementary Figure 4S). We conclude that contraction causes on average 10x reduction in sarcomere volume. It is remarkable that such large reduction in volume is not visible in DIC images of contracted myofibrils. Perhaps cross-linking keeps the overall structure intact, while some internal collapse of myofilaments takes place.

The above experiments show the necessity of normalization when comparing degree of disorder. The order is best represented by probability distribution of orientations. Probability distribution are plots of polarization values vs. the number of times that a given orientation occurs during a 20-second experiment. A thin probability distribution indicates that orientations of the lever arms are narrowly distributed, i.e. that the cross-bridges are uniformly oriented. Conversely, thick probability distribution indicates that orientations of the lever arms are widely distributed i.e. that cross-bridges are disordered. Figure 7 shows typical normalized probability distribution of contracting myofibrils. In contraction the lever arms undergo power-stroke cycles. The dispersion of orientations arises from changes of orientations of LC1 and from stochastic noise. The dispersion could also be fitted by two Gaussians as reported in [27], where it was suggested that one Gaussian represented cross-bridges in the pre-power stroke state and the second one cross-bridges in the post-power stroke state. In the present chapter, however, in order to derive a single FWHM parameter from the distributions we fitted all histograms by a single Gaussian. The distributions in relaxation were similar. Here, the myosin heads are free to oscillate, driven by thermal fluctuations. Normalization involves multiplying FWHM of weaker signal by the square root of the ratio between larger and smaller signals.

The normalized FWHM's of 27 experiments are summarized in Table 1. The difference between the effective FWHM's of contracting and relaxed myofibrils was large enough to suggest that it was not due to the difference in random sampling (t = -8.532 with 53 degrees of freedom, P = <0.001).

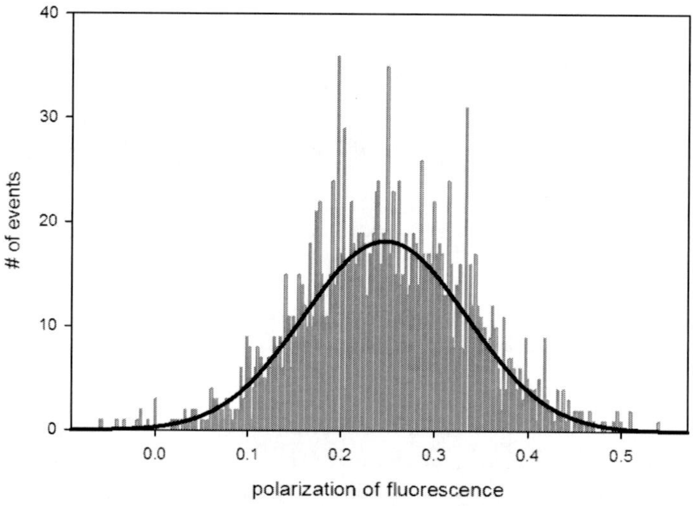

Figure 7. Probability distribution of orientations of cross-bridge lever arms during contraction of myofibrils. All the probability distributions were fit by a single Gaussian $y=a \exp[-0.5(x-x_0/b)^2]$. The goodness of fit was assessed by χ^2 in the Origin program.

Table 1. Effective FWHM values and their SD in all 27 experiments.

State	Effective FWHM	SD of Effective FWHM	PF of the peak
Contracting	0.121	0.06	0.2884± 0.0552
Relaxing	0.230	0.043	0.2861± 0.0255

We conclude that cross-bridges are more organized during contraction then during relaxation. The same result was obtained when the measurements were done on a different confocal microscope (ISS Alba).

CONCLUSION

The experiments revealed that cross-bridge lever arms rotated during contraction. This has been demonstrated before in *in vitro* in solution experiments [39], by X-ray [40, 41], by AFM [42], by Electron Tomography [43], by laser trap [44] and in single molecule experiments in myosin II [5, 6], myosin V [9], myosin VI [45-47] and in myosin X [11]. However, this is not at all obvious *in vivo* because isometric condition could have prevented cross-bridges from reaching neighboring actin target zones while detached. Moreover, muscle is an exceptionally dense environment [48] and the exclusion volume effects [49] must play a significant role during contraction. In muscle force could have been developed by a mechanism that does not require conformational changes of lever arm.

Autocorrelation experiments demonstrated that lever arms rotated asynchronously in every half-sarcomere. If they did not, the mesoscopic polarization signal and its auto-correlation function would have been periodic (autocorrelation of periodic signal is periodic, [35]). This finding is not surprising in view of the fact that synchronization would have imposed periods in which all the cross-bridges in a half-sarcomere would be dissociated from actin causing extension by the neighbors. On the other hand, synchrony could have increased efficiency: the myofilament lattice imposes 143 Å spacing between cross-bridges. The lateral extension associated with a power stroke may be as large as 100 Å, and so synchronization of cross-bridges would prevent them from "bumping" into each other and decreasing efficiency of contraction.

It is well known that in rigor myosin heads [22, 50-52] and myosin lever arms [53-55] are well ordered. But recently, it has been proposed that under certain conditions, relaxed cross-bridges may be ordered as well. Electron microscopic studies revealed that the helically ordered arrangement of the myosin heads characteristic of the relaxed state was lost upon phosphorylation of the regulatory light chain (RLC) in the thick filaments isolated from striated muscles of tarantula [56], Limulus [57] and rabbit psoas muscle [58]. Recent X-ray diffraction work showed that phosphorylation of RLC caused a change in mass distribution of cross-bridges as they moved farther from the surface of thick filaments to become closer to the thin filament [59]. CryoEM work from the laboratory of Craig et al. using the three-dimensional reconstruction of tarantula myosin filaments demonstrated the mechanism by which phosphorylation can regulate myosin activity [60]. Electron Paramagnetic Resonance (EPR) work of Cooke and his collaborators using nucleotide-analog spin label probes showed that the myosin heads are highly ordered in the relaxed fibers, when all the heads are dephosphorylated, and that this perfectly ordered structure of myosin cross-bridges disappears with phosphorylation of RLC [61]. This prompted a proposal of anew, super-relaxed state (SRX) in muscle existing in unphosphorylated skeletal [62] and cardiac [63] muscle fibers. It was shown that the SRX state corresponds to the highly ordered array of myosin heads in the absence of phosphorylation [64]. The results presented here do not argue against this hypothesis because we have not checked the degree of phosporylation of muscle.

The extrusion of solvent from contracting myofibrils deserves a comment. Solvent extrusion from filament lattice was proposed earlier as a mechanism of muscle contraction. Szent-Györgyi proposed that contraction is brought about by ATP-induced shedding of the water envelope surrounding the S2 segment of myosin in rigor [65]. Nuclear magnetic relaxation times of water protons in living skeletal frog muscle increased during isometric contraction [66]. This led to the suggestion that water in muscle is organized [67-69] and aligned along the myofilaments, and that the state of the intracellular water changes with physiological conditions [70]. Oplatka and collaborators suggested that hydrolysis of ATP causes directional flow of water away from the Z-discs forcing buildup of pressure in the center of a sarcomere, thus causing sliding of filaments [71]. A similar mechanism was proposed for the cytoplasmic streaming in *physarum polycephallum* [72], thrombostenin [73] and actomyosin solutions [74]. Our results show that solvent is massively extruded. In the presence of sarcolemma, such extrusion could lead to the pressure buildup in the center of sarcomere. Interestingly, in demembranated system, the solvent is flowing continuously (Figure 4S).

No attempt was made here to translate polarization values to the absolute orientation of transition dipoles with respect to the myofibrillar axis. Such translation is critically dependent on the model of arrangement of cross-bridges. For example, using the model of Tregear and Mendelson [31], which assumes that a fraction α of the cross-bridges are arranged helically along the long axis of muscle and the rest are arranged randomly and that the cross-bridges execute only polar motions (and not, more realistically, combination of polar and azimuthal motions [75]), the range of angles corresponding to a given range of polarizations can be very broad for small values of α.

We labeled LC1 with the red dye, SeTau-647-mono-maleimide (SeTau). SeTau has several important advantages over a single isomer of tetramethylrhodamine-5-iodoacetamide dihydroiodide used earlier [27]. Most importantly, SeTau is excited in the red and thus reduces the contribution of autofluorescence [76]. It is well suited for excitation with 635 nm diode lasers, has large Stokes shift (44 nm), has much higher photostability than Cy5 or Alexa647, has a high extinction coefficient (230,000), and it has several times longer fluorescent lifetime than Cy5 or Alexa647 or SETA dyes. A long fluorescence lifetime is critical in polarization measurements: if a dye has a short lifetime then it has a high intrinsic polarization. In this case, there is a significant contribution of unbound fraction of fluorophores to the observed polarization of fluorescence. A change of polarization caused by cross-bridge rotation may constitute only a small fraction of total polarization of fluorescence. In summary, red organic dyes have short lifetimes, high extinction coefficient and small Stoke's shift. The UV and visible dyes have lower extinctions and longer lifetimes but easily photobleached. SeTau is the unique dye with high excitation in red and relatively long lifetime.

To exclude a possibility of systematic errors associated with PicoQuant MicroTime 200 microscope, we have carried out measurements using a completely different optical system – ISS Alba microscope. Alba differs from PQ in that it does not use Time-Correlated Single Photon Counting electronics in Time-Tagged Time-Resolved mode where each photon is recorded individually. Rather, the sample is illuminated by a Continuous Wave (CW) laser. A myofibril is imaged by objective scanning, and the fluorescence emission is split by a 50-50 beam splitter before passing through separate pinholes for each respective detector. Analyzers are inserted before each detector. Using this system we have obtained similar results: The mean FWHM of 18 experiments in contraction and relaxation were 0.112 ± 0.020 and 0.128 ± 0.028 and not statistically different.

We have considered four possible artifacts and conclude that all are unlikely:

The fluctuations are caused by translation of a whole myofibril, not by rotations of a cross-bridge. This is unlikely because we measured correlation of polarization values, which is sensitive only to the rotational motions. The control experiments showed no evidence of sarcomere twisting during contraction.

Effect of EDC. In principle, it is possible that EDC affects the functional properties of myofibrils and the narrowness of distribution during contraction is due to cross-linking. To asses this, we compared the distribution before and after cross-linking rigor myofibrils. The goodness of fit, assessed by χ^2, showed that the distributions were the same [27]. We think that the lack of effect was due to the fact that cross-linking occurred between the myosin heads and actin, and between LMM tails, both of which are far away from LC1. Moreover, on the average only <20% heads are cross-linked [77, 78].

Reversibility. To ensure the shape of the distribution observed during contraction was not an artifact induced by the contraction protocol itself, we examined the orientational distributions in rigor before and after the contraction of the myofibrils. Both distributions were identical, within the experimental error [27].

Photobleaching. The effect of photobleaching was tested by increasing laser intensity to induce more severe outcome. Supplementary Fig 5S-A shows the time course of intensity decay. In these control experiments, the intensity decreased by ~70% during 20-second time period. Figure 5S-B shows that even under these extreme conditions both orthogonal intensities decreased at the same rate and so the mean value of the polarization was unchanged.

ACKNOWLEDGMENTS

Supported by grants from the NIH R01AR048622 and R01HL090786 grant N N202 112340 from the Polish National Science Center.

REFERENCES

[1]   M. A. Geeves and K. C. Holmes, *Adv. Protein Chem.* 71(24), 161 (2005).

[2]   A. Houdusse and H. L. Sweeney, *Curr. Opin. Struct. Biol.* 11(2), 182 (2001).

[3]   M. A. Ferenczi, S. Y. Bershitsky, N. Koubassova, V. Siththanandan, W. I. Helsby, P. Panine, M. Roessle, T. Narayanan and A. K. Tsaturyan, *Structure* 13(1), 131 (2005).

[4]   H. Qian, S. Saffarian and E. L. Elson, *Proc. Natl. Acad. Sci. U.S.A.*, 99(16), 10376 (2002).

[5]   D. M. Warshaw, E. Hayes, D. Gaffney, A. M. Lauzon, J. Wu, G. Kennedy, K. Trybus, S. Lowey and C. Berger, *Proc. Natl. Acad. Sci. U.S.A.*, 95(14), 8034 (1998).

[6]   M. E. Quinlan, J. N. Forkey and Y. E. Goldman, *Biophys. J.* 89(2), 1132 (2005).

[7]   H. Lu, M. Y. Ali, C. S. Bookwalter, D. M. Warshaw and K. M. Trybus, *Traffic* 10(10), 1429 (2009).

[8]   A. Yildiz, M. Tomishige, R. D. Vale and P. R. Selvin, *Science* 303(5658), 676 (2004).

[9]   J. N. Forkey, M. E. Quinlan, M. A. Shaw, J. E. Corrie and Y. E. Goldman, *Nature* 422(6930), 399 (2003).

[10]   H. Lu, G. G. Kennedy, D. M. Warshaw and K. M. Trybus, *J. Biol. Chem.*, 285(53), 42068 (2010).

[11]   Y. Sun, O. Sato, F. Ruhnow, M. E. Arsenault, M. Ikebe and Y. E. Goldman, *Nat. Struct. Mol. Biol.* 17(4), 485 (2010).

[12]   D. Toomre and J. Bewersdorf, *Annu. Rev. Cell. Dev. Biol.* 26, 285 (2010).

[13]   S. W. Hell and J. Wichmann, *Optics Lett.* 19, 780 (1994).

[14]   T. A. Klar, S. Jakobs, M. Dyba, A. Egner and S. W. Hell, *Proc. Natl. Acad. Sci. U.S.A.*, 97(15), 8206 (2000).

[15]   S. C. Hopkins, C. Sabido-David, J. E. Corrie, M. Irving and Y. E. Goldman, *Biophys. J.* 74(6), 3093 (1998).

[16]   C. G. Dos Remedios, R. G. Millikan and M. F. Morales, *J. Gen. Physiol.* 59, 10 (1972).

[17]   C. G. Dos Remedios, R. G. Yount and M. F. Morales, *Proc. Natl. Acad. Sci. U.S.A.*, 69, 2542 (1972).

[18]   T. P. Burghardt, J. E. Charlesworth, M. F. Halsetad, J. E. Tarara and K. Ajtai, *Biophys. J.* 24, 24 (2006).

[19]   Y. E. Goldman, *Cell*, 93, 1 (1998).

[20]   M. F. Morales, *Protein Science*, 4, 130 (1995).

[21]   C. L. Berger, J. S. Craik, D. R. Trentham, J. E. Corrie and Y. E. Goldman, *Biophys. J.* 68, 78S (1995).

[22]   C. L. Berger, J. S. Craik, D. R. Trentham, J. E. Corrie and Y. E. Goldman, *Biophys. J.* 71, 3330 (1996).

[23]   J. Borejdo, D. Ushakov, R. Moreland, I. Akopova, Y. Reshetnyak, L.D. Saraswat, K. Kamm and S. Lowey, *Biochemistry*, 40, 3796 (2001).

[24]   D. S. Ushakov, V. Caorsi, D. Ibanez-Garcia, H. B. Manning, A. D. Konitsiotis, T. G. West, C. Dunsby, P. M. French and M. A. Ferenczi, *J. Biol. Chem.* 286(1), 842 (2011).

[25]   C. Herrmann, C. Lionne, F. Travers and T. Barman, *Biochemistry*, 33(14), 4148 (1994).

[26]   A. K. Tsaturyan, S. Y. Bershitsky, R. Burns and M. A. Ferenczi, *Biophys. J.* 77(1), 354 (1999).

[27]   K. Midde, R. Luchowski, H. K. Das, J. Fedorick, V. Dumka, I. Gryczynski, Z. Gryczynski and J. Borejdo, *Biophys. J.* 100(4), 1024 (2011).

[28]   S. T. Hess and W. W. Webb, *Biophys. J.* 83(4), 2300 (2002).

[29]   V. Buschmann, B. Kramer and Koberlink, *Quantitive FCS: Determination of confocal volume by FCS and bead scanning with Micro Time 200.* PicoQuant, Application note Quantitative FCS v. 1.1, 2009.

[30]   J. Borejdo, D. Szczesna-Cordary, P. Muthu and N. Calander, *Biochemistry,* 49, 5269 (2010).

[31]   R. T. Tregear and R. A. Mendelson, *Biophys. J*, 15, 455 (1975).

[32]   D. Magde, E. L. Elson and W. W. Webb, *Biopolymers*, 13(1), 29 (1974).

[33]   E. L. Elson, *Annu. Rev. Phys. Chem.*, 36, 379 (1985).

[34]   E. L. Elson, *Introduction to FCS.* Short Course on Cellular and Molecular Fluorescence, ed. Z. Gryczynski. Vol. 2, Fort Worth: UNT. pp. 1-10, (2007).

[35] R. Bracewell, *The Fourier Transform and Its Applications*, McGraw-Hill, New York (1965).

[36] E. L. Elson and D. Magde, *Biopolymers*, 13, 1 (1974).

[37] P. Mettikolla, N. Calander, R. Luchowski, I. Gryczynski, Z. Gryczynski, J. Zhao, D. Szczesna-Cordary and J. Borejdo, *J. Theor. Biol.*, 284, 71 (2011).

[38] E. L. Elson, *J. Biomed. Opt.*, 9(5), 857 (2004).

[39] D. J. Jacobs, D. Trivedi, C. David and C. M. Yengo, *J. Mol. Biol.*, 407(5), 716 (2011).

[40] J. H. Brown, V. S. Kumar, E. O'Neall-Hennessey, L. Reshetnikova, H. Robinson, M. Nguyen-McCarty, A. G. Szent-Györgyi and C. Cohen, *Proc. Natl. Acad. Sci. U.S.A.*, 108(1), 114 (2011).

[41] H. Minoda, T. Okabe, Y. Inayoshi, T. Miyakawa, Y. Miyauchi, M. Tanokura, E. Katayama, T. Wakabayashi, T. Akimoto and H. Sugi, *Biochem. Biophys. Res. Commun.*, 405(4), 651 (2011).

[42] N. Kodera, D. Yamamoto, R. Ishikawa and T. Ando, *Nature*, 468(7320), 72 (2010).

[43] S. Wu, J. Liu, M. C. Reedy, R. T. Tregear, H. Winkler, C. Franzini-Armstrong, H. Sasaki, C. Lucaveche, Y. E. Goldman, M. K. Reedy and K. A. Taylor, *PLoS One*, 5(9), e12643 (2010).

[44] J. Sleep, A. Lewalle and D. Smith, *Proc. Natl. Acad. Sci. U.S.A.*, 103(5), 1278 (2006).

[45] J. G. Reifenberger, E. Toprak, H. Kim, D. Safer, H. L. Sweeney and P. R. Selvin, *Proc. Natl. Acad. Sci. U.S.A.*, 106(43), 18255 (2009).

[46] J. A. Spudich, *Curr. Biol.*, 18(2), R68 (2008).

[47] Y. Sun, H. W. Schroeder, 3rd, J. F. Beausang, K. Homma, M. Ikebe and Y. E. Goldman, *Mol. Cell.*, 28(6), 954 (2007).

[48] C. R. Bagshaw, *Muscle Contraction*, Chapman & Hall, London, 1982.

[49] A. P. Minton, *J. Biol. Chem.*, 276(14), 10577 (2001).

[50] D. D. Thomas and R. Cooke, *Biophys. J.* 32, 891 (1980).

[51] J. Borejdo, O. Assulin, T. Ando and S. Putnam, *J. Mol. Biol.* 158, 391 (1982).

[52] E. M. Ostap, V. A. Barnett and D. D. Thomas, *Biophys. J.*, 69(1), 177 (1995).

[53] N. Ling, C. Shrimpton, J. Sleep, J. Kendrick-Jones and M. Irving, *Biophys. J.*, 70, 1836 (1996).

[54] S. C. Hopkins, C. Sabido-David, U. A. van der Heide, R. E. Ferguson, B. D. Brandmeier, R. E. Dale, J. Kendrick-Jones, J. E. Corrie, D. R. Trentham, M. Irving and Y. E. Goldman, *J. Mol. Biol.*, 318(5), 1275 (2002).

[55] C. Sabido-David, B. Brandmeier, J. S. Craik, J. E. Corrie, D. R. Trentham and M. Irving, *Biophys. J.*, 74(6), 3083 (1998).

[56] R. Craig, R. Padron and J. Kendrick-Jones, *J. Cell Biol.*, 105(3), 1319 (1987).

[57] R. J. Levine, P. D. Chantler, R. W. Kensler and J. L. Woodhead, *J. Cell Biol.*, 113(3), 563 (1991).

[58] R. J. Levine, R. W. Kensler, Z. Yang and H. L. Sweeney, *Biophys. J.*, 68(4 Suppl), 224S (1995).

[59] B. A. Colson, M. R. Locher, T. Bekyarova, J. R. Patel, D. P. Fitzsimons, T. C. Irving and R. L. Moss, *J. Physiol.*, 588(Pt 6), 981 (2010).

[60] L. Alamo, W. Wriggers, A. Pinto, F. Bartoli, L. Salazar, F.Q. Zhao, R. Craig and R. Padron, *J. Mol. Biol.*, 384(4), 780 (2008).

[61] N. Naber, R. Cooke and E. Pate, *EPR Spectroscopy Shows Oriented Myosin Heads in Relaxed Muscle Fibers*. Biophysical Soc. 55th Annual Meeting, p26, (2011).

[62] M. A. Stewart, K. Franks-Skiba, S. Chen and R. Cooke, *Proc. Natl. Acad. Sci. U.S.A.*, 107(1), 430 (2010).

[63] P. Hooijman, M. A. Stewart and R. Cooke, *Biophys. J.*, 100(8), 1969 (2011).

[64] R. Cooke, *Biophys. Rev.*, 3(1), 33 (2011).

[65] A. Szent-Györgyi, *Proc. Natl. Acad. Sci. U.S.A.*, 71(9), 3343 (1974).

[66] C. B. Bratton, A. L. Hopkins and J. W. Weinberg, *Science*, 147, 738 (1965).

[67] F. W. Cope, *Biophys. J.*, 9(3), 303 (1969).

[68] C. F. Hazlewood, B. L. Nichols and N. F. Chamberlain, *Nature*, 222(5195), 747 (1969).

[69] R. Damadian, *Science*, 193(4253), 528 (1976).

[70] T. Yamada, *Cell Mol. Biol.* (Noisy-le-grand), 47(5), 925 (2001).

[71] A. Oplatka, H. Gadasi, R. Tirosh, Y. Lamed, A. Muhlrad and N. Liron, *J. Mechanochem. Cell Motil.*, 2(4), 295 (1974).

[72] R. Tirosh, A. Oplatka and I. Chet, FEBS Lett., 34(1), 40 (1973).

[73] I. Cohen, R. Tirosh and A. Oplatka, Pflugers Arch., 352(1), 81 (1974).

[74] A. Oplatka and R. Tirosh, *Biochim. Biophys. Acta*, 305(3), 684 (1973).

[75] T. P. Burghardt and K. Ajtai, *Biochemistry*, 33, 5376 (1994).

[76] J. R. Lakowicz, *Principles of Fluorescence Spectroscopy*, Springer, (2006).

[77] P. F. Flicker, R. A. Milligan and D. Applegate, *Adv. Biophys.*, 27, 185 (1991).

[78] W. C. Cooper, L. R. Chrin and C. L. Berger, *Biophys. J.*, 78(3), 1449 (2000).

ABBREVIATIONS

ACF	AutoCorrelation Function
APD	Avalanche PhotoDiode
DIC	Differential Interference Contrast
DTT	Dithiothreitol
DV	Detection Volume
ECV	Elliptical Confocal Volume
EDC	1-ethyl-3-[3-(dimethylamino)propyl]carbodiimide
FWHM	Full Width at Half Maximum
FLC	Fluorescence Lifetime Correlation
HS	Half Sarcomere
LC1	Myosin Alkaline Light Chain 1
LMM	Light Meromyosin
PF	Polarization of Fluorescence
PSF	Point Spread Function
QD	Quantum Dots
RLC	Regulatory Light Chain
S1	Myosin Subfragment-1
S2	Myosin Subfragment-2
SeTau	SeTau-647-mono-maleimide
SeTau-LC1	Myosin Alkaline Light Chain 1 labeled with SeTau

STED	Stimulated Emission Detection
SMD	Single Molecule Detection
TIRF	Total Internal Reflection Fluorescence
TFP	TriFluoroPiperazine

SUPPLEMENTARY MATERIAL

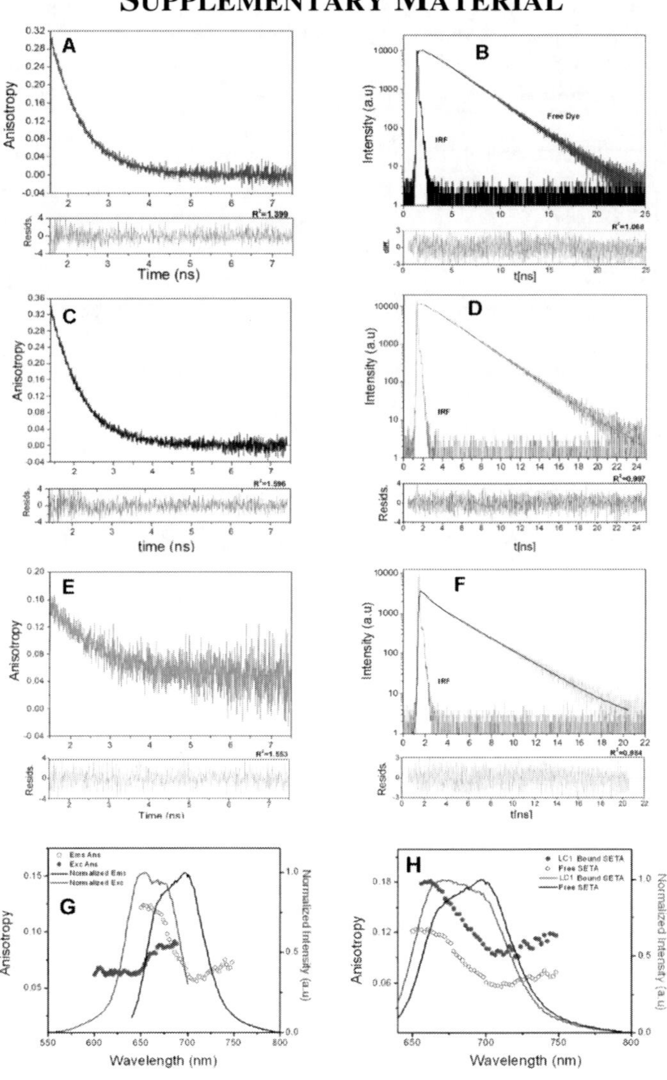

Figure 1S. SeTau-LC1 is immobilized by a myosin lever arm. To ensure that the orientation of the transition dipole of the fluorophore reflects the orientation of the lever arm, we measured the decay of anisotropy of SeTau-LC1 exchanged into myofibrils. Anisotropy is defined as $r = (I_\parallel - I_\perp)/(I_\parallel + 2I_\perp)$. The decay of anisotropy of free SeTau is shown in panel **A**. The decay was best fit by a single exponential curve $r(t) = R_{INF} + R_1 \cdot \exp(-t/\theta)$, where $R_{INF}=0$ is the anisotropy at infinite time, $R_1 = 0.301$ is the initial anisotropy and Θ is the rotational correlation time = 0.776 ns. The anisotropy decay comprised entirely of the fast decay. **B**: The lifetime of free SeTau (2.449 ns). (**C**): The decay of anisotropy of SeTau-LC1. The decay was best fit by a single exponential curve $r(t) = R_{INF} + R_1 \cdot$

exp(-t/θ), where R_{INF}=0 is the anisotropy at infinite time, R_1 = 0.335 is the initial anisotropy and Θ is the rotational correlation time = 0.784 ns. **D**: The lifetime of SeTau bound to LC1 (2.474 ns). The anisotropy decay comprised entirely of the fast decay suggesting that the dye was not immobilized by LC1. **E**: The decay of SeTau-LC1 bound to myofibrillar myosin was quite different: it was best fit by the exponential curves r(t) =R_{INF}+ R_1 · exp(-t/θ$_1$)+ R_2 · exp(-t/θ$_2$) with R_{INF}=0.050, R_1 = 0.047, R_2 = 0.106, θ$_1$ = 223 ns, θ$_2$ = 1.268 ns. The fast decay comprised 47% of the positive anisotropy components, indicating that 53% of the dye was immobilized by myosin. The mobile fraction decayed with the correlation time characteristic of the dye alone. **F**: The lifetime of SeTau-LC1 bound to myofibrillar myosin. The decay consists of fast (0.513 ns) and slow (2.695 ns) components. **G, H**: Absorption (red lines), emission (black lines), absorption anisotropy (red circles) and emission anisotropy (black circles) of Free (G) and LCV1 bound Setau. Anisotropies suggest that the fluorophore resides in two different environments.

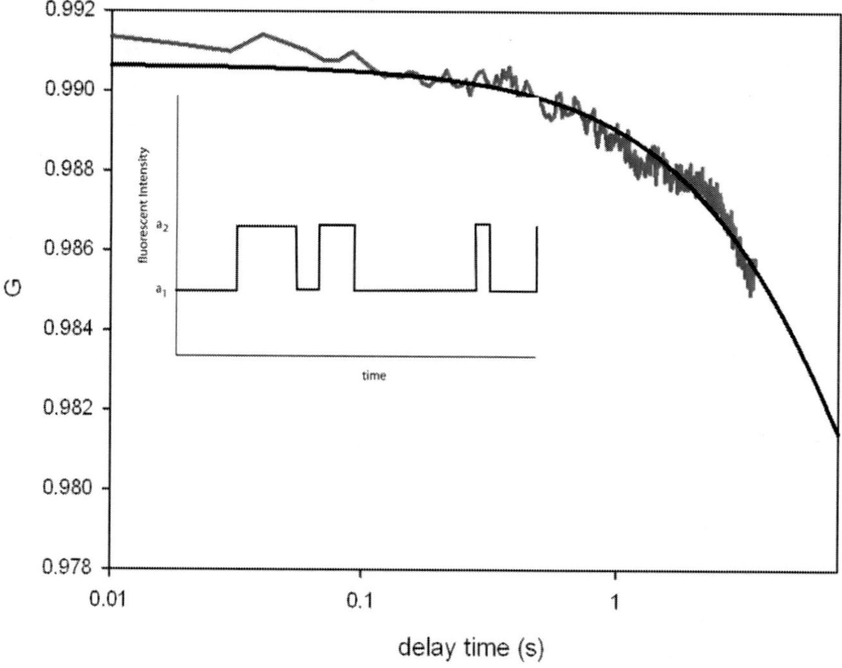

Figure 2S. For 2 steps process such as shown in insert, the correlation function is:

$$R_2\left(t\right)=\frac{\left(a_1 k_2+a_2 k_1\right)^2}{\left(k_1+k_2\right)^2}+k_1 k_2 \frac{\left(a_1-a_2\right)^2}{\left(k_1+k_2\right)^2}\exp\left(-\left(k_1+k_2\right)t\right) \qquad (1)$$

[1]. Assuming a$_1$=0, a$_2$=1 we get

$G(\tau) = k_1^2/ (k_1+k_2)^2 + k_1 k_2/(k_1+k_2)^2 \exp(-(k_1+k_2)t)$. Typical autocorrelation function during contraction (e.g. text Fig 5, shown here in green in non-normalized form) is best fit by a single 3 parameter exponential fit y=y$_o$+ae^{-bt} fit (black) with y$_o$=0.972, a=0.0186 and b=0.0912 s^{-1}. This translates to k$_1^2$/ (k$_1$+k$_2$)^2 = 0.972, k$_1$k$_2$/(k$_1$+k$_2$)^2 = 0.0186, k$_1$+k$_2$ = 0.0912 s^{-1} and k$_1$ = 10.82 s^{-1}, k$_2$ = 0.21 s^{-1}

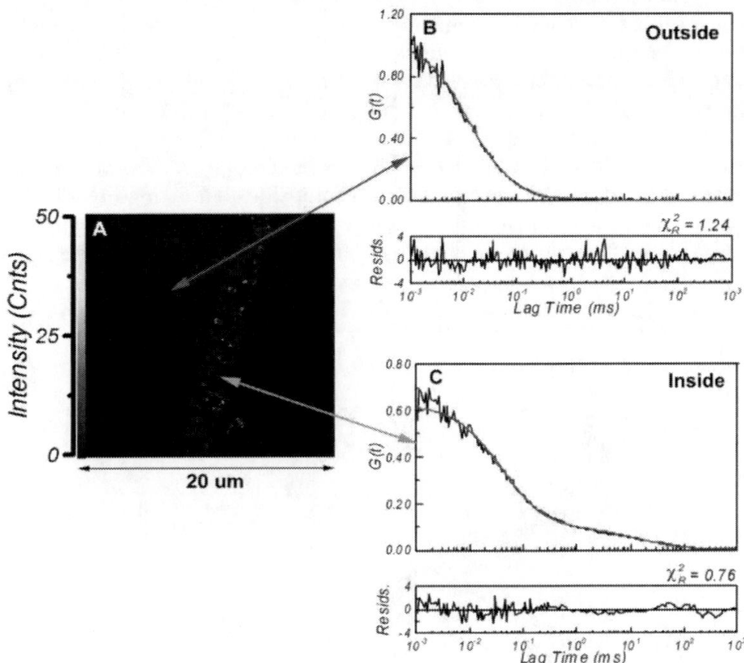

Figure 3S. Estimating the extent of solvent extrusion. 10 nM of fluoresent tracer hydrazine was added to myofibrils bathed in rigor solution and fluorescence lifetime correlation (FLC) functions of hydrazine fluorescence was measured in order to calculate the number of hydrazine molecules inside myofibril (a measure of the volume) in rigor. (**A**): image and (**B, C**): FLC in rigor myofibril. The correlation function is similar outside (**B**), data collected from the area indicated by the red arrow, a single decay, $\chi2$=1.24, the diffusion coefficient D=470 μm^2/s, transit time τ=0.0132 ms, average number of molecules in the DV <N>=0.97) and inside (**C**), data collected from the area indicated by the green arrow, two decays, $\chi2$=0.76, D_1=0 μm^2/s, D_2=160 um^2/s, τ_1=15.288 ms, τ_2=0.0386 ms, <N>=1.6). The first and second diffusions contributed 14.8 and 85.2% to the total intensity, respectively. The fact that diffusion of hydrazine inside myofibril is similar to diffusion outside myofibril indicates that hydrazine is moving freely within myofibrillar lattice. The average number of hydrazine molecules inside myofibril was 1.6. Data collected for 60 s. Sarcomere length 2.5 μm.

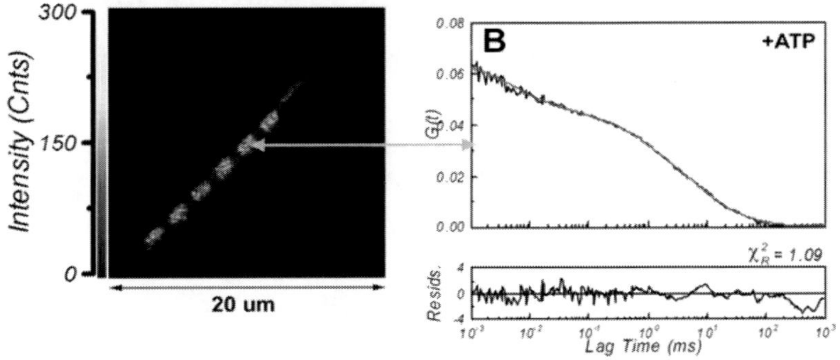

Figure 4S. Estimating the extent of solvent extrusion, continued. A: typical intensity image of cross-linked myofibril after replacing rigor with contracting solution. The fluorescent image is formed by inclusion of hydrazine in contracting solution (there no other fluorescent dye present). Contraction (with associated solvent extrusion) obviously exposed surfaces of some sarcomeric proteins to which hydrazine could now bind. Nevertheless, a majority of the dye is free to move as shown in panel **B**: FLC of hydrazine fluorescence in area of contracted myofibril indicated by green arrow. FLC function is well fitted ($\chi 2$=1.09) with the three exponent model. The average diffusion coefficients were 0.7, 7 and 1100 μm^2/s, respectively. The average transit times were 9.041, 0.900 and 0.005 ms. The first, second and third diffusions contributed 6.1, 45.7 and 48.2% of the intensity, respectively. Thus nearly half of the tracer is mobile with diffusion coefficient 8 times larger than in rigor. This acceleration of motion of tracer is most likely due to streaming of solvent inside contracting myofibril. We conclude that there is solvent inside contracted myofibril and that cross-bridges are not inhibited to rotate. Note that the number of solvent molecules is ~1.6 in rigor (from Figure 2S) and ~17 molecules (from this figure, N=1/G(0)), i.e. the volume collapsed ~10 times.

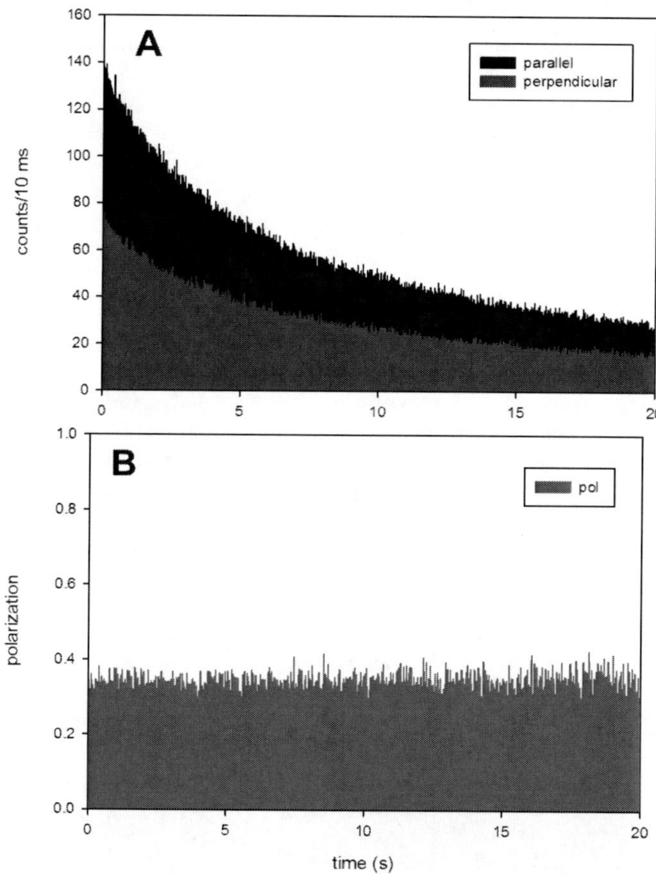

Figure 5S. Photobleaching induced by illuminating myofibril with strong exciting light. While the orthogonal intensities (red and blue) show significant decay, the polarization of fluorescence was constant.

REFERENCES TO SUPPLEMENTARY MATERIAL

[1]. P. Mettikolla, N. Calander, R. Luchowski, I. Gryczynski, Z. Gryczynski, J. Zhao, D. Szczesna-Cordary and J. Borejdo, *J. Theor. Biol.* 284, 71-81 (2011).

In: Skeletal Muscle
Editor: Mark Willems

ISBN: 978-1-62417-271-7
© 2013 Nova Science Publishers, Inc.

Chapter 2

MECHANISMS AND CONSEQUENCES OF SKELETAL MUSCLE AGING

Ju Li and Irina M. Conboy[*]
Department of Bioengineering,
University of California, Berkeley, California, US

ABSTRACT

Muscle aging is associated with a decline in skeletal muscle mass and function, which leads to a decline in the strength and agility of older individuals. The age-related muscle loss and impaired regeneration capability of muscle are due to changes of both muscle stem cells and the systemic environment that are responsible for muscle maintenance and repair. This chapter presents an overview of the physiological, cellular and molecular changes during the process of muscle aging and related diseases. In particular, the signal pathways that either regulate the behavior of aged muscle stem cells or modulate the aging muscle niche elements are summarized. Advancements in understanding the mechanism of muscle aging will facilitate therapeutic strategies to cure muscle aging related diseases and to improve muscle function in aged individuals.

INTRODUCTION

Aging is associated with a series of dysfunctional elements in the musculoskeletal system, including fragile bones, redistributed fat, loss of tendon elasticity, and significantly diminished mass and functionality of skeletal muscle. Age-related skeletal muscle dysfunction, named sarcopenia, is the pathological consequence of losing muscle mass and contractibility. In addition to the muscle atrophy process, there is also a remarkable decline in the regenerative capability of muscle with age, which may indeed contribute to the progressive loss of this tissue, but which definitely leads to delayed recovery from muscle

---

[*] Corresponding author: Irina M. Conboy. E-mail address: iconboy.berkeley.edu.

injuries [1]. Therefore, skeletal muscle aging has a large impact on the quality of life for the elderly and draws more and more attention for public health.

In order to promote healthy aging, it is important to learn the mechanism of muscle aging and to identify the factors that regulate this process. Since muscle stem cells are the major population responsible for muscle regeneration as well as tissue homeostasis, the question arises as to whether the age-associated decline of skeletal muscle function is due to a diminished regenerative response of muscle stem cells.

As with other types of stem cells, muscle stem cells are susceptible to different types of damage that are known to accumulate with time, including telomere shortening, reactive oxidative species (ROS)-induced damage, and genetic mutations.

However, these intrinsic changes in the muscle stem cells alone do not fully account for the extreme phenotype of age-related muscle dysfunction. Muscle-specific stem cells reside adjacent to muscle fibers and are surrounded by various tissues (connective, vascular, nerve and immune), which form a dynamic signaling environment of cross talk between cells via molecules that they display on cell surface and that they secrete. The productive cross talk between muscle stem cells and their niches (specifically studied are differentiated tissue and blood circulation) change with time in ways that become inhibitory for regenerative responses.

In this chapter, the influences of various intrinsic and environmental factors on skeletal muscle aging are discussed. In particular, we focus on the age-related changes in cellular signal transduction.

A better understanding of the mechanisms that lead to the age-specific decline in tissue regeneration will facilitate the development of therapeutic interventions that can delay or even reverse the onset of organismal aging.

1. PHYSIOLOGY OF MUSCLE AGING

The aging process leads to muscle atrophy, which is measured by the decreases of muscle mass and fiber cross section area (CSA). Early studies found an average of 10% reduction of total muscle mass at 50 years of age and 40% by 80 years by measuring muscle CSA of different muscles in different age groups [2]. Since the histological studies of CSA are limited by the availability of human muscle samples, in further studies, muscle mass in a greater population was measured by the non-invasive methods, such as X-ray absorptiometry (DEXA) or magnetic resonance imaging (MRI). Both DEXA and MRI results confirmed the age-associated decline in muscle mass. Moreover, when different genders are considered, men have significantly greater skeletal muscle mass than women but exhibit greater losses of muscle mass with aging [3, 4]. Age-associated muscle atrophy has been explained by myonuclear domain theory, which states that the cytoplasmic volume controlled by each myonucleus is constant [5]. With age, there is a decrease in the number of myonuclei, which leads to a transient increase in myonuclear domain and in turn triggers myofiber atrophy to restore the original myonuclear domain through decreasing volume and observed CSA [6].

With aging, there is a decline of muscle function, which is caused not only by the decrease in muscle mass but also by reduced innervation. One assessment of muscle function is measured by contractile strength normalized to CSA, and it declines with advancing age

[7]. This age-associated decline of muscle contractibility may be due to changes in neural drive [8], decreases of contractile proteins [9], and infiltration of non-contractile fatty and fibrous tissues within and between myofibers [10]. At the cellular level, aging of muscle is accompanied by a reduced rate of biosynthesis of myofibrillar proteins, which is one of the main causes of the decline in muscle mass and contractibility [8]. Since skeletal muscle is a very stable tissue with very low protein turnover rate (1-2% per week), the decline of muscle regenerative capability does not immediately translate into a lack of functional tissue; however with time, the poor performance of muscle stem cells contributes to the onset and progression of sacropenia in the old [11]. Skeletal muscle is a highly specialized tissue, composed of bundles of post-mitotic myofibers. The remarkable regenerative capacity of skeletal muscle is dependent on a special type of muscle-specific stem cell, named a satellite cell [12]. In resting muscles, satellite cells reside adjacent to muscle fibers and stay in a quiescent state [13]. Remarkably, in response to muscle injury or attrition, these cells are able to produce new muscle tissue throughout the adult life by activation to proliferate, differentiation into new muscle fibers and self-renewal [14]. Regrettably, aged satellite cells exhibit reduced mitotic activity in culture and a lack of myogenic responses *in vivo* [15]. The aging process may affect satellite cells through either decreasing cell number or reducing cell activity. There are different opinions about whether the number of satellite cells changes with age. Some results show a decline in the number of satellite cells with age [16-19], whereas others find no significant differences [15, 20-23]. The satellite cell content varies among muscle fiber types, different species, the ages and health status of examined subjects. As a result, the different results may be due to variations in the muscles examined or the quantitation methodologies. While there is a lack of consensus on the age-induced changes in satellite cell number, it has been hypothesized that in addition to the debate over cell number, it is clear that satellite cell activity is also an essential regulator of muscle aging. Satellite cells from aged muscle display an impaired ability to activate and proliferate in response to injury, which leads to a lack of muscle regenerative capacity and a delay in recovery from muscle traumas [15, 23, 24]. Moreover, aged satellite cells have been demonstrated to be susceptible to differentiating into unproductive tissue lineage pathways (such as fat or connective tissue) [25-27] and to be prone to apoptosis [28]. As a result, even though aged muscle might contain similar numbers of satellite cells as young muscle, the ability of aged satellite cells for muscle regeneration becomes impaired.

2. MUSCLE STEM CELL AGING

Satellite cells, the muscle-resident stem cells, are responsible for post-natal muscle growth and muscle regeneration. They are activated by pathological or physiological stimuli, such as degeneration of muscle fibers due to disease, injury or exercise. Their main function is to form new functional muscle fibers while replenishing the stem cell pool. With age, satellite cells go through a series of changes that impair their function, leading to the abandonment of muscle tissue maintenance and repair.

2.1. Telomere and Telomerase

Telomere length and telomerase activity are closely related to the aging process. Telomeres are repetitive DNA sequences at the end of chromosomes that protect chromosome ends from degradation and maintain genomic stability, and the length of the telomere is regulated by an enzyme, telomerase [29]. Lack of a functional telomere or telomerase can accelerate some pathological changes associated with aging. For example, Werner syndrome is caused by mutations in Werner RecQ helicase, a DNA helicase/exonuclease, which is involved in repair of double strand DNA breaks and synthesis of lagging strand telomeres. Loss of telomeres in the lagging strand causes an accelerated aging phenotype with premature tissue senescence and genomic instability [30]. With respect to tissue-specific progenitor cells, age-associated telomere shortening likely contributes to the impairment of cell proliferation, migration, and self-renewal [31]. It has been hypothesized that the telomeres of satellite cell become shortened with age, as they do in other tissues with high turnover [29]. While adult skeletal muscle has a very low turnover, during neonatal and early post-natal years, skeletal muscle undergoes rapid growth [32]. Hence, theoretically, the active mitotic cycles that satellite cells go through during growth could decrease their telomere length. However, it has been shown that the telomere length of muscle fibers in young and old mice is not proportional to age (and are of approximately equal length) and that both young and old satellite cells have telomerase activity and hence are able to restore the length of telomeres when these stem cells divide.

While it has been proposed that telomere length of satellite cell can be used as a mitotic clock that records the maximal number of division the satellite cells can achieve *in vitro*, it was found that the maximal doubling capacity of satellite cells decreases during childhood, yet remains stable after adulthood, suggesting no further shortening of telomere during aging [33, 34]. In fact, it has been confirmed that there is no significant difference of telomere length between young and aged satellite cells [35, 36]. Therefore, there may not be a direct linkage between telomere length and satellite cell aging. Notably, telomere shortening is a reversible process. Telomerase has been shown to be able to elongate short telomeres and reset the mitotic clock. Immortalized myogenic C2C12 cells, which are derived from mouse satellite cells, have extremely long telomeres and an associated high level of telomerase activity [37]. During *in vitro* culture of satellite cells, both telomere length and telomerase activity decreases with time [37], and telomere shortening during *in vitro* culture is accompanied by a decreased capacity for regeneration when cells are transplanted [38]. However, it is unclear whether this phenotype is the result of culture, and it may not reflect normal aging of satellite cells. Additionally, some data suggest that telomeres in skeletal muscle can be regarded as dynamic structures regulated by their genetics and environment. For example, there are variations in the telomere length among donors from the same age group, which may be due to different genetic background, health states, and physical characteristics of the donors [35, 36]. Overall, there is no definitive data suggesting that the decline in telomerase activity of aged satellite cells contributes to the age-specific differences in muscle regeneration.

2.2. Oxidative Damage

Oxidative damage to macromolecules is another key effector of cell aging. Reactive oxygen species (ROS) are among the major contributors to damage of cellular components such as RNA, DNA, proteins, and lipids. The degree of ROS-induced damage might be different between the post-mitotic cells and stem cells. Since post-mitotic cells like myofibers have a more constant nuclear and protein content, ROS-induced damage can accumulate with time. In contrast, dividing organ stem cells repair ROS-induced damage more effectively than post-mitotic cells. As most satellite cells stay quiescent in adults and do no frequently proliferate, ROS-induced damage can accumulate in these muscle stem cells. Additionally, age-induced oxidative stress causes significant damage to myofibers, which have high metabolic activity; this leads to apoptosis of myonuclei and ultimately muscle atrophy [39]. The main oxidative stress comes from energy metabolism. A high caloric intake increases oxidative stress by generating more free radicals and weakening oxidative defense [40], and caloric restriction is thought to be able to lower the oxidative damage, possibly delaying the aging process.

It has been demonstrated that ROS-induced damage can accumulate in mitochondria of skeletal muscle with age, resulting in increases in iron content and oxidative damage to mitochondrial DNA [41]. ROS-induced mutations in mitochondrial DNA, increased mitochondrial permeability, and increased protein aggregation have all been found in skeletal muscle from aged rodents and humans [42, 43]. This progressive loss of mitochondrial function with age results in a loss of muscle oxidative capacity, contributing to apoptosis of myonuclei, and in turn reduced sizes of myofibers [44]. Since there are fewer mitochondria in type II fibers than in type I fibers, type II fibers have a greater level of oxidative stress and are more sensitive to mitochondrial dysfunction. As a result, type II fibers are preferentially lost during aging, resulting in a fast-to-slow transition of fiber types [44]. Cumulatively, these changes cause the poor performance of old muscle tissue and might be also responsible for the age-specific decline in the regenerative capacity of muscle stem cells.

2.3. Genetic Mutation and Epigenetic Modification

Not only is mitochondrial DNA damaged by oxidative stress, but also genomic DNA is sensitive to ROS. Several types of diseases with premature aging phenotypes are due to mutation of genes responsible for DNA modification and damage repair, including Cockayne syndrome, Hutchison-Gilford progeria syndrome, and Werner syndrome [45]. However, there is no evidence as to whether age-associated satellite cell defects are due to genetic DNA changes. Age-related differences in gene expression and proteomics profile have been reported in human and mouse satellite cells [46-48]. Expression of several genes encoding for matrix proteins and elements of cellular signal transduction are changed during aging. Furthermore, early cross-age transplantation experiments showed that the regeneration potential of whole muscle grafts depends only on the age of the host rather than the age of the donor. Satellite cells from old rats are able to regenerate as effectively as those from young animals when transplanted into a young host [49]. Cross-age parabiotic paring experiments also showed that the regeneration capacity of old satellite cells could be restored when old mice shared their circulatory systems with young animals [22]. Thus, age-related changes of satellite cells are reversible when they are in a more optimal environment. These results support the hypothesis that age-induced damage to satellite cells themselves is quickly reversible by "youthful" modifications of the organismal niche of these adult stem cells.

3. Environmental Aging

Individual muscle fibers are surrounded by various types of tissues, including the connective tissues, motor neurons, adipose tissue, and blood vessels. The performance of skeletal muscle is regulated by a combination of the local and systemic environments. Productive regenerative responses of old satellite cells can be restored by either transplantation to young muscle or exposure to the young blood circulation [22, 49]. Therefore, the extrinsic environment instead of the intrinsic satellite cell characteristics may be the key determinant of the health status of skeletal muscle in young versus old mammals.

3.1. Exracellular Matrix (ECM)

Muscle fibers are surrounded by interstitial connective tissue composed of extracellular matrix (ECM) components [50]. ECM consists of fibrous proteins (primarily collagen, laminin, fibronectin, tenascin, and vitronectin), embedded in a mixture of proteoglycans [51]. The collagen fiber network bound to proteoglycan molecules maintains the matrix strength and elasticity. With age, the relative amount of proteoglycan becomes reduced due to decreased synthesis and increased degradation. These changes alter the properties of skeletal muscle making it more susceptible to injury and attrition. Another function of proteoglycan is to interact with various growth factors and their receptors through sulfate chains. This interaction is required for growth factor-regulated satellite cell activity and muscle regeneration [52, 53]. For example, in healthy muscle, perlecan and heparin sulfate proteoglycan bind to fibroblast growth factor 2 (FGF2) and hepatocyte growth factor (HGF), respectively, resulting in a sequestered growth factor pool that can be activated for signaling and induce satellite cell responses when muscle is injured [54-56]. Similarly, during muscle regeneration, decorin and biglycan bind to TGF-β and diminish the availability of this inhibitory growth factor, thus releasing satellite cells from quiescence [57, 58].

Gene expression profiling shows that several genes encoding ECM fiber proteins are downregulated during muscle aging, suggesting that they may be involved in age-specific changes in maintenance of muscle integrity and a proper satellite cell niche [59]. Collagen is the major ECM fibrous protein of skeletal muscle. The strength and stiffness of the matrix comes from the degree of cross-linking of collagen fibers. With age, the number of collagen cross-links increases, which makes the matrix too stiff for optimal muscle function. Fibrillin, another ECM component, plays a central role in maintaining the satellite cell niche by interacting with TGF-β [60], and the TGF-β pathway is responsible for age-related satellite cell dysfunction [62]. A fibrillin gene mutation is associated with Marfan syndrome, characterized by muscle weakness and the inability to respond to growth and exercise [61]. With age, fibrillin gene expression is downregulated in skeletal muscle [60]. Thus the decrease in fibrillin may contribute to muscle aging.

ECM composition and structure play an essential role in regulating satellite cell activity and muscle function. To improve myoblast transplantation, several engineering approaches have been used to optimize myoblast viability and myogenicity with the help of synthetic scaffolds. An aligned collagen matrix was used in one experiment to mimic the muscle structure *in vitro* and was grafted *in vivo* to improve muscle function [63]. In another approach, a three-dimensional porous collagen scaffold was used to deliver myogenic precursor cells to dystrophic mice. The muscle cells showed decreased apoptosis, increased proliferation and improvement of regeneration efficiency when transplanted in engineered scaffolds [64]. Satellite cells were delivered to injured muscle in three-dimensional matrix scaffolds with sustained release of growth factors, such as HGF, FGF2 and VEGFA, and the long-term survival and migration of satellite cells were dramatically improved [65, 66]. These approaches prove the importance of the ECM in regulating satellite cell activity and suggest novel therapeutic interventions for sarcopenia and genetic myopathies.

With respect to muscle degenerative disorders, the dystrophin protein at the plasma membrane of muscle fibers connects the cytoskeletal network with the ECM, and as such it is required for skeletal muscle function. Mutations in the dystrophin gene result in several muscular dystrophy diseases, for example, Duchenne's muscular dystrophy (DMD) and Becker's muscular dystrophy (BMD) [67]. DMD patents suffer from progressive muscle loss and satellite cell exhaustion [68]. Both DMD patients and a dystrophic mice model (mdx mice) show premature aging phenotypes and develop more severe muscle atrophy later in life. Particularly, mdx mice are healthy and maintain their mobility at an early age; however, the decline in tissue regenerative capability with age causes loss of muscle mass, and this is dramatically exacerbated by a transgenic mutation in telomerase that accelerates aging and stem cell loss [69].

3.2. Anabolic Hormones

As mentioned before, satellite cells from old mice can be rejuvenated when they share the circulatory system with young mice [22]. There must be some systemic environmental factors responsible for age-associated satellite cell defects. It has been shown that sarcopenia is associated with a decline in the circulating levels of muscle anabolic hormones such as testosterone, dehydroepiandrosterone (DHEA), growth hormone, and IGF-I [70]. Among these hormones, the GH/IGF-I axis is the main regulator of skeletal muscle mass. Transgenic mice overexpressing or deficient of several components of GH/IGF-I axis exhibit increased or reduced muscle mass, respectively. Skeletal muscle expresses at least two IGF-I isoforms, IGF-IEa and mechano-growth factor (MGF), derived from alternative splicing of the same transcript. IGF-I modulates muscle mass in response to hormonal and mechanical stimulation via IGF-IEa and MGF, respectively [71]. The GH/IGF-I axis impinges on muscle aging as well. Low-dose of GH can improve muscle function during aging, while a combination of GH supplementation and resistance exercise can to a slight degree reverse sarcopenia [72].

3.3. Local and Systemic Signaling Pathways

A recent cross-age parabiotic pairing experiment demonstrates the function of extrinsic factors in the satellite cell niche which regulate satellite cell activity and modulate muscle function [22]. These findings suggested that age-specific changes in several signaling pathways are involved with muscle aging, including Notch, Wnt and TGF-β. These pathways cooperate to regulate satellite cell behavior as a network system.

3.3.1. Notch Pathway

Notch signal is activated when the membrane-anchored ligand Delta interacts with the transmembrane Notch receptor, which then undergoes a series of cleavages resulting in the intracellular domain of Notch (NICD) translocating to the nucleus. NICD acts as a transcription factor that regulates target genes such as Hey and Hes, which encode proteins involved in many cellular behaviors, including myogenesis [73]. During muscle aging, old satellite cells express functional Notch receptor, but through lack of the Notch ligand Delta fail to activate Notch signaling. Diminished activation of Notch impairs regeneration of young muscle, whereas forced Notch activation rejuvenates old muscle [15].

Therefore, Notch signaling is a key determinant of the age-associated decline of regenerative potential of skeletal muscle. Several Notch ligands, receptors and antagonists are found to be asymmetrically located during satellite cell division. Notch receptors are highly expressed in the quiescent satellite cells, whereas Notch ligands and Notch antagonists are preferentially located in the myogenic precursors [74, 75]. The asymmetric localization of Notch signaling components, such as Numb, in dividing muscle progenitor cells suggests that the Notch pathway may be involved in satellite cell fate determination. Thus, insufficient Notch activation in the aged muscle may impair muscle regeneration through both interrupting satellite cell division and skewing self-renewal.

3.3.2. Wnt Pathway

Wnt is another key pathway involved in muscle aging. The age-associated decline of muscle regeneration capacity is accompanied by an increase of connective tissue [72]. Satellite cells from old muscle tend to convert from a myogenic to a fibrogenic lineage, which is mediated by the old systemic environment [22].

It was found that this lineage conversion in the aged satellite cells is associated with premature activation of canonical Wnt signaling, which plays an important positive role during the later stages of myogenesis, namely to promote terminal differentiation of myogenic progenitor cells into fusion-competent myoblasts and myotubes [76]. Addition of an exogenous Wnt activator or inhibitor can alter satellite cell identity through promoting or revering the myogenic-to-fibrogenic transition and by altering the myogenic proliferation and differentiation rates [27, 76].

3.3.3. TGF- β Pathway

The TGF-β superfamily is the major negative regulator of muscle size and satellite cell activity [77]. There is an elevated level of TGF-β1 in the local and systemic niches of aged satellite cells, which interferes with muscle regeneration through stimulation of cyclin-dependent kinase (CDK) inhibitors in satellite cells [78]. Attenuation of TGF-β1 in aged mice can restore the regenerative capability of satellite cells and rejuvenate the aged muscle. Myostatin, also called growth and differentiation factor 8 (GDF8), is another member of the TGF-β superfamily.

A natural deficiency or loss-of-function genetic mutation in myostatin in many species leads to an increase in skeletal muscle mass, which is referred as double-muscling phenotype [79, 80]. Further studies indicate that myostatin expressed by satellite cells plays an essential role in maintaining a balance among proliferation, differentiation, and self-renewal of muscle stem cells. Myostatin negatively regulates satellite cell function through withdrawal of satellite cells from the cell cycle, leading them to either a quiescent state or a terminally differentiated state [81, 82]. It has been shown that serum myostatin levels increase with age, while elevated serum myostatin levels correlate with age-related sarcopenia [83].

3.3.4. Cross Talk between Signal Pathways

The aged satellite cell niche is regulated by multiple local and systemic environmental factors that interact with each other. A balance between positive and negative signaling pathways is critical to maintain the tissue state. Notch and TGF-β1 compete with each other to control satellite cell proliferation during muscle regeneration. High level of TGF-β1 in aged muscle or exogenously added TGF-β1 in young muscle causes a prompt upregulation of CDK inhibitors, whereas forced activation of Notch blocks any TGF-β-induced effect [84]. However, TGF-β1 does not inhibit the expression or activation of Notch signaling components. Likewise, activation of Notch does not diminish TGF-β1/pSmad3 levels either. Thus, the age-specific activation of Notch and TGF-β1 seems to initiate independently of each other and reaches a balance that controls the regenerative competence of muscle stem cells. Another example of balance between different signal pathways is Notch and Wnt signaling interacting with each other to control muscle homeostasis [76]. In response to muscle injury, activated Notch promotes proliferation of satellite cells and inhibits their differentiation by inhibiting Wnt, allowing committed muscle precursors to be generated and expanded at the injury site. The subsequent decline of Notch and activation of Wnt ensures the differentiation of muscle precursors into new muscle fibers. In this process, GSK3b, an intracellular messenger of Wnt and many other pathways, plays a crucial role in switching signaling from Notch to Wnt or vice versa [76].

3.3.5. Other Pathways in Muscle Aging

In addition to the three pathways mentioned above, numerous signal transductions have been found to play a role in the regulation and aging of adult myogenesis. For example, an important mitotic signaling extracellular signal-regulated kinase (ERK) has been shown to positively regulate proliferation and differentiation of muscle progenitor cells and yet decline in aged skeletal muscle. Many soluble factors from youthful environment rejuvenate aged muscle stem cell through activating the ERK signal pathway [85].

The Nuclear factor kappa B (NF-κB) transcription factor is also involved in the progression of age-related muscle atrophy. Aged skeletal muscle has high local and circulating level of inflammatory factors, including TNF-α, IL-1 and IL-6, which activate NF-κB signaling in satellite cells [86-88]. Activation of the NF-κB signal in turn regulates muscle mass and regenerative capacity through its target genes: MuRF1, MyoD, cyclin D, ubiquitin-conjugated enzyme, and a proteasome subunit [89, 90]. Blockage of NF-κB activity is able to reverse the aging phenotype by altering the global gene expression program in skin tissue

[89]. Moreover, it has been reported that exercise and calorie restriction are capable of delaying muscle aging through downregulating NF-κB activation in older individuals [91].

The calcineurin (Cn)-Nuclear factor of activated T cell (NFAT) pathway plays an essential role for skeletal muscle growth and aging as well. Calcineurin (Cn) is a heteodimeric serine/threonine phosphatase composed of a calmodulin-binding catalytic A subunit and a Ca^{2+}-binding regulatory B subunit.

In response to calcium increase, activated Cn stimulates dephosphorylation, nuclear translocation, and activation of NFAT, which in turn activates NFAT-dependent gene expression [92]. In resting skeletal muscle where intracellular calcium levels are low, the Cn-NFAT signal is inactivated. Under certain physical and pathological conditions, elevated intracellular calcium levels activate the Cn signal, which stimulates muscle hypertrophy through positively regulating NFAT signal, and inhibiting the myostatin signal [93].

3.4. Neighboring Tissues

Skeletal muscle is surrounded by various tissues, which modulate muscle function through either direct interaction or by indirect secretory molecules. Heterochronic muscle transplantation has demonstrated that old muscle is less capable of supporting muscle regeneration than young muscle due to alteration in the neighboring tissues, including neurons and muscle innervation and the immune cells for wound clearance and support for the regenerative process [49].

3.4.1. Neural System

Functional activity of skeletal muscle is regulated by the motor neurons that innervate it and trigger contractive forces. The age-related loss of muscle force is mainly due not only to the loss of muscle mass but also to the loss of motor neurons and poor re-innervation of newly formed muscle fibers [94]. Aging can affect the proportion of motor unit types, innervation rates, and the mechanical properties of neuro-muscular junctions. For example, with advancing age, fast-type motor neurons are preferentially lost, leaving slow-type motor neurons, which leads to more progressive denervation and apoptosis of fast-twitch muscle fibers [5, 95]. The fast-twitched fibers can generate greater force for muscle contraction, but they are more susceptible to muscle fatigue. The fast-to-slow transition of fiber type with age requires muscle to use energy more efficiently to produce persistent forces but at the loss of burst contractile activity.

Together, these age-related changes tend to decrease the diversity of motor units and reduce the force generation spectrum. Motor neurons innervate muscle fibers via a specific structure, the neuromuscular junction (NMJ). Satellite cells tend to locate close to the NMJ [96]; and since muscle denervation leads to satellite cell activation either directly or via the death of denervated muscle fibers [97], the NMJ structure may be required to maintain a functional pool of satellite cells. Regrettably, during muscle aging, there is an alteration of the NMJ structure, which results in motor neuron degeneration and muscle atrophy [98].

3.4.2. Immune System

The immune system is the primary responder to muscle injury. Immune responses include infiltration of leukocytes, chemokine secretion, macrophage recruitment and wound clearance [72]. Aged neutrophils display an impaired capability to secrete chemokines, recruit macrophages, and clean the damaged tissues, as compared with young cells [99]. Macrophage infiltration and wound clearance have been implicated in the age-associated defects in muscle maintenance and repair [100]. Co-culture of minced muscle with functional bone marrow cells before implantation results in faster muscle regeneration compared with the muscle alone or muscle mixed with bone marrow with impaired phagocytic activity [100].

However, the mechanism of how the age-induced changes in the immune system affect skeletal muscle function and/or regeneration is still unclear. One possible explanation is that macrophages may regulate satellite cells through indirect paracrine or endocrine factors. Aged muscle has elevated local and serum levels of cytokines, such as TNF-a, IL-1, and IL-6 [86]. High levels of TNF-a in aged muscle activate nuclear factor kappa B (NF-kB), which in turn activates apoptosis signaling [88]. Elevated levels of IL-6 with age is associated with a high risk of muscle wasting, which is indicated by the loss of muscle strength instead of muscle mass [101].

Thus, age-related alterations in the immune system may affect satellite cells and myofibers through changing the cytokine levels in the muscle niche and via differences in the clearance of dead muscle tissue, as well as the age-specific increase in the inflammatory response.

CONCLUSION

Similar to other tissues, the function of skeletal muscles and associated satellite cells become impaired by the aging process. Since the regulation of satellite cells is well understood, skeletal muscle is a good model for studying the mechanism of tissue aging. The key age-specific intrinsic culprits of muscle atrophy and satellite cell dysfunction are unlikely to include telomere shortening but might include ROS-induced damage to macromolecules and might exacerbate the differences in rates of biosynthesis of nucleic acids and proteins. The intrinsic changes of muscle stem cells are transient and reversible, but the extrinsic factors from the aged local and systemic environments impose a stable inhibition on satellite cell regenerative performance. Tissues surrounding satellite cells influence their function through direct interaction, secretion of different matrix proteins or growth factors, innervation and clearance of damaged myofibers. Notch, Wnt and TGF-β are the three major pathways that play a role in the age-imposed dysfunction of satellite cell. The cross talk between these signal transductions changes during aging in ways that interfere with muscle homeostasis and regeneration. Future efforts will be focused on translating our understanding of the mechanisms of muscle aging into developing therapeutic approaches for the rejuvenation of mammalian tissues.

ACKNOWLEDGMENTS

We would like to thank Michael J. Conboy for editing this chapter. This work was supported by a Keck Fellowship to IMC.

REFERENCES

[1] T. J. Doherty, *J. Appl. Phys.* 95, 1717 (2003).

[2] W. J. Evans, *J. Gerontol. A. Biol. Sci. Med. Sci.* 50 Spec. No, 5 (1995).

[3] D. Gallagher, M. Visser, R. E. De Meersman, D. Sepulveda, R. N. Baumgartner, R. N. Pierson, T. Harris, and S. B. Heymsfield, *J. Appl. Phys.* 83, 229 (1997).

[4] I. Janssen, S. B. Heymsfield, Z. M. Wang, and R. Ross, *J. Appl. Phys.* 89, 81 (2000).

[5] D. L. Allen, R. R. Roy and V. R. Edgerton, *Muscle Nerve* 22, 1350 (1999).

[6] A. S. Brack and T. A. Rando, *Stem Cell Rev.* 3, 226 (2007).

[7] N. A. Lynch, E. J. Metter, R. S. Lindle, J. L. Fozard, J. D. Tobin, T. A. Roy, J. L. Fleg, and B. F. Hurley, *J. Appl. Phys.* 86, 188 (1999).

[8] G. N. Klein, A. F. Mannion, M. M. Panjabi, and J. Dvorak, *Eur. Spine J.* 10, 141 (2001).

[9] D. L. Hasten, J. Pak-Loduca, K. A. Obert, and K. E. Yarasheski, *Am. J. Phys. Endo. Metab.* 278, E620 (2000).

[10] J. A. Kent-Braun, A. V. Ng, and K. Young, *J. Appl. Phys.* 88, 662 (2000).

[11] H. Schmalbruch and D. M. Lewis, *Muscle Nerve* 23, 617 (2000).

[12] S. B. Charge and M. A. Rudnicki, *Physiol Rev.* 84, 209 (2004).

[13] A. Mauro, *J. Biophys. Biochem. Cytol.* 9, 493 (1961).

[14] P. Seale and M. A. Rudnicki, *Dev. Biol.* 218, 115 (2000).

[15] I. M. Conboy, M. J. Conboy, G. M. Smythe, and T. A. Rando, *Science* 302, 1575 (2003).

[16] A. S. Brack, H. Bildsoe and S. M. Hughes, *J. Cell Sci.* 118, 4813 (2005).

[17] G. Shefer, D. P. Van de Mark, J. B. Richardson, and Z. Yablonka-Reuveni, *Dev. Biol.* 294, 50 (2006).

[18] C. A. Collins, P. S. Zammit, A. P. Ruiz, J. E. Morgan, and T. A. Partridge, *Stem Cells* 25, 885 (2007).

[19] G. Shefer, G. Rauner, Z. Yablonka-Reuveni, and D. Benayahu, *PLoS One* 5, e13307 (2010).

[20] S. M. Roth, G. F. Martel, F. M. Ivey, J. T. Lemmer, E. J. Metter, B. F. Hurley, and M. A. Rogers, *Anat. Rec.* 260, 351 (2000).

[21] H. C. Dreyer, C. E. Blanco, F. R. Sattler, E. T. Schroeder, and R. A. Wiswell, *Muscle Nerve,* 33, 242 (2006).

[22] I. M. Conboy, M. J. Conboy, A. J. Wagers, E. R. Girma, I. L. Weissman, and T. A. Rando, *Nature,* 433, 760 (2005).

[23] A. J. Wagers and I. M. Conboy, *Cell,* 122, 659 (2005).

[24] E. Schultz and B. H. Lipton, *Mech. Ageing Dev.* 20, 377 (1982).

[25] J. M. Taylor-Jones, R. E. McGehee, T. A. Rando, B. Lecka-Czernik, D. A. Lipschitz, and C. A. Peterson, *Mech. Ageing Dev.* 123, 649 (2002).

[26] G. Shefer, M. Wleklinski-Lee and Z. Yablonka-Reuveni, *J. Cell Sci.* 117, 5393 (2004).

[27] A. S. Brack, M. J. Conboy, S. Roy, M. Lee, C. J. Kuo, C. Keller, and T. A. Rando, *Science,* 317, 807 (2007).

[28] S. S. Jejurikar, E. A. Henkelman, P. S. Cederna, C. L. Marcelo, M. G. Urbanchek, and W. M. Kuzon, Jr., *Exp. Gerontol.* 41, 828 (2006).

[29] J. M. Wong and K. Collins, *Lancet.* 362, 983 (2003).

[30] L. Crabbe, R. E. Verdun, C. I. Haggblom, and J. Karlseder, *Science,* 306, 1951 (2004).

[31] I. Spyridopoulos, Y. Erben, T. H. Brummendorf, J. Haendeler, K. Dietz, F. Seeger, C. K. Kissel, H. Martin, J. Hoffmann, B. Assmus, A. M. Zeiher, and S. Dimmeler, *Arterioscler. Thromb. Vasc. Biol.* 28, 968 (2008).

[32] M. Murphy and G. Kardon, *Curr. Top. Dev. Biol.* 96, 1 (2011).

[33] V. Mouly, A. Aamiri, A. Bigot, R. N. Cooper, S. Di Donna, D. Furling, T. Gidaro, V. Jacquemin, K. Mamchaoui, E. Negroni, S. Perie, V. Renault, S. D. Silva-Barbosa, and G. S. Butler-Browne, *Acta. Physiol. Scand.* 184, 3 (2005).

[34] S. Di Donna, V. Renault, C. Forestier, G. Piron-Hamelin, D. Thiesson, R. N. Cooper, E. Ponsot, S. Decary, R. Amouri, F. Hentati, G. S. Butler-Browne, and V. Mouly, *Neurol. Sci.* 21, S943 (2000).

[35] S. Decary, V. Mouly, C. B. Hamida, A. Sautet, J. P. Barbet, and G. S. Butler-Browne, *Hum. Gene Ther.* 8, 1429 (1997).

[36] V. Renault, L. E. Thornell, P. O. Eriksson, G. Butler-Browne, and V. Mouly, *Aging Cell,* 1, 132 (2002).

[37] M. S. O'Connor, M. E. Carlson and I. M. Conboy, *Biotechnol. Prog.* 25, 1130 (2009).

[39] J. L. Hagen, D. J. Krause, D. J. Baker, M. H. Fu, M. A. Tarnopolsky, and R. T. Hepple, *J. Gerontol. A. Biol. Sci. Med. Sci.* 59, 1099 (2004).

[40] F. S. Facchini, N. W. Hua, G. M. Reaven, and R. A. Stoohs, *Free Radic. Biol. Med.* 29, 1302 (2000).

[41] T. Hofer, E. Marzetti, J. Xu, A. Y. Seo, S. Gulec, M. D. Knutson, C. Leeuwenburgh, and E. E. Dupont-Versteegden, *Exp. Gerontol.* 43, 563 (2008).

[42] J. Wanagat, Z. Cao, P. Pathare and J. M. Aiken, *FASEB J.* 15, 322 (2001).

[43] L. Combaret, D. Dardevet, D. Bechet, D. Taillandier, L. Mosoni, and D. Attaix, *Curr. Opin. Clin. Nutr. Metab. Care* 12, 37 (2009).

[44] K. E. Conley, S. A. Jubrias, and P. C. Esselman, *J. Physiol.* 526 Pt 1, 203 (2000).

[45] G. A. Garinis, G. T. van der Horst, J. Vijg, and J. H. Hoeijmakers, *Nat. Cell Biol.* 10, 1241 (2008).

[46] S. Bortoli, V. Renault, E. Eveno, C. Auffray, G. Butler-Browne, and G. Pietu, *Gene,* 321, 145 (2003).

[47] I. Piec, A. Listrat, J. Alliot, C. Chambon, R. G. Taylor, and D. Bechet, *FASEB J.* 19, 1143 (2005).

[48] A. Scime, J. Desrosiers, F. Trensz, G. A. Palidwor, A. Z. Caron, M. A. Andrade-Navarro, and G. Grenier, *Mech. Ageing Dev.* 131, 9.

[49] B. M. Carlson and J. A. Faulkner, *Am. J. Physiol.* 256, C1262 (1989).

[50] M. D. Grounds, L. Sorokin and J. White, *Scand. J. Med. Sci. Sports* 15, 381 (2005).

[51] D. T. Woodley, C. N. Rao, J. R. Hassell, L. A. Liotta, G. R. Martin, and H. K. Kleinman, *Biochim. Biophys. Acta* 761, 278 (1983).

[52] A. Langsdorf, A. T. Do, M. Kusche-Gullberg, C. P. Emerson, Jr., and X. Ai, *Dev. Biol.* 311, 464 (2007).

[53] J. Villena and E. Brandan, *J. Cell Physiol.* 198, 169 (2004).

[54] J. Larrain, J. Alvarez, J. R. Hassell, and E. Brandan, *Exp. Cell Res.* 234, 405 (1997).

[55] R. Tatsumi and R. E. Allen, *Muscle Nerve,* 30, 654 (2004).

[56] S. Yamada, N. Buffinger, J. DiMario, and R. C. Strohman, *Med. Sci. Sports Exerc.* 21, S173 (1989).

[57] J. C. Casar, B. A. McKechnie, J. R. Fallon, M. F. Young, and E. Brandan, *Dev. Biol.* 268, 358 (2004).

[58] R. Droguett, C. Cabello-Verrugio, C. Riquelme, and E. Brandan, *Matrix Biol.* 25, 332 (2006).

[59] D. D. Cornelison, *J. Cell Biochem.* 105, 663 (2008).

[60] A. Scime, A. Z. Caron and G. Grenier, *Front. Biosci.* 14, 3012 (2009).

[61] R. D. Cohn, C. van Erp, J. P. Habashi, A. A. Soleimani, E. C. Klein, M. T. Lisi, M. Gamradt, C. M. ap Rhys, T. M. Holm, B. L. Loeys, F. Ramirez, D. P. Judge, C. W. Ward, and H. C. Dietz, *Nat. Med.* 13, 204 (2007).

[62] M. E. Carlson and I. M. Conboy, *Aging Cell,* 6, 371 (2007).

[63] W. Yan, S. George, U. Fotadar, N. Tyhovych, A. Kamer, M. J. Yost, R. L. Price, C. R. Haggart, J. W. Holmes, and L. Terracio, *Tissue Eng.* 13, 2781 (2007).

[64] S. Carnio, E. Serena, C. A. Rossi, P. De Coppi, N. Elvassore, and L. Vitiello, *J. Tissue Eng. Regen. Med.* 5, 1.

[65] E. Hill, T. Boontheekul and D. J. Mooney, *Tissue Eng.* 12, 1295 (2006).

[66] B. Blumenthal, P. Golsong, A. Poppe, C. Heilmann, C. Schlensak, F. Beyersdorf, and M. Siepe, *Artif. Organs* 34, E46.

[67] R. M. Lovering, N. C. Porter and R. J. Bloch, *Phys. Ther.* 85, 1372 (2005).

[68] M. A. Luz, M. J. Marques and H. Santo Neto, *Braz. J. Med. Biol. Res.* 35, 691 (2002).

[69] A. Sacco, F. Mourkioti, R. Tran, J. Choi, M. Llewellyn, P. Kraft, M. Shkreli, S. Delp, J. H. Pomerantz, S. E. Artandi, and H. M. Blau, *Cell,* 143, 1059 (2010).

[70] A. W. van den Beld and S. W. Lamberts, *Novartis. Found. Symp.* 242, 3 (2002).

[71] G. Goldspink, *Physiology (Bethesda)* 20, 232 (2005).

[72] M. D. Grounds, *Ann. N. Y. Acad. Sci.* 854, 78 (1998).

[73] S. Artavanis-Tsakonas, M. D. Rand and R. J. Lake, *Science,* 284, 770 (1999).

[74] I. M. Conboy and T. A. Rando, *Dev. Cell* 3, 397 (2002).

[75] S. Fukada, A. Uezumi, M. Ikemoto, S. Masuda, M. Segawa, N. Tanimura, H. Yamamoto, Y. Miyagoe-Suzuki, and S. Takeda, *Stem Cells,* 25, 2448 (2007).

[76] A. S. Brack, I. M. Conboy, M. J. Conboy, J. Shen, and T. A. Rando, *Cell Stem Cell,* 2, 50 (2008).

[77] A. M. Solomon and P. M. Bouloux, *J. Endocrinol.* 191, 349 (2006).

[78] M. E. Carlson, M. J. Conboy, M. Hsu, L. Barchas, J. Jeong, A. Agrawal, A. J. Mikels, S. Agrawal, D. V. Schaffer, and I. M. Conboy, *Aging Cell* 8, 676 (2009).

[79] A. C. McPherron, A. M. Lawler and S. J. Lee, *Nature,* 387, 83 (1997).

[80] A. C. McPherron and S. J. Lee, *Proc. Natl. Acad. Sci. US* 94, 12457 (1997).

[81] C. McFarlane, A. Hennebry, M. Thomas, E. Plummer, N. Ling, M. Sharma, and R. Kambadur, *Exp. Cell Res.* 314, 317 (2008).

[82] M. Manceau, J. Gros, K. Savage, V. Thome, A. McPherron, B. Paterson, and C. Marcelle, *Genes Dev.* 22, 668 (2008).

[83] K. E. Yarasheski, S. Bhasin, I. Sinha-Hikim, J. Pak-Loduca, and N. F. Gonzalez-Cadavid, *J. Nutr. Health Aging* 6, 343 (2002).

[84] M. E. Carlson, M. Hsu and I. M. Conboy, *Nature,* 454, 528 (2008).

[85] I. M. Conboy, H. Yousef and M. J. Conboy, *Aging (Albany NY)* 3, 555 (2011).

[86] R. Roubenoff, *Curr. Opin. Clin. Nutr. Metab. Care* 6, 295 (2003).

[87] S. J. Lees, K. A. Zwetsloot and F. W. Booth, *Aging Cell,* 8, 26 (2009).

[88] A. J. Dirks and C. Leeuwenburgh, *J. Nutr. Biochem.* 17, 501 (2006).

[89] A. S. Adler, S. Sinha, T. L. Kawahara, J. Y. Zhang, E. Segal, and H. Y. Chang, *Genes Dev.* 21, 3244 (2007).

[90] H. Li, S. Malhotra and A. Kumar, *J. Mol. Med. (Berl.)* 86, 1113 (2008).

[91] T. Phillips and C. Leeuwenburgh, *FASEB J.* 19, 668 (2005).

[92] N. Al-Shanti and C. E. Stewart, *Biol. Rev. Camb. Philos. Soc.* 84, 637 (2009).

[93] R. N. Michel, S. E. Dunn and E. R. Chin, *Proc. Nutr. Soc.* 63, 341 (2004).

[94] J. A. Faulkner, L. M. Larkin, D. R. Claflin, and S. V. Brooks, *Clin. Exp. Pharmacol. Physiol.* 34, 1091 (2007).

[95] A. A. Vandervoort, *Muscle Nerve,* 25, 17 (2002).

[96] P. S. Zammit, T. A. Partridge and Z. Yablonka-Reuveni, *J. Histochem. Cytochem.* 54, 1177 (2006).

[97] E. Schultz, *Anat. Rec.* 190, 299 (1978).

[98] M. R. Deschenes, M. A. Roby, M. K. Eason, and M. B. Harris, *Exp. Gerontol.* 45, 389.

[99] G. S. Ashcroft, S. J. Mills and J. J. Ashworth, *Biogerontology,* 3, 337 (2002).

[100] S. I. Zacks and M. F. Sheff, *Muscle Nerve,* 5, 152 (1982).

[101] T. Tsujinaka, J. Fujita, C. Ebisui, M. Yano, E. Kominami, K. Suzuki, K. Tanaka, A. Katsume, Y. Ohsugi, H. Shiozaki, and M. Monden, *J. Clin. Invest.* 97, 244 (1996).

Chapter 3

MITOCHONDRIAL RESPIRATORY CHAIN UNCOUPLING, OXIDATIVE STRESS AND SKELETAL MUSCLE ENERGETICS

*T.N. Tran[1,2], E. Noll[1,2], O. Collange [1,2], J. Bouitbir [1,3], A. L. Charles[1,3], M. Hengen[1,2], M. Kindo[14], J. Zoll [1,3], P. Diemunsch[1,2] and B. Geny[1, 3]**

[1] Equipe d'Accueil 3072, Institut de Physiologie, Faculté de Médecine, Université de Strasbourg, Cedex, France
[2] Pôle Anesthésie Réanimation Chirurgicale, SAMU, Hôpitaux Universitaires de Strasbourg, Strasbourg, Molière France
[3] Service de Physiologie et d'Explorations Fonctionnelles, Pôle de Pathologie Thoracique, Hôpitaux Universitaires, CHRU de Strasbourg, Cedex, France
[4] Service de Chirurgie Cardio-Vasculaire, Pôle d'Activité Médico-Chirurgicale Cardio-Vasculaire, Nouvel Hôpital Civil, Hôpitaux Universitaires de Strasbourg, Strasbourg, France

ABSTRACT

Skeletal muscles need ATP to generate mechanical work. This fuel is obtained by the mitochondrial coupling of the respiratory chain that creates a high electrochemical gradient of protons through the inner membrane and the ATP synthase complex, which uses this energy to phosphorylate ADP into ATP. Oxidative phosphorylation also generates reactive oxygen species (ROS), involved in the regulation of many physiological processes. However, whatever its origin, ROS overproduction leads to oxidative stress that can be a signal pathway to apoptosis and other cellular impairments, resulting into organs dysfunctions. The mitochondrial respiratory chain uncoupling is a phenomenon allowing the return of protons into the matrix without ATP production. This proton leak lowers the membrane potential across the inner membrane and increases the mitochondrial respiration rate. Major mitochondrial uncoupling appears deleterious and

---

* E-mail address: Bernard.GENY@chru-strasbourg.fr

increases ROS production. On the other hand, mild mitochondrial uncoupling could be an approach to promote lifespan extension since it is a highly effective intervention to prevent the formation of ROS. Mitochondrial coupling can be appreciated principally *in vitro* by measuring the acceptor control ratio (ratio of oxygen consumption in phosphorylating and non-phosphorylating states) that shows the efficiency of the mitochondrial coupling or by the simultaneous assessment of membrane potential and mitochondrial respiration. The proton leak is shown from the measurement of the membrane potential together with the respiration rate in non-phosphorylating state. Uncoupling proteins (UCPs) favor mitochondrial uncoupling. UCP1, physiologically observed in small hibernating animals is localized in brown adipose tissue and is involved in cold-induced thermogenesis. The role of the other UCPs, principally UCP2 (ubiquitous) and UCP3 (almost exclusively in skeletal muscle), is more controversial. As they are activated in extreme conditions (e.g. fasting, intensive exercise, high fat diet...) they could correspond to a protective mechanism against oxidative stress and its consequences. Finding therapeutic approaches aiming to induce mild mitochondrial uncoupling appears interesting. The overexpression of UCP3 in transgenic mice protects these animals against diabetes and obesity. It also increases life expectancy. Thus, knock out of the transcriptional intermediary factor 2 (TIF2) enhance overexpression of UCP3 and could be a promising pharmacological target. Further, in clinical practice, induction of mild uncoupling has already been obtained by using drugs such as isoflurane. It allowed organ protection through pharmacological preconditioning. Increasing knowledge on mild uncoupling in skeletal muscle appears therefore useful in order to decrease oxidative stress, thereby reducing one of the most common origins of metabolic dysfunctions involved in many pathologies.

INTRODUCTION

To support and move the skeleton, skeletal muscles need energy, which is stored in adenosine triphosphate (ATP) molecules. ATP is generated in the mitochondria by the process of oxidative phosphorylation. Mitochondria are ubiquitous in the cells cytosol with high concentrations in muscle cells because of the high energetic requirements of muscles.

Structure of Mitochondria

Mitochondria are membrane-enclosed organelles. They consist of two membranes: the outer and the inner membranes. The outer membrane is a relatively simple phospholipid bilayer that is permeable to nutrient molecules, ions, ATP and ADP molecules thanks to the porins. Indeed, those protein structures allow the passage of molecules of 10 kDa in weight. The inner membrane ordered in folds is electively permeable to oxygen, carbon dioxide and water and contains the electron transport chain, ATP synthetase, and transport proteins. The folds are organized into layers called cristae, which increase total surface area of the inner membrane. The space between the outer and inner membranes is called the intermembrane space. The inner membrane enclosed the mitochondrial matrix. Those two spaces have an important role in oxidative phosphorylation.

Functions of Mitochondria

The major mitochondrial function is the energy conversion from cellular respiration. The other main functions are control of apoptosis-programmed cell death, mitochondrial membrane potential preservation, cell differentiation, growth and development, calcium ions storage and calcium signaling, steroid synthesis, and reactive oxygen species production and clearance. In this chapter, we are going to approach only the energy conversion and the ROS production.

COUPLING OF MITOCHONDRIAL RESPIRATORY CHAIN

The process of aerobic cellular respiration starts in the cytosol and finishes in the mitochondrial inner membrane. It can be summarized in a global equation:

$$C_6H_{12}O_6 + 6O_2 + \sim36\ ADP + \sim36\ Pi \rightarrow 6CO_2 + 6H_2O + \sim36\ ATP + heat$$

Energy conversion is a chemical process of releasing energy stored in glucose.

The first step is the glycolysis in the cytoplasm. The glycolytic pathway is regulated by many cytoplasmic enzymes and ends in the transformation of a glucose molecule into two molecules of pyruvate. Products of glycolysis are two ATP molecules, two pyruvate molecules and two NADH (nicotinamide adenine dinucleotide) molecules.

$$Glucose + 2\ NAD^+ + 2\ ADP + 2\ Pi \rightarrow 2\ pyruvate + 2\ NADH + 2\ H^+ + 2\ ATP + 2\ H2O.$$

Then, the aerobic respiration takes place into the mitochondria. It is a complex mechanism using the oxygen as the final electron acceptor of respiration. First, the pyruvate decarboxylation converts pyruvate to acetyl-CoA and CO2. Second, the acetyl-CoA is used in the citric acid cycle (Krebs Cycle) where it is fully oxidized to carbon dioxide and water, producing yet more NADH but also FADH2 (reduced flavin adenine dinucleotide), which store the released energy. Third, NADH and FADH2 are oxidized to NAD^+ to FAD by the electron transport chain (ETC). The ETC is consists of many complexes across the mitochondrial inner membrane. Three of them are protons pumps: complexes I, III and IV. Indeed, the redox reactions through the complexes I and III are coupled to the extrusion of protons from the matrix. This process generates a "hydrogen ion gradient" called proton motive force (pmf) across the inner membrane of the mitochondria. At last, the pmf is used to produce a large amount of ATP in a process called oxidative phosphorylation. Indeed, the ATP synthase complex uses energy of the high electrochemical gradient of protons through the inner membrane to phosphorylate ADP into ATP and can changes the pmf depending on its phosphorylation activity.

The pmf has two components, ΔpH (the pH gradient across the mitochondrial inner membrane) and $\Delta\psi m$ (mitochondrial membrane potential, the difference in electrical potential between the cytoplasm and the mitochondrial matrix). The pmf can be appreciated by the equation:

$$Pmf = \Delta\psi m - 61{,}5\log_{10}\Delta pH$$

Two Nobel Prizes in Chemistry were awarded for works on the energy conversion within the mitochondria. The first one was awarded in 1978 to Peter Mitchell for having described the energy transfer [1], the second one in 1997 to Paul D. Boyer and John E. Walker for elucidation of the enzymatic mechanism underlying the synthesis of ATP [2, 3].

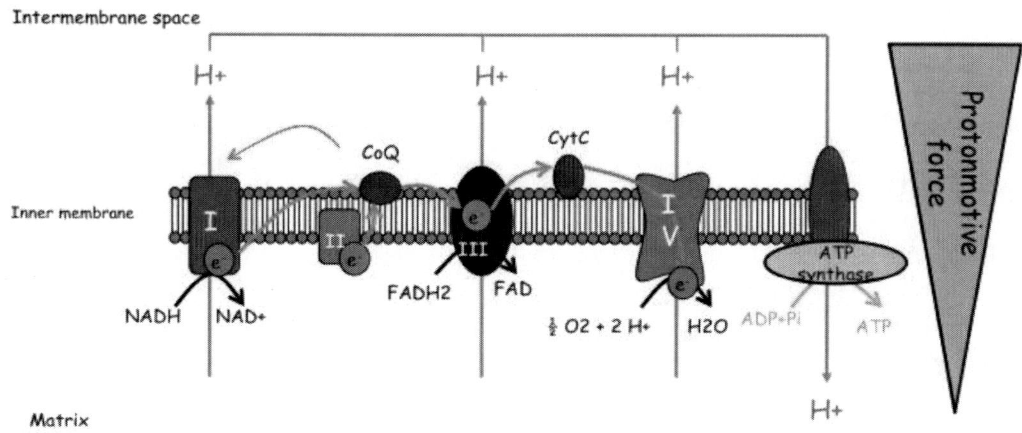

Figure 1. The electron transport chain across the mitochondrial inner membrane.

MITOCHONDRIAL ROS PRODUCTION

It has been shown that under physiological conditions some electrons in the mitochondrial electron transport chain do not form a water molecule with the oxygen as their final electron acceptor but produce radical oxygen species (ROS), especially superoxide and hydrogen peroxide [4,5]. Mitochondria have many antioxidants skills to scavenge their own ROS production: free radicals can be neutralized by mitochondrial enzymatic systems (superoxide dismutase, catalase and glutathione peroxidase), trace elements (copper, zinc, iron, selenium) or vitamins A, C and E. The few ROS released from mitochondria may explain the different lifespan between various species with the same metabolic rates [6] exposing the free radical hypothesis of aging. ROS are involved in the regulation of many physiological processes and may participate in many signaling and regulation pathways. However, under various cell stresses, such as ischemia–reperfusion [7, 8], hypoxia–reoxygenation [9], diabetes mellitus [10], lipotoxicity [11 ,12], or treatment with toxic agents [13], mitochondrial ROS can be overproduced, exceeding the antioxidant capacities. Subsequently, they can be rapidly released into cytoplasm, where they may have damaging effects leading to oxidative stress and cell injury. This oxidative stress contributes to a broad spectrum of diseases and pathological conditions.

MITOCHONDRIAL MILD UNCOUPLING

What is a Mild Uncoupling?

Mitochondrial membrane potential and ROS production are correlated [14]. Small increases in the membrane potential induce ROS formation, whereas slight decreases can substantially diminish the production of these reactive intermediates. A mild uncoupling slightly decreases the membrane potential that can curtail ROS production without greatly diminishing the efficiency of oxidative phosphorylation [15]. More precisely, mitochondrial uncoupling allows the return of protons from the intermembrane space into the matrix without ATP production. First observed by Nicholls in 1974 [16], this proton leak lowers the membrane potential across the inner membrane and increases the mitochondrial respiration rate [17]. This leak could be basal, depending of the composition of fatty acids of the inner membrane [18] or could be induced by proteins localized through the inner membrane: the uncoupling proteins. Discovered by Nicholls in 1978, the first uncoupling protein UCP1, also called thermogenin is physiologically observed in small hibernating animals [19, 20]. It is localized in brown adipose tissue and involved in cold-induced thermogenesis. The role of the other UCPs, principally UCP2 (ubiquitous) and UCP3 (almost exclusively in skeletal muscle) is more controversial. These are activated in extreme conditions (e.g. fasting, intensive exercise and high fat diet), so they could be a protective mechanism against oxidative stress and their consequences [21]. However, their activation also impairs whole-body efficiency by a decrease of the membrane potential and an increase of the mitochondrial rate. The outcome of this reaction is the decline of ROS production according to the work of Korshunov [14]. This mechanism can be a key of longevity. Indeed, mitochondrial uncoupling could be an approach to promote lifespan extension due to its ability to prevent the formation of ROS [22, 23].

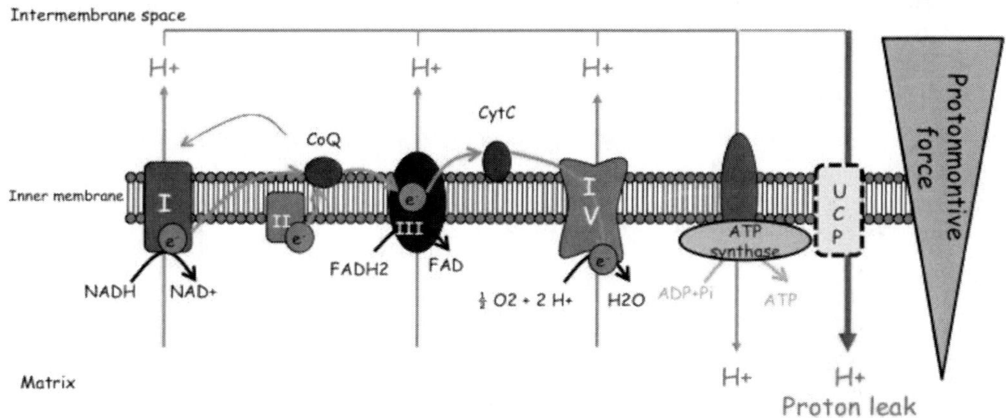

Figure 2. The electron transport chain and the proton leak through the uncoupling protein.

MEASUREMENTS OF MITOCHONDRIAL COUPLING/UNCOUPLING

Mitochondrial coupling can be appreciated principally *in vitro* by measuring the acceptor control ratio (ratio of oxygen consumption in phosphorylating and non-phosphorylating states) that shows the efficiency of the mitochondrial coupling or by simultaneous assessment of the membrane potential and mitochondrial respiration. The proton leak is shown from the measurement of the membrane potential together with the respiration rate in non-phosphorylating state.

Oxygraph Recordings

An oxygraph apparatus can measure oxygen uptake of mitochondria depending of stimulation of the different complexes of the electron transport chain [24,25]. After preparation, isolated mitochondria are incubated in an oxygen electrode in an isotonic medium containing substrate and phosphate. The addition of ADP causes an activation of ATP synthase complex, which converts ADP into ATP. This activation called "state 3" can be appreciated by the oxygen consumption measured by the electrode. "State 3" is then compared to "state 4" which is the mitochondrial oxygen consumption measured after all the ADP has been phosphorylated. The ratio [state 3 rate] : [state 4 rate] is called the respiratory control ratio and appreciates the efficiency of the mitochondrial coupling. When permeabilized fibers are used instead of isolated mitochondria, endogenous ATPases prevent the establishment of state 4. The "state 2" is then used. This state is the basal oxygen consumption under glutamate and malate substrates. The ratio [state 3 rate] : [state 2 rate] is called the acceptor control ratio (ACR).

Simultaneous Assessment of Membrane Potential and Mitochondrial Respiration

TPP^+ (tetraphenylphosphonium ion) or $TPMP^+$ (triphenyl-methyl-phosphonium ion) probes can be used to measure $\Delta\psi m$. The mitochondrial uptaking of those lipophilic membrane-permanent cations (C^+) is depending on the pmf and can be monitored by an external macro-electrode [26,27]. $\Delta\psi m$ is then measured applying the Nernst equation:

$$\Delta\psi m = 61.5 \; \log_{10} [C^+_i] \, / \, [C^+_e]$$

where 61,5 is calculated according to the universal gas constant R, the absolute temperature T, the Faraday constant F and the valency z of the lipid-soluble ions (z = 1 for tetraphenylphosphonium cation, TPP+), and C^+I is the concentration of the lipid-soluble ions in the mitochondrial matrix, and C^+e the concentration in the medium. $\Delta\psi m$ is less sensitive than respiration rates to show mitochondrial dysfunctions but their simultaneous assessment allows bringing to light mitochondrial uncoupling. Indeed, respiration rates can be artificially underestimated because of an artefactual decrease of $\Delta\psi m$ due to mitochondria isolation. Some apparatus can determine $\Delta\psi m$ in parallel with respiration, usually with oxygen and

TPMP$^+$ electrodes inserted into the same incubation. This technique appreciates respiration rates depending on $\Delta\psi$m and then shows a proton leak, which is an increased respiration rate to maintain a normal membrane potential. Respiration rate or membrane potential can be modulated by addition of substrates or complexes inhibitors, i.e. the addition of UCP inhibitors as GDP recouples the mitochondrial respiration [20].

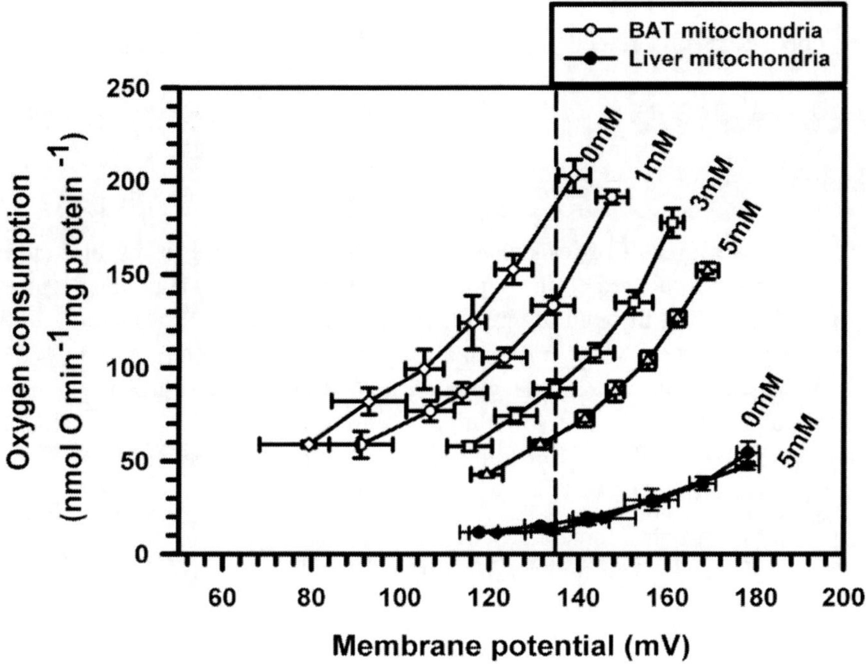

© 2007 the American Physiological Society.

Figure 3. Dependence of proton leak rate (measured as the respiration rate driving proton leak) in the presence of oligomycin on membrane potential of isolated brown adipose tissue BAT (○) and liver (●) mitochondria of cold-acclimated E. myurus in the presence of 0, 1, 3, and 5 mM GDP. Values are ± SE from five independent preparations. The broken line indicates the highest common potential (135.3 mV). From Mzilikazi et al. [20].

IS THE MITOCHONDRIA UNCOUPLING BENEFICIAL OR DELETERIOUS?

Mitochondrial membrane potential and ROS production are correlated [14]. Small increases in the membrane potential induces ROS formation, whereas slight decreases can substantially diminish the production of these reactive intermediates. A mild uncoupling slightly decreases the membrane potential that can curtail ROS production without greatly diminishing the efficiency of oxidative phosphorylation [15]. A complete uncoupling between the respiratory chain and the production of ATP is lethal. The aim of oxidative phosphorylation is indeed to produce ATP that is the energy-rich intermediate used by cells to power a large repertoire of reactions and that is absolutely necessary for life [28]. For example, the chemical mitochondrial uncoupler 2,4-dinitrophenol (DNP) which had initially shown potential benefits on energy balance has been proven to be lethal [29]. On the other hand, mild mitochondrial uncoupling is probably useful for living organism, allowing for example cells to switch off rapidly for high-efficiency mode of ATP production that does not require huge increases in oxygen demand [15]. We actually do not know how much uncoupling can be beneficial or deleterious at the scale of the cell or at the entire organism. Thus, mitochondrial uncoupling is associated with some pathogenesis like Type 2 diabetes and, on the opposite, with protective processes limiting the deleterious effect of oxidative stress.

Type 2 Diabetes

Type 2 diabetes is a result of resistance to the actions of insulin in muscle, fat and liver tissues, and defective insulin secretion by pancreatic beta cells. The beta cells sense glucose through its metabolism and through the resultant increase in the ATP/ADP ratio and the rise in this ratio is considered to be an important coupling signal between glucose sensing and insulin secretion. UCP2 decreases the generation of ATP from glucose metabolism in pancreatic islets and impairs glucose-stimulated insulin secretion (GSIS) [30]. Thus, UCP2 is considered as a negative regulator of insulin secretion and as a link between obesity, beta-cell dysfunction and Type 2 diabetes [28].

Thermogenesis

In addition to the brown adipose tissue, skeletal muscle has been shown to contribute to adaptive thermogenesis. In this view, mitochondrial uncoupling is postulated to be a hypothetical target to elevate the whole-body energy expenditure [31]. UCP3 is probably involved in hyperthermia induced with MDMA (3,4-methylenedioxymethamphetamine, "ecstasy") and finally, mice without UCP3 are protected to this condition.

Mild Uncoupling and Oxidative Stress

Reactive oxygen species (ROS) play dual roles within aerobic cells, serving either as signaling molecules in a wide range of normal cellular processes or as oxidizing agents that can potentially trigger cell death [15]. Mitochondrial ROS overproduction is thought to be associated with many neurological degenerative diseases (Alzheimer and Parkinson diseases), obesity, diabetes, cardiovascular disease, aging and inflammation (ischemia reperfusion, sepsis). To capitalize the signaling effect of ROS (0.2-2% of O_2 consumption [32]) and to avoid their toxic effects, aerobic cells have to keep these singlet electron carriers within tolerable limits. To this end, the antioxidative enzymes and low-molecular-weight compounds overpower the excess ROS molecules generated by aerobic respiration. Some antioxidative systems can be considered unproductive, as they are energetically "expensive". They are indeed reliant on both ATP and NADPH, produced

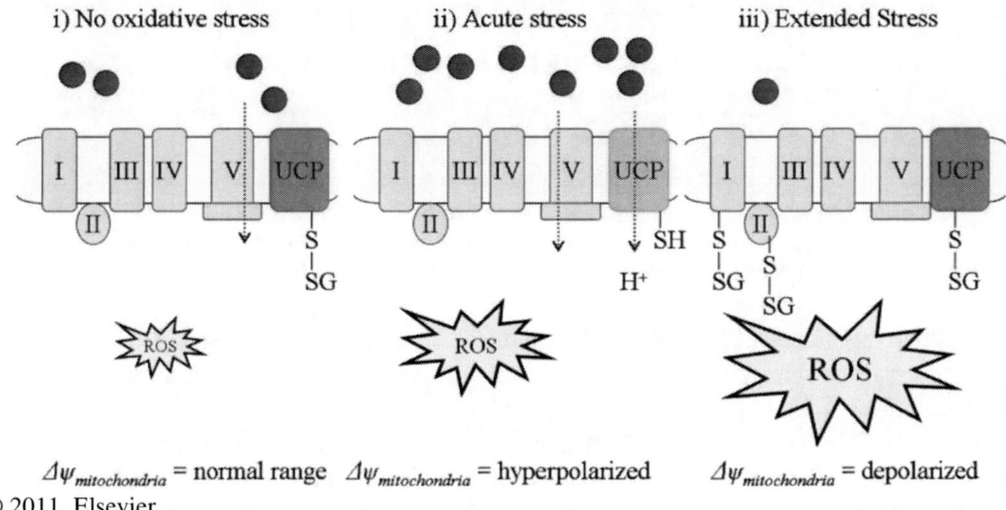

Figure 4. Hypothetical scheme depicting the various levels of the regulation of mitochondrial ROS emission. Filled red circles represent protons. (i) Under normal conditions, mitochondrial electron transfer through the respiratory complexes to the terminal electron acceptor O_2 is efficient (with minimal ROS production from the respiratory complexes), thus maintaining Δψm and limiting ROS emission from the electron transport chain. UCP are inactivated. (ii) Slight increases in Δψm and mitochondrial ROS emission lead to the activation of UCP and to the induction of proton leak. The respiration increase restores Δψm and lowers mitochondrial ROS emission. (iii) A major uncoupling inhibits complex V, decreases Δψm and increases ROS production. From Mailloux and Harper [15].

during mitochondrial respiration, which is the major source of ROS in the aerobic cell [33]. To this end, "mild" uncoupling of the mitochondrial oxidative phosphorylation system represents an acute, yet energetically costly mechanism to decrease ROS production [15]. "Short circuiting" the mitochondrial oxidative phosphorylation, the mild uncoupling may represent the first line of defense against oxidative stress. Mitochondrial membrane potential and ROS production are correlated. Small increases in the membrane potential induce ROS formation, whereas slight decreases can substantially diminish the production of these reactive intermediates. A mild uncoupling slightly decreases the membrane potential that can curtail ROS production without greatly diminishing the efficiency of oxidative phosphorylation [15]. In others terms, a mild uncoupling would reduce the efficiency of energy conversion and the ROS production without compromising intracellular high-energy phosphate levels [34].

ISCHEMIA REPERFUSION

Mitochondria uncoupling has been shown to be either protective or deleterious during ischemia reperfusion [35]. Mice without UCP2 have less tissue injury after cerebral ischemia. However, artificial overexpression of UCP2 in the brain is associated with decreased cell death in cultured neurons [36]. The timing and the degree of mitochondrial uncoupling are crucial to target therapeutic effects during ischemia-reperfusion.

THERAPEUTIC APPROACHES AIMING TO INDUCE MILD MITOCHONDRIAL UNCOUPLING

Local Anesthetics & Mitochondrial Uncoupling

Papa et al. first described the amines local anesthetics (LA) uncoupling potency in 1972 [37]. Dibucaine exerted on bovine-heart isolated mitochondria an increase in proton influx from the inner membrane space to the mitochondria matrix. This caused a proton turnover leading to an increased energetic dissipation and a decreased oxidative phosphorylation. Amines local anesthetics are classified into ester or amide with respect to the linkage compound between the substituted aromatic ring and the tertiary amine. The aromatic ring gives the molecule hydrophobic properties and the tertiary amine hydrophilic properties. Garlid and Nakashima showed that hydrophobic amines uncouple mitochondria in presence of anions through an electro-neutral anion uptake or an electrophoretic proton transport [38]. The lipophilic LA uncoupling was also described by Dabadie et al. and involved a protonophoric pathway with bupivacaine (a worldwide used LA in human clinical setting) [39]. On opposite, ropivacaine, another clinical worldwide used LA, was found to be probably have a less uncoupling effect on mitochondria due to his reduced lipophilicity [40]. The protonophoric mechanism for uncoupling the mitochondria seems to be more a channel mechanism than a mobile carrier mechanism [41]. LA also uncouples via direct perturbation of the proton pumping complex [42]. This uncoupling without modification of the membrane proton permeability is called « decoupling ».

Non Steroidal Anti-Inflammatory Drugs and Mitochondrial Uncoupling

Mitochondrial effects of Piroxicam have been studied focused on uncoupling properties. On isolated rat liver mitochondria, piroxicam has a dose-dependent decrease on the respiratory control ratio [43]. There is probably a protonophoric mechanism underlying this observation. The effect was observed with a concentration higher than the therapeutic use of this oxicam group molecule.

Volatile Anesthetics & Mitochondrial Uncoupling.

A significant part of the data concerning volatile anesthetics consequences upon the mitochondrial function comes from studies focused on the anesthetic preconditioning mechanism. On isolated rat ventricular cardiomyocytes, sevoflurane and desflurane increased flavoproteins oxidation as a compensatory mechanism for mitochondrial uncoupling [44]. The same uncoupling effect was observed with isoflurane [45]. Hanley et al. observed that at a concentration of 2 MAC (minimal alveolar concentration), the activity of complex I was reduced by about 20% in the presence of halothane and isoflurane and by about 10% in the presence of sevoflurane [46]. The attenuation of complex I by isoflurane decreases ROS release and protects myocardium during reperfusion [47]. The mitochondrial uncoupling was kept constant after the volatile anesthetic removal, suggesting a "memory" effect upon mitochondrial physiology after exposure. This memory effect is consistent with the time course of the anesthetic preconditioning protocols in use in experimental settings [48]. Mitochondrial K-ATP channels are probably involved in the isoflurane mediated uncoupling but this is still matter of debate [49].

Propofol and Mitochondrial Uncoupling

2,6-Diisopropylphenol (or propofol) modifies the inner membrane proton permeability resulting in a decrease of the transmembrane potential but without consequences on the coupling of the electron flux between the respiratory chain and the ATP synthetase [50]. Advances in mitochondrial pharmacology of 2,6-diisopropylphenol showed that it has a mild uncoupling effect in non-phosphorylating mitochondria via a proton leak through the proton channel of the ATPase [51]. This phenomenon does not persist under sustained ATPase activity. However, large doses of propofol have been associated with the 'propofol infusion syndrome' (PRIS), which is a metabolic disorder with a fatal outcome. An acute refractory bradycardia leading to asystole is associated with a metabolic acidosis, rhabdomyolysis, hyperlipidaemia, and enlarged or fatty liver. According to many authors, PRIS could be the result of a severe mitochondrial uncoupling by inhibiting the electron transport chain (complexes II and IV, cytochrome C) leading to a failure of ATP production and to metabolic acidosis [52].

Experimental Transgenic Uncoupling

Progresses in molecular biology have led to the development of transgenic mice with high or low expression of mitochondrial uncoupling protein (UCP). For example, over expression of UCP1 in mice muscle and heart lead to a phenotype with low body weight, high food intake and energy expenditure [53] as a consequence of mitochondrial uncoupling. Experimental transgenic uncoupling has other consequences upon mitochondrial function: UCP1 overexpression is protective upon heart ischemia-reperfusion [54]. After the initiation of reperfusion, the respiratory chain restarts immediately but without a simultaneous restart of ATP intake via muscular contraction, leading to high membrane potential and ROS production. UCP1 induced proton leak in this situation is probably the explanation for the protective effect observed with this experimental uncoupling. Experiments on UCP2 and UCP3 are more controversial. If overexpression of UCP3 in mice lead to increased basal mitochondrial respiration and diminishes the age-induced increase in ROS production under activation conditions [55], the same research group could demonstrate any beneficial effect, nor uncoupled respiration, nor ROS production, nor acute lipotoxicity and nor oxidative damage in isolated skeletal muscle mitochondria from UCP3-ablated mice [56]. Nevertheless, UCP3 seems to have an important role in exercise-induced oxidative stress. Several studies have shown its upregulation during prolonged exercise to lower ROS production [57,58]. Furthermore, a recent study has shown that mice, in which the transcriptional regulator TIF2 was selectively ablated in skeletal muscle myofibers at adulthood, present a mitochondrial uncoupling in skeletal muscle myocytes by overexpression of UCP3. This uncoupling protected them from decreased muscle oxidative capacities induced by sedentariness, delayed the development of Type 2 diabetes, and attenuated high-caloric-diet-induced obesity. Thus, modulation of the expression of this co-regulator represents an indirect target to interact on UCP function [59]. Experimental transgenic uncoupling seems to be a promising way of understanding the consequences of mitochondrial physiopathology.

CONCLUSION

Since its first description in 1974, the mitochondrial uncoupling is the object of many studies. The beneficial effect of "mild uncoupling" is very interesting for therapeutics of many mitochondrial diseases. Moreover, many drugs used routinely in anesthesia are able to induce mitochondrial uncoupling. But the mechanisms to control mitochondrial uncoupling are still unresolved. Despite the identification of UCP2 and UCP3 in 1997, their role is still controversial because of conflicting results about their regulation, their activation, and their effects. Furthermore, we actually do not know how much uncoupling can be beneficial or deleterious. By the way, the Mitochondrial Free Radical Theory of Aging (MFRTA) of Harman [60] is more and more criticized [61,62]. But the endogenous upregulation of UCP in oxidative stress situations provides major evidence for the implication of mild uncoupling in ROS regulation and is a promising way to understand mitochondrial physiopathology, even if more studies are needed.

REFERENCES

[1]   P. Mitchell, *Nature*, 191, 144 (1961).
[2]   P. D. Boyer, W. E. Kohlbrenner, D. B. McIntosh, L. T. Smith and C. C. O'Neal. *Ann. N. Y. Acad. Sci.* 402, 65 (1982).
[3]   J. P. Abrahams, A. G. W. Leslie, R. Lutter and J. E. Walker, *Nature*, 370(6491), 621 (1994).
[4]   J. St-Pierre, J. A. Buckingham, S. J. Roebuck and M. D. Brand, *J. Biol. Chem.* 277(47), 44784 (2002).
[5]   R. G. Hansford, B. A. Hogue and V. Mildaziene, *J. Bioenerg. Biomembr.* 29(1), 89 (1997).
[6]   R. S. Sohal, I. Svensson, B. H. Sohal and U. T. Brunk, *Mech Ageing Dev.* 49(2), 129 (1989).
[7]   A. L. Charles, A. S. Guilbert, J. Bouitbir, P. Goette-Di Marco, I. Enache, J. Zoll, F. Piquard and B. Geny, *Brit. J. Surg.* 98(4), 511 (2011).
[8]   I.I. Pipinos, V. G. Sharov, A. D. Shepard, P. V. Anagnostopoulos, A. Katsamouris, A. Todor A, K.A. Filis and H. N. Sabbah, *J. Vasc. Surg.* 38(4), 827 (2003).
[9]   L. Zuo L and T. L. Clanton, *Am. J. Physiol. Cell Physiol.* 289(1), C207 (2005).
[10]   A. Bravard, C. Bonnard, A. Durand, M. A. Chauvin, R. Favier, H. Vidal and J. Rieusset, *Am. J. Physiol. Endocrinol. Metab.* 300(3), E581 (2011).
[11]   J. Hoeks J, M. K. C. Hesselink and P. Schrauwen, *Exp. Gerontol.* 41(7), 658 (2006).
[12]   L. V. Yuzefovych, V. A. Solodushko, G. L. Wilson and L. I. Rachek, *Endocrinology* 153(1), 92 (2012).

[13] F. M. Cerqueira, F. R. M. Laurindo and A. J. Kowaltowsk, *PLoS ONE* 6(3), e18433 (2011).

[14] S. S. Korshunov, V. P. Skulachev and A. A. Starkov, *FEBS Lett.* 416(1), 15 (1997).

[15] R. J. Mailloux and M. E. Harper, *Free Radic. Biol. Med.* 51(6), 1106 (2011).

[16] D. G. Nicholls, *Eur. J. Biochem.* 50(1), 305 (1974).

[17] M. D. Brand, *Biochim. Biophys. Acta.* 1018(2-3), 128 (1990).

[18] E. M. Fontaine, M. Moussa, A. Devin, J. Garcia, J. Ghisolfi, M. Rigoulet and X. M. Leverve, *Biochim. Biophys. Acta.* 1276(3), 181 (1996).

[19] D. G. Nicholls, V. S. Bernson and G. M. Heaton, *Experientia Suppl.* 32, 89 (1978).

[20] N. Mzilikazi, M. Jastroch, C. W. Meyer and M. Klingenspor, *Am. J. Physiol - Regul. Integr. Comp. Physiol.* 293(5), R2120 (2007).

[21] T. C. Esteves and M. D. Brand, *Biochim. Biophys. Acta* 1709(1), 35 (2005).

[22] D. Dikov, A. Aulbach, B. Muster, S. Dröse, M. Jendrach, J. Bereiter-Hahn, *Exp. Gerontol.* 45(7-8), 586 (2010).

[23] M. D. Brand, *Exp. Gerontol.* 35(6-7), 811 (2000).

[24] E. Gnaiger, *Respir. Physiol.* 128(3), 277 (2001).

[25] A. V. Kuznetsov, V. Veksler, F. N. Gellerich, V. Saks, R. Margreiter and W. S. Kunz, *Nat Protoc.* 3(6), 965 (2008).

[26] N. Kamo, M. Muratsugu, R. Hongoh and Y. Kobatake, *J. Membr. Biol.* 49(2), 105 (1979).

[27] M. D. Brand, Measurement of mitochondrial proton motive force. In: Bioenergetics: a practical approach. G. C. Brown and C. E. Cooper, 1996 pp39–62.

[28] S. Krauss, C. Y Zhang, B. B. Lowell, *Nat. Rev. Mol. Cell Biol.* 6(3), 248 (2005).

[29] J. Bartlett, M. Brunner and K. Gough, *Emerg. Med J.* 27(2), 159 (2010).

[30] C. Y Zhang, G. Baffy, P. Perret, S. Krauss, O. Peroni, D. Grujic, T. Hagen, A. J. Vidal-Puig, O. Boss, Y. B. Kim, X. X. Zheng. M. B. Wheeler, G. I. Shulman, C. B. Chan and B. B. Lowell, *Cell* 105(6), 745 (2001).

[31] S. A. van den Berg, W. van Marken Lichtenbelt, K. Willems van Dijk and P. Schrauwen, *Curr. Opin. Clin. Nutr. Metab. Care* 14(3), 243 (2011).

[32] E. Cadenas and K. J. A. Davies, *Free Radic. Biol. Med.* 29(3–4), 222 (2000).

[33] J. F. Turrens. *J. Physiol.* 552(2), 335 (2003).

[34] F. M. Cunha, C. C. Caldeira da Silva, F. M. Cerqueira and A. J. Kowaltowski, *Curr. Drug Targets* 12(6), 783 (2011).

[35] B. Cannon, I. G. Shabalina, T. V. Kramarova, N. Petrovic and J. Nedergaard, *Biochim Biophys Acta* 1757(5–6), 449 (2006).

[36] G. Mattiasson, M. Shamloo, G. Gido, K. Mathi, G. Tomasevic, S. Yi, C. H. Warden, R. F. Castilho, T. Melcher, M. Gonzalez-Zulueta, K. Nikolich and T. Wieloch, *Nat. Med.* 9(8), 1062 (2003).

[37] S. Papa, F. Guerrieri, S. Simone and M. Lorusso, *J. Bioenerg.* 3(6), 553 (1972).

[38] K. D. Garlid and R. A. Nakashima, *J. Biol. Chem.* 258(13), 7974 (1983).

[39] P. Dabadie, P. Bendriss, P. Erny and J-P Mazat, *FEBS Lett.* 226(1), 77 (1987).

[40] A. Floridi, M. D. Padova, R. Barbieri and E. Arcuri, *Biochem Pharmacol.* 58(6), 1009 (1999).

[41] X. Sun and K. D. Garlid. *J. Biol. Chem.* 267(27), 19147 (1992).

[42] K. van Dam, Y. Shinohara, A. Unami, K. Yoshida and H. Terada, *FEBS Lett.* 277(1–2), 131 (1990).

[43] C. L. Salgueiro-Pagadigorria, J. Constantin, A. Bracht, E. A. Nascimentoand and E. L. Ishii-Iwamoto. *Comp. Biochem. Physiol. C Pharmacol. Toxicol. Endocrinol.* 113(1), 93 (1996).

[44] F. Sedlic, D. Pravdic, M. Ljubkovic, J. Marinovic, A. Stadnicka and Z.J. Bosnjak, *Anesth Analg.* 109(2), 405 (2009).

[45] M. Ljubkovic, Y. Mio, J. Marinovic, A. Stadnicka, D. C. Warltier, Z. J. Bosnjak and M. Bienengraeber, *Am. J. Physiol Cell Physiol.* 292(5), C1583 (2007).

[46] P. J. Hanley, J. Ray, U. Brandt and J. Daut, *J. Physiol.* 544(Pt 3), 687 (2002).

[47] N. Hirata, Y. H. Shim, D. Pravdic, N. L. Lohr, P.F. Pratt, D. Weihrauch, J. R. Kersten, D. C. Warltier, Z. J. Bosnjak and M. Bienengraeber, *Anesthesiology* 115, 531 (2011).

[48] C. J. McLeod, A. P. Jeyabalan, J. O. Minners, R. Clevenger, R. F. Hoyt and M. N. Sack, *Circulation* 110(5), 534 (2004).

[49] M. Murata, M. Akao, B. O'Rourke and E. Marbán, *Circ. Res.* 89(10), 891 (2001).

[50] D. Branca, M. S. Roberti, P. Lorenzin, E. Vincenti and G. Scutari, *Biochem. Pharmacol.* 42(1), 87 (1991).

[51] M. Rigoulet, A. Devin, N. Avéret, B. Vandais and B. Guérin, *Eur. J. Biochem.* 241(1), 280 (1996).

[52] P. C. A. Kam and D. Cardone, *Anaesthesia* 62(7), 690 (2007).

[53] E. Couplan, C. Gelly, M. Goubern, C. Fleury, B. Quesson, M. Silberberg, E. Thiaudiere, P. Mateo, M. Lonchampt, N. Levens, C. De Montrion, S. Ortmann, S. Klaus, M. D. Gonzalez-Barroso, A. M. Cassard-Doulcier, D. Ricquier, A. X. Bigard, P. Diolez and F. Bouillaud, *J. Biol. Chem.* 277(45), 43079 (2002).

[54] J. Hoerter, M. D. Gonzalez-Barroso, E. Couplan, P. Mateo, C. Gelly, A. M. Cassard-Doulcier, P Diolez and F. Bouillaud, *Circulation* 110(5), 528 (2004).

[55] M. Nabben, J. Hoeks, J. J. Briedé, J. F. C. Glatz, E. Moonen-Kornips, M. K. C. Hesselink and P. Schrauwen, *FEBS Lett.* 582(30), 4147 (2008).

[56] M. Nabben, I. G. Shabalina, E. Moonen-Kornips, D. van Beurden, B. Cannon, P. Schrauwen, J. Nedergaard and J. Hoeks, *Biochim. Biophys. Acta* 1807(9), 1095 (2011).

[57] N. Jiang, G. Zhang, H. Bo, J. Qu, G. Ma, D. Cao, L. Wen, S. Liu, L. L. Ji and Y. Zhang, *Free Radic. Biol.Med.* 46(2), 138 (2009).

[58] E. J. Anderson, H. Yamazaki and P. D. Neufer, *J. Biol. Chem.* 282(43), 31257 (2007).

[59] D. Duteil, C. Chambon, F. Ali, R. Malivindi, J. Zoll, S. Kato, B. Geny, P. Chambon and D. Metzger, *Cell Metab.* 12(5), 496 (2010).

[60] D. Harman, *J Gerontol.* 11(3), 298 (1956).

[61] A. Sanz and R. K. A. Stefanatos, *Curr. Aging Sci.* 1(1), 10 (2008).

[62] M. Ristow and S. Schmeisser, *Free Radic. Biol.Med.* 51(2), 327 (2011).

In: Skeletal Muscle
Editor: Mark Willems

ISBN: 978-1-62417-271-7
© 2013 Nova Science Publishers, Inc.

Chapter 4

THE EMERGING ROLE OF MITOCHONDRIA IN INFLAMMATORY MYOPATHIES

Alain Meyer[a,b], Joffrey Zoll[a,b], Anne Laure Charles[a],*
François Piquard[a,b], Béatrice Lannes[c],
Jacques-Eric Gottenberg[d], Jean Sibilia[d] and Bernard Geny[a,b]

[a]Equipe d'Accueil 3072, Institut de Physiologie,
Faculté de Médecine, Université de Strasbourg, Cedex, France.
[b]Service de Physiologie et d'Explorations Fonctionnelles, Pôle de Pathologie Thoracique,
Hôpitaux Universitaires, CHRU de Strasbourg, Cedex, France
[c]Département de pathologie, Hôpitaux Universitaires de Strasbourg, Strasbourg, France
[d]Service de Rhumatologie, Pôle de Médecine Interne-Rhumatologie-Nutrition-
Endocrino-Diabétologie Hôpitaux Universitaires de Strasbourg,
Strasbourg, France

ABSTRACT

Inflammatory myopathies (IM) are rare systemic diseases which all share weakness and inflammation of skeletal muscles. While most investigations have focused on immune mechanisms underlying these diseases, it becomes increasingly clear that non-immune processes take part into IM pathogenesis. One of the most original is the role played by mitochondria. In daily practice, mitochondrial abnormalities are frequently observed in muscle histological examination of patients with IM. Moreover, defects in respiratory chain enzymes and mitochondrial DNA (mtDNA) deletion suggest that an authentic acquired mitochondrial muscle disease occur during IM. Given the mtDNA sensibility to oxidative damage, this may be the illustration of an increase in mitochondrial oxidative stress during IM. Indeed, surprising links between inflammation and mitochondrial processes are emerging. Pro-inflammatory cytokines and signaling through some innate immune receptors, which are present in IM patient muscles, have recently been shown to increase mitochondrial reactive oxygen species (mROS) generation. In turn, evidence has merged that mitochondria are surprising key modulators

---

* Email address: alain.meyer1@chru-strasbourg.fr

of inflammation. Especially, mROS potentiate the synthesis of pro-inflammatory cytokines known to play important roles in IM pathogenesis. On the other hand, numerous mitochondrial components, released during cells injury, are potential agonists of innate immune receptors which are expressed in IM patients muscle. Thus, an original crosstalk between inflammatory and mitochondrial processes may exist during IM, in which mitochondria may be centrally positioned hubs in the regulation of inflammation. Specific antioxidant strategies may therefore represent promising adjuvant therapeutic options to be explored in IM management. Little but encouraging data concerning these treatments support this view.

INTRODUCTION

The inflammatory myopathies (IM) are a group of rare systemic diseases who all share weakness and inflammation of skeletal muscles. They mainly encompass polymyositis (PM), dermatomyositis (DM), sporadic inclusion body myositis (sIBM) and myositis associated to other conditions [1]. These muscular diseases are of particular importance because they represent the largest group of acquired and treatable myopathies in both children and adult. IM are believed to be autoimmune diseases and most investigations have focused on immune mechanisms underlying these diseases (Table 1).

Adaptive Immunity in IM Pathogenesis

The frequent detection of myositis-specific auto-antibodies in IM patients' serum [2] and the presence of B cells and T cells in IM patients muscle biopsy [3] support the involvement of adaptive immunity in IM pathogenesis. Moreover, it has been shown that both B cells and T cells undergo maturation and clonal expansions during IM [4,5], which suggest an antigen driven process. This could occur within the muscle itself, through an original lymphoid organization, which is capable of antigen-stimulated clonal maturation of antibody-producing plasma cells [6]. The importance of these adaptive immune processes is finally highlighted by the efficacy of therapeutic strategies targeting B cells with rituximab [7] or T cells with abatacept [8].

Innate Immunity in IM Pathogenesis

Innate immune processes are emerging players of IM pathogenesis.

Pattern recognition receptors (PRRs) are the hallmark of innate immunity. These germ-line encoded receptors initiate the immune response by sensing the presence of microorganisms through the recognition of structures conserved among microbial species, which are called pathogen-associated molecular patterns (PAMPs). Recent evidence indicates that PRRs are also responsible for recognizing endogenous molecules released from damaged cells, termed damage associated molecular patterns (DAMPs) [9].

Toll like receptors (TLRs) are the best characterized PRRs. It has been shown that the expression levels of TLR2, TLR4 and TLR9 in muscle tissues are significantly higher in patients with IM compared with those in controls [10,11] and correlate with the expressions of inflammatory cytokines [10]. TLR3 and TLR7 have been also recently found in muscle tissues of IM patients but not in non-inflammatory muscle tissues [11,12].

Members of the family of NOD-like Receptors (NLRs) are cytoplasmic PRRs that assemble into large multiprotein complexes, termed inflammasome, which activate interleukin-1 (IL-1) [13]. The involvement of this cytokine in IM suggests that the inflammasome probably play a role in IM pathogenesis but this has not been explored yet.

MDA-5 is a cytoplasmic PRR that is involved in the recognition of virus. Interestingly, autoantibodies targeting MDA-5 has been found in patients with dermatomyositis suggesting that this PRR may also be involved in IM pathogenesis [14].

Professional innate immune cells are present within muscle of IM patients. Especially, whereas in normal control muscles, few immature dendritic cells (DC) have been detected, both immature and mature DC have been found in muscle of IM patients [15]. These cells play a crucial role in initiating immunity through the activation of naive T cells by antigen presenting and production of large amounts of type I interferons.

Pro-Inflammatory Cytokines in IM

The role of immune processes is finally illustrated by the involvement of cytokines that form an integrated signaling network regulating both innate and adaptive immune responses.

Type 1 (α and β) and type 2 (γ) interferons have been shown to be overexpressed in serum and muscle samples of patients with IM [16,17]. These cytokines are notably thought to up-regulate the major histocompatibility complex I (MHC-I) on muscle fibers, a key event in IM initiation [18]. Type 2 interferon also promotes the differentiation of T lymphocyte helper into Th1 or Th17 cells. Cases reports of IM following interferons therapies [19] illustrate the important of these cytokines in IM initiation.

IL-1α IL-1β and the related cytokine IL-18 are other pro-inflammatory cytokines that have been localized in muscles of IM patients [17, 20-22]. High levels of IL-1Ra and IL-18 have been found in the serum of patients with IM, [21,23]. IL-1 family cytokines are known to increase the expression of adhesion molecules on endothelial cells enabling transmigration of leukocytes, to stimulate immune cells activation and to participate in Th1 and Th17 immune response polarization. IL-1 receptor antagonist (anakinra) efficacy in the treatment of IM supports the key role of this cytokines in IM physiopathology [24].

IL-6 has been referred as the first identified "myokine" as skeletal muscle is a major source of this cytokine [25]. Increased serum levels of IL-6 and increased levels of IL-6 messenger RNA in the muscles of patients with IM have been observed [26,27]. Functionally, IL-6 exercises a chemotactic activity and activates both innate and adaptive immune cells. It promotes Th17 polarization of the immune response. Accordingly, IL-6-deficient mice show impaired inflammatory response in a model of myosin-induced experimental myositis [28] and IL-6 blockade had therapeutic effects in a murine model of IM [29].

Increased presence of tumor necrosis factor α(TNF-α) and its soluble receptors have been shown by immunostaining and PCR analysis in muscle biopsy specimens from patients with IM [30]. This cytokine is known as a strong activator of inflammation. The association between DM and the -308A TNF polymorphism [31] and the efficiency of etanercept, an antagonist of TNF-α in IM also illustrate the importance of this cytokine in IM pathogenesis [32].

Muscle Fiber: An Original Immune Cell in IM

One of the most surprising aspects of immune process in IM is the role played by muscle fibers themselves. During IM, muscle fibers acquired original propriety that makes them behave like professional immune cells.

Muscle fiber expresses PRRs and secretes proinflammatory cytokines. Muscle cells constitutionally express TLR4 [33] and can express TLR3, TLR7 and TLR9 during IM [11, 12]. Immature myoblast precursors stimulated through TLR3 may represent a local source of IFN-β [34]. Muscle cells are also able to secrete IL-6 [25], TNF-α and IL-1 [35].

Muscle fibers express co-stimulatory molecules. Whereas muscle fibers do not normally express MHC, during IM, these cells express both HLA class I [3] and II [36]molecule that may render these cells able to present antigenic peptides to the TCR of CD8 and CD4 T-cells. Moreover, muscle fibers of IM patients also express co-stimulatory molecules of the B7-family members and establish direct cell-to-cell contact with auto-invasive CD8 positive T cells through the ligands these molecules that could contribute to complete activation of T cells [37].

Table 1. PRRs and cytokines involved in IM

PRRs	Localisation	Notable effects
TLR2, TLR4	Extracellular membrane	Pro-inflammatory cytokine secretion
TLR3, TLR7, TLR9	Endosome	Type I interferons secretion
MDA-5	Cytoplasm	Chemokines secretion
NALP3	Cytoplasm	IL-1 secretion
Cytokines	Origin	Notable effects
INF-α/β	Plamacytoid dendritic cells Muscle fibres	Up regulation of CMH I and II
INF-γ	T cells	Up regulation of CMH I and II Adaptative immune responsepolarization
IL-1	Muscle fibers Inflammatorycells Endothelialcells	Chemotaxism Expression of adhesionmolecule Innate and adaptative immune cells activation
IL-6	Muscle fibers Inflammatory cells	Adaptative immune responsepolarization
TNF-α	Muscle fibers Inflammatory cells	Innate immune cell activation Pro-inflammatoty cytokine secretion

Acquired mitochondrial myopathy in IM (Table 2)

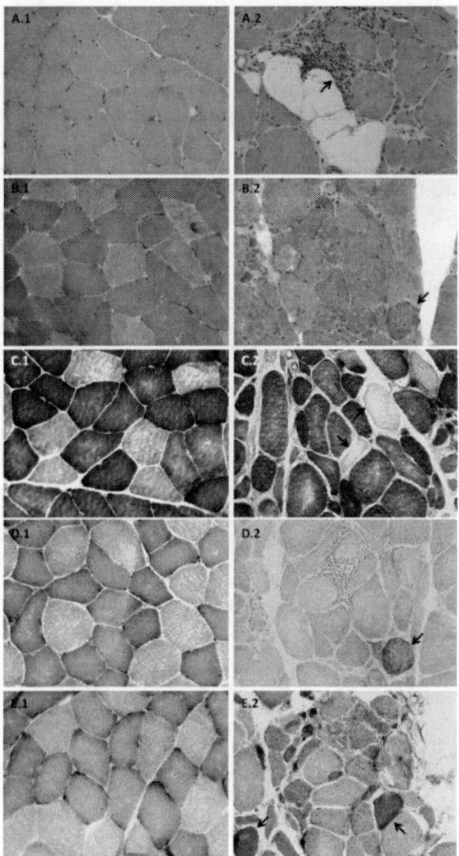

Figure 1: Muscle biopsy of (1) a control patient; (2) a sIBM patient with defects of respiratory chain enzyme complexes III and IV and multiple mtDNA deletion. *A. Haematoxylin and eosin staining.* A.1: Fibers have normal size and shape, fiber's nuclei are located in subsarcolemma areas, endomysium is barely visible, blood vessels are found in perimysium. A.2: Some muscle fibers are atrophic and show round-shaped with internal nuclei,endomysium is focally enlarged with an area of inflammatory cells (arrow). Also note the presence of adipose tissue. *B. Gomori trichrome.* B.1: Nuclei are stained red, mitochondria appear as very small dots that are invisible at this magnification. B.2: A fiber (arrow) is disrupted and red-stained especially in peripheral area (ragged-red fiber). *C. Cytochrome oxidase (COX) staining.* C.1:Normal fibers are homogeneously stained with an intensity that depends on fiber type. C.2: Two fibers (arrow) displayed absent COX activity (COX-negative fibers). *D. NADH tetrazoliumreductase (NADH-TR) staining.*D.1: Normal fibers are homogeneously stained with an intensity that depends on fiber type. D.2: One fiber (arrow) displayed increased and irregular staining. *E. Succinic dehydrogenase (SDH) staining.*E.1: Normal fibers are homogeneously stained with an intensity that depends on fiber type. E.2 Some fibers (arrows) displayed increased and irregular staining.

In spite of the strong evidence for immune processes involvement in IM and despite the success of therapeutics that target the actors of immunity, some data argue for a potential role of non-immune mechanisms in these diseases [38]. First, the degree of inflammation in skeletal muscle does not consistently correlate with the severity of the structural changes observed in the muscle fibers or with the severity of the clinical disease. Second, structural changes in the muscle fibers occur even in the absence of any inflammatory cells in muscle. Third, some IM patients do not respond even to potent anti-inflammatory therapy. The identification of some of these non-immune processes has just begun [39]. One of the most original may be the role played by mitochondria.

In vivo data. Using phosphorus magnetic resonance spectroscopy (^{31}P-MRS), Cea et al [40] found that IM patients have impaired post-exercise muscle oxidative metabolism which was similar to that found in a group of 10 patients with a primary mitochondrial disorder. Interestingly, there was no correlation between the magnetic resonance spectrometry abnormality and the degree of inflammation or fatty infiltration of muscle as measured by quantitative MRI suggesting that impaired post-exercise muscle oxidative metabolism was not directly immune mediated or secondary to muscle atrophy in these patients [40].

Histological data (Figure 1). In daily practice, mitochondrial disorders are evaluated on muscle biopsies, using modified Gomoritrichrome, cytochrome oxidase and succinate dehydrogenase (SDH) stains. These disorders are suggested when muscle fibers have a ragged red appearance on Gomoritrichrome, intense irregular abnormal SDH or NADPH-TR staining and negative COX staining [41]. Interestingly, such changes have been described in patients with different forms of idiopathic myositis including sIBM, PM, DM and juvenile myositis [41-43]. Using electron microscopy, abnormality in ultrastructure of mitochondria has been reported as well [44,45].

Muscle biochemistry findings. Arenas et al [46] found that 11 of 13 consecutive patients with IM had muscle carnitine insufficiency and 4 of 11 had muscle defects of respiratory chain enzyme complexes that suggest that an authentic mitochondrial muscle disease may arise in IM.

Findings from molecular biology. Using PCR analysis, Molnar et al [45] found multiple mitochondrial DNA (mtDNA) deletions, between 300 bp and 6600 bp in size, in muscle of patients with IM including sIBM and PM. The deletions were heteroplasmic and the degree of the heteroplasmy varied between 30–80%. The localizations of the deletions were notably the cytochrome b (a component of respiratory chain complex III) region, the COX III region and the "common deletion" [47] region. Hereditary disorders with multiple mitochondrial DNA deletions have been associated with mutations in the nuclear genes encoding thymidine phosphorylase (ECGF1) [48], adenine nucleotide translocator1 (ANT1) [49], twinkle (C10orf2) [50], and mitochondrial DNA polymerase gamma (POLG1) [51]. Oldfors et al [52] have studied three of these genes in five patients with sIBM characterized by a high proportion COX-deficient fiber and multiple mitochondrial DNA deletions in muscle. No mutations were identified [52] suggesting that mitochondrial myopathy during IM may rely on acquired mitochondrial DNA deletion.

Mitochondrial abnormalities in IM: do they reflect an increase in mitochondrial oxidative stress? The origin of these mtDNA deletions is still unclear. Interestingly, mtDNA is physically associated with the inner mitochondrial membrane, where the majorities of the cellular reactive oxygen species (ROS) are generated. Moreover, mtDNA has particular properties that render it very sensitive to oxidation damages [53, 54]. This raised the hypothesis that mtDNA deletions acquired during in IM may reflect an increased mitochondrial ROS (mROS) production in IM muscle. As all 13 proteins encoded by mitochondrial genome are regulatory subunits of the oxidative phosphorylation complexes, mtDNA integrity is essential for mitochondrial function. Thus, mtDNA deletions can in turn lead to an increase in mitochondrial oxidative stress through a feed forward loop between mtDNA deletion and mROS production (Figure 2C).

Table 2. acquired mitochondrial myopathy in IM

Muscle histology	
- *Gomoritrichrome staining*	Ragged red fibers
- *COX staining*	COX-negative fibers
- *SDH staining*	Intense irregular SDH activity
- *Electronic microscopy*	Increase in number and decrease in size of mitochondria
Muscle biochemistry	
- *Respiratory chain enzyme activity*	Complex I, III and/or IV deficiency
- *Carnitine*	Carnitine deficiency
Muscle *in vivo* analysis	
31P-MRS	Impaired post-exercise muscle oxidative metabolism
Muscle molecular biology analysis	
mtDNA	mtDNA deletion (300-6600pb) Heteroplasmic degree: 30-80%
Inter-nuclear communication genes (*ANT1, twinkle , POLG1*)	No mutation detected

Cross talk between inflammation and mitochondria

Inflammation Enhances mROS Production

There is growing evidence that an inflammatory micro-environment could contribute to an increase of mitochondrial oxidative stress through various pathways (Figure 2). Involvement of some PRRs, which are expressed in muscles of patients with IM, results in enhanced mROS production (Figure 2A). It has been shown that signaling via TLR2 and TLR4 enhance mROS generation in macrophages. This response involves translocation of the adaptor TRAF6 to mitochondria, where it interacts with the protein ECSIT [55] that is implicated in mitochondrial respiratory chain assembly [56] suggesting that the heighted mROS generation result from direct perturbation in the mitochondrial respiratory chain. Involvement of TLR4 could also increase mROS generation through an alternative pathway that is mitochondrial uncoupling. Uncoupling protein 2 (UCP2) is a molecule that is thought to decrease mitochondrial superoxide generation through a reduction of electron leak from oxidative phosphorylation complexes [57]. Emre et al. [58] showed that stimulation of murine bone marrow-derived macrophages by the TLR4 ligand LPS down-regulate UCP2 through the JNK and p38 pathways.

mROS production is common to many activators of the NLRP3 inflammasome [59]. It has been shown that inflammasome stimulation lead to NLRP3 being redistributed to the perinuclear space where it co-localizes with endoplasmic reticulum and mitochondria organelle clusters relocation close to mitochondria suggesting that mROS contribute to the augmentation of oxidative stress in response to inflammasome activation.

Pro-inflammatory cytokines environment may also represent a source of mROS during IM (Figure 2B). Several cytokines that are present in muscles of patients with IM are able to enhanced mROS generation. In macrophages, INF-γ has been shown to trigger an ERRα- and PGC1β-dependent transcriptional program that promotes mitochondrial function and mROS production during bacterial infection [60]. Recently, TNF-α has also been shown to promotes mROS generation in a process that involves ROS modulator 1 (Romo1), a protein located in the mitochondria which recruits Bcl-XL to reduce the mitochondrial membrane potential [61]. Finally, IL-1β, and to a lesser extend IL-6 are capable to deregulate enzymatic antioxidant defenses in mitochondria that could lead to an accumulation of H_2O_2 [62].

mROS Contribute To Immune Processes

mROS have emerged as a surprising key modulator of numerous inflammatory processes that are known to take part in IM pathogenesis (Figure 2A).

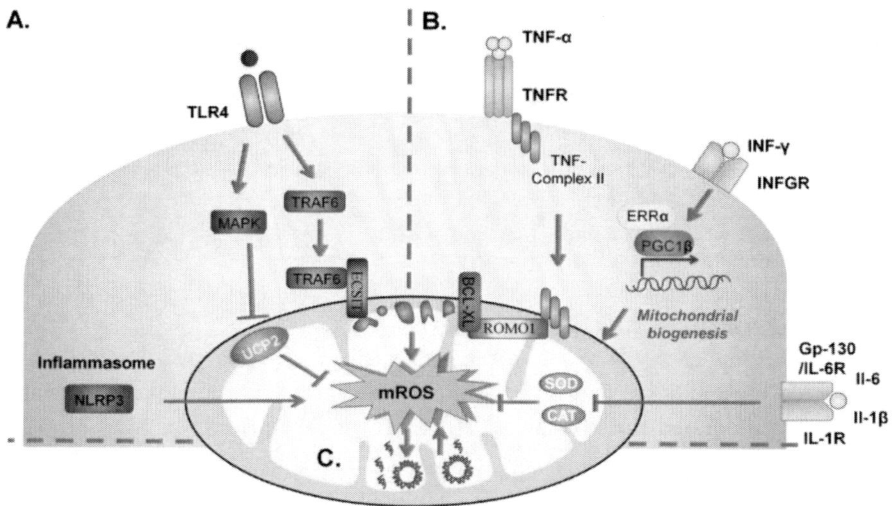

Figure 2. Potential mechanisms of mROS generation and mtDNA damage during IM. *A. PRRs-induced mROS generation:* TLR4 activation interferes both with respiratory chain assembly, through TRAF6/ECSIT involvements and with mitochondrial coupling, through MAPK dependent inhibition of UCP2. NLRP3 inflammasome activation also increases mROS through unknown mechanism. *B. Cytokine-induced mROS generation:* In response to TNF-α, TNF complex II, (which consists of receptor-interacting protein 1, TNF receptor-associated protein with death domain, TNF receptor-associated factor 2, Fas-associate death domain protein, and pro-caspase-8) binds to the C-terminus of Romo1 located in the mitochondria which in turn recruits Bcl-XL to reduce the mitochondrial membrane potential. INFG receptor activation triggers an ERRα- and PGC1β-dependent transcriptional program that promotes mitochondrial biogenesis. IL-1β and IL-6 inhibits expression of Cu/Zn superoxide dismutases (SOD) and catalase (CAT) expressions. *C. Positive feed forward loop between mROS generation and mtDNA damage:* mROS may induced mtDNA damage that in turn increase mitochondrial respiratory chain dysfunctions and mROS generation.

mROS contribute to inflammation. Several studies have highlighted that mROS potentiate pro-inflammatory cytokines secretion. Endotoxin-induced fever, IL-4 and IL-6 increases are lower in transgenic mice for human both UCP2 and UCP3 [63]. Disruption of the UCP2 gene in mice results in increase of mROS production and heightened activation of both NF-κB (64) and MAPK signaling pathways [58] that consequently augments pro-inflammatory cytokines, including IL-1β, IL-6, TNF-α, INF-γ, in response to TLR4 ligand both *in vivo* [64] and *in vitro* [58]. The elevated basal activity NF-κB in UCP2$^{-/-}$macrophages can be blocked by mROS inhibitors [64] suggesting that mitochondrial oxidative stress is directly involved in this enhanced inflammatory phenotype.

Similarly, treatment with rotenone (a potent mROS generator) strongly enhances IL-1β secretion in THP1 [65] and in LPS/ATP stimulated macrophage [66], as well as IFN-β in MEF cells stimulated with the MDA-5 ligand poly I:C [67]. Rotenone induced pro-inflammatory cytokines secretion is blocked by the adding of mROS inhibitors [65-66].

The mechanism by which mROS can enhance pro-inflammatory cytokines synthesis is unanswered. In THP-1, rotenone-induced IL-1 secretion may be mediated by the dissociation of TXNIP from thioredoxin in a ROS sensitive manner that allowed it to bind and activate NLRP3 [65].

CAN mROS CONTRIBUTE TO AUTOIMMUNITY?

In animal models, immunized UCP2-deficient mice developed significantly higher disease scores than littermate controls. A polymorphism in the mitochondrial genome, G-to-T at position 7778 alters mitochondrial performance, increasing H_2O_2 production and susceptibility to multiple autoimmune diseases including collagen-induced arthritis, spontaneous nephritis and autoimmune pancreatitis [68]. The influence of increase in mROS generation in the susceptibility to IM models has not been assessed yet.

In human, The UCP2-866G allele is correlated with lower levels of UCP2 and susceptibility to multiple sclerosis suggesting that mROS may play a role in autoimmunity arising [69]. However, the correlation between polymorphism of gene encoding mito-chondrial component and IM has not been assessed. Some data suggest a potential role of oxidative stress in the modulation of lymphocyte activation. In ESb-LT lymphoma cells, micromolar concentrations of hydrogen peroxide rapidly induced activation of the transcription factor NF-κB and significantly increased T cell proliferation when applied for a short period under reducing conditions. These data indicate that ROS may act as an important competence signal in T- lymphocytes [70].

An interesting still open question is whether ROS could influence immunogenicity of auto-antigens that are thought to be implicated in IM [71]. Van Dooren et al found that oxidative stress does induce modifications of histidyl-tRNAsynthetase that affect its tRNAaminoacylation activity but does not modify its immunoreactivity [72].

Mitochondrial Components Are Potential Triggers for the Innate Immunity

Because of their bacterial origin, mitochondria share several features with bacteria and might contain sequestered DAMPs that could trigger innate immune responses during pathological insult such inflammation and necrosis (73), two processes that occur in muscle during IM (Figure 2B).

mtDNA as a ligand of TLR9.mtDNA contains unmethylatedCpG motifs that resemble bacterial CpG DNA, which is a potent activator of TLR9. Collins et al. demonstrated that *In vivo* injection of purified mtDNA into the joints of miceinduces inflammation and arthritis, and that *in vitro* exposure of splenocytes to mtDNA, but not nuclear DNA, augments TNF-α secretion [74]. Zhang et al [73,75] showed that PNN activation with purified mtDNA induced is mediated by TLR9.

N-formylpeptides as ligands of formylpeptide-receptors. Like bacteria, mitochondria initiate protein synthesis with an N-formylmethionine residue. This molecular pattern is sensed by the formyl-peptide receptors (FPRs) that are expressed by neutrophils and activate the innate immune response. It has been showed that mitochondrial N-formyl peptide induced PNN chemotaxis [76,77], oxidative burst [78] and cytokine secretion [79] through FPR1 involvement. Moreover, the combination of mitochondrial N-formyl peptides and mitochondrial transcription factor A (TFAM) dramatically increased IL-8 release from monocytes [79].

ATP as a ligand of P2RX7. Mitochondria are a major source of ATP. Extracellular ATP is known to contribute to NLRP3 activation via the purinergic receptor P2RX7. Iyer et al. have shown that necrotic cells produced by pressure disruption, hypoxic injury, or complement-mediated damage were capable of activating the NLRP3 inflammasome and that this activation was triggered in part through ATP produced by mitochondria sensed by P2RX7 [80]. Similary, McDonald et al [77] showed that ATP released from necrotic cells activated the NLRP3 inflammasome to generate an inflammatory microenvironment that alerted circulating neutrophils to adhere within liver sinusoids. Necrosis, complement mediated damage and hypoxemia are all injuries that exist in IM patients muscle [81].

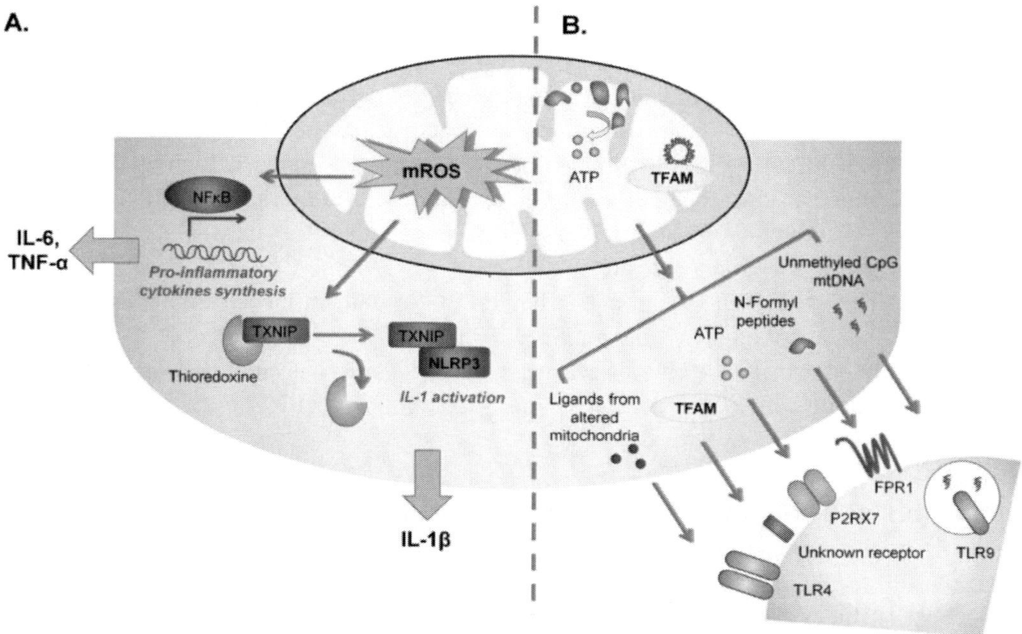

Figure 3.Potential role of mitochondrial changes in inflammation during IM. *A. mROS-potentiated pro-inflammatory cytokines secretion:*mROS increase IL-6 and TNF-α secretion through heighted NF-κB activation and IL-1β secretion through TXNIP dependent activation of NLRP3. *B. mtDAMPs-induced activation of innate immunity.*UnmethyledCpG motif of mtDNA, N-formyl peptides, ATP, TFAM and uncharacterized ligands peculiar to altered mitochondria are sequestered DAMPs that can be released

during cell injuries and trigger innate immunity through respectively TLR9, formylpeptide receptor 1 (FRP1), P2RX7, unknown receptor and TLR4.

Extracts of dysfunctional mitochondria as ligands of TLR4. Finally, mitochondria extracts obtained from hydrogen peroxide dysfunctionalised cells contain endogenous high-affinity human Toll-like receptor 4 (TLR4) ligands and induce TLR4-mediated inflammatory reactions in human THP-1 cells. Interestingly, inflammatory response was much stronger than that induced by the extracts of normal mitochondria suggesting that ROS-induced mitochondrial dysfunction enhanced immunogenicity of mitochondria components [82].

Do Mitochondrial Changes Play a Primary Role in Autoimmunity and IM Pathogenesis?

As explained above, some mitochondrial changes have been linked to the emergence of autoimmunity *in vivo,* both in animal models and in human. Nevertheless, there is little data to support the view that such changes play a primary role in IM emergence.

Primary mitochondrial disorders due to genetic abnormalities rarely show muscle inflammation. One patient with MELAS had muscle histopathology with endomysial inflammation and MHC class I up regulation by muscle fibers. However, the MELAS-related mutation, which was detected in other tissues, was not present in muscle, making it likely that the association of the two disorders was coincidental [83].

Nevertheless, toxic mitochondrial myopathy secondary to zidovudine therapy in HIV patients very often displays inflammation. Thus, mitochondrial dysfunction may facilitate muscle inflammation when an immune trigger, such as HIV, is present. Interestingly, this myopathy may respond only to zidovudin discontinuation in some patient, only to immunomodulation in others and to both in others, indifferently of the presence of muscle inflammation [84].

TARGETING OXIDATIVE STRESS: A THERAPEUTIC OPTION IN IM?

Because of the evidence of interconnection between inflammation, ROS level and mitochondrial dysfunction, anti-oxidant drugs could represent an adjuvant strategy in IM treatment. *In vitro* data have suggested that antioxidant treatments can act as cytokine secretion inhibitors through their scavenger activity on mROS. Tal et al [67] showed that propyl-gallate inhibits type 1 INF released in MEF cells stimulated by poly I:C and that N-acetyl-cystein was able to blockade poly I:C induced type 1 INF secretion in autophagy-deficient MEF cells. Similarly, the mitochondrial specific anti-oxidant Mito-TEMPO inhibits IL-1β secretion in macrophages stimulated with ATP and LPS [66] and the antioxidant PDTC blocked the activation of NF-κB by LPS [85]

Antioxidant drugs can also interfere with alternative pathways of inflammation. ICAM1 up-regulation by muscular endothelial cells is though to play a pivotal role in the recruitment of lymphocytes within the muscle (86). It has been showed that phenanthroline,disulfiram, vitamine C, vitamine E and N-acetyl-cysteine can inhibit ICAM-1 expression in endothelial cells activated with cytokines [87].

In human, antioxidant treatments have been poorly assessed in IM. However, several cases reports found a beneficial of vitamin E therapy in IM [88-91]. A more recent work reported the efficacy of high-dose vitamin C therapy in some cases of sIBM (92).

CONCLUSION

Inflammation does not account of all IM aspects. Especially, there is now evidence that an acquired mitochondrial myopathy can occurred during IM and in turn, mitochondrial change may potentiate muscle inflammation. This may rely on a surprising crosstalk between immune and mitochondrial processes in which, mitochondria may be centrally positioned hubs in the regulation of inflammation. Pharmacological antioxidant strategies and exercise may therefore represent adjuvant therapeutic options in IM treatment.

REFERENCES

[1]   A. L. Mammen, *Nat. Rev. Neurol.* 7, 343 (2011).

[2]   J. Sibilia, E. Chatelus, A. Meyer, J. E. Gottenberg, C. Sordet and J. Goetz, *Presse Med.* 39, 1010 (2010).

[3]   D. Dimitri, *Presse Med.* 38, 1141 (2009).

[4]   E. M. Bradshaw, A. Orihuela, S. L. McArdel, M. Salajegheh, A. A. Amato, D. A. Hafler, S. A. Greenberg, K. C. O'Connor, *J. Immunol.* 178, 547 (2007).

[5]   M. Salajegheh, G. Rakocevic, R. Raju, A. Shatunov, L. G. Goldfarb and M. C. Dalakas, *Neurology* 69, 1672 (2007).

[6]   M. Salajegheh, J. L. Pinkus, A. A. Amato, C. Morehouse, B. Jallal, Y. Yao and S. A. Greenberg, *Muscle Nerve* 42(4), 576 (2010).

[7]   M. Couderc, J. E. Gottenberg, X. Mariette, E. Hachulla, J. Sibilia, O. Fain, A. Hot, M. Dougados, L. Euller-Ziegler, P. Bourgeois, C. Larroche, A. Tournadre, Z. Amoura, B. Mazières, P. Arlet, M. De Bandt, T. Schaeverbeke and M. Soubrier, *Rheumatology (Oxford)* 50, 2283 (2011).

[8]   J. L. Musuruana and J. A. Cavallasca, *Joint Bone Spine* 78, 431 (2011).

[9]   O. Takeuchi and S. Akira, *Cell* 140, 805 (2010).

[10]   G. T. Kim, M. L. Cho, Y. E. Park, W. H. Yoo, J. H. Kim, H. J. Oh, D. S. Kim, S. H. Baek, S. H. Lee, J. H. Lee, H. Y. Kim and Kim SI, *Clin. Rheumatol.* 29, 273 (2010).

[11]   C. Cappelletti, F. Baggi, F. Zolezzi, D. Biancolini, O. Beretta, M. Severa, E. M. Coccia, P. Confalonieri, L. Morandi, M. Mora, R. Mantegazza and P. Bernasconi, *Neurology* 76, 2079 (2011).

[12]   A. Tournadre, V. Lenief and P. Miossec, *Arthritis Rheum* 62, 2144 (2010).

[13]   C. Bryant and K. A. Fitzgerald, *Trends Cell Biol.* 19, 455 (2009).

[14]   R. Nakashima, Y. Imura, S. Kobayashi, N. Yukawa, H. Yoshifuji, T. Nojima, D. Kawabata, K. Ohmura, T. Usui, T. Fujii, K. Okawa and T. Mimori, *Rheumatology (Oxford)* 49, 433 (2010).

[15] G. Page, G. Chevrel and P Miossec, *Arthritis Rheum.* 50, 199 (2004).

[16] S. A. Greenberg, *Arthritis Res. Ther.* 12 Suppl 1, S4 (2010).

[17] J. Schmidt, K. Barthel, A. Wrede, M. Salajegheh, M. Bähr and M. C. Dalakas, *Brain* 131, 1228 (2008).

[18] K. Nagaraju, N. Raben, L. Loeffler, T. Parker, P. J. Rochon, E. Lee, C. Danning, R. Wada, C. Thompson, G. Bahtiyar, J. Craft, R. Hooft Van Huijsduijnen and P. Plotz, *Proc. Natl. Acad. Sci. U. S. A.* 97, 9209 (2000).

[19] A. Aouba, S. Georgin-Lavialle, B. Terrier, L. Guillevin and F. J. Authier, *Joint Bone Spine* 78, 94 (2011).

[20] G. S. Baird and T. J. Montine, *Arch. Pathol. Lab. Med.* 132, 232 (2008).

[21] M. Tucci, C. Quatraro, F. Dammacco and F. Silvestris, *Clin. Exp. Immunol.* 146, 21 (2006).

[22] M. Tucci, C. Quatraro, F. Dammacco and F. Silvestris, *Ann. N. Y. Acad. Sci.* 1107, 184 (2007).

[23] K. Son, Y. Tomita, T. Shimizu, S. Nishinarita, S. Sawada and T. Horie, *Intern. Med.* 39, 128 (2000).

[24] A. Furlan, C. Botsios, A. Ruffatti, S. Todesco and L. Punzi, *Joint Bone Spine* 75, 366 (2008).

[25] B. K. Pedersen, *Med. Sci. Sports Exerc.* 44(3), 392 (2012).

[26] H. Lepidi, V. Frances, D. Figarella-Branger, C. Bartoli, A. Machado-Baeta and J. F. Pellissier, *Neuropathol. Appl. Neurobiol.* 24, 73 (1998).

[27] C. Gabay, F. Gay-Croisier, P. Roux-Lombard, O. Meyer, C. Maineti, P. A. Guerne, T. Vischer and J. M. Dayer, *Arthritis Rheum.* 37, 1744 (1994).

[28] F. Scuderi, F. Mannella, M. Marino, C. Provenzano and E. Bartoccioni, *J. Neuroimmunol.* 176, 9 (2006).

[29] N. Okiyama, T. Sugihara, Y. Iwakura, H. Yokozeki, N. Miyasaka and H. Kohsaka, *Arthritis Rheum.* 60, 2505 (2009).

[30] [P. Efthimiou, *Semin. Arthritis Rheum.* 36, 168 (2006).

[31] V. P. Werth, J. P. Callen, G. Ang and K. E. Sullivan, *J. Invest. Dermatol.* 119(3), 617 (2002).

[32] Muscle Study Group. *Ann. Neurol.* 70(3), 427 (2011).

[33] W. J. Lee, *Immune Netw.* 11, 223 (2011).

[34] A. Tournadre, V. Lenief, A. Eljaafari-Dany, P. Miossec, *Arthritis Rheum.* 64, 533 (2012).

[35] B. De Paepe, K. K. Creus and J. L. De Bleecker, *Curr. Opin. Rheumatol.* 21, 610 (2009).

[36] C. W. Keller, C. Fokken, S. G. Turville, A. Lünemann, J. Schmidt, C. Münz and J. D. Lünemann, *J. Biol. Chem.* 286, 3970 (2011).

[37] K. Murata and M. C. Dalakas, *Am. J. Pathol.* 155, 453 (1999).

[38] A. Henriques-Pons and K. Nagaraju, *Curr. Opin. Rheumatol.* 21, 581 (2009).

[39] I. Loell and I. E. Lundberg, *J. Intern. Med.* 269, 243 (2011).

[40] G. Cea, D. Bendahan, D. Manners, D. Hilton-Jones, R. Lodi, P. Styles and D. J. Taylor, *Brain.* 125, 1635 (2002).

[41] A. A. Varadhachary, C. C. Weihl and A. Pestronk, *Curr. Opin. Rheumatol.* 22, 651 (2010).

[42] P. Chariot, E. Ruet, F. J. Authier, D. Labes, F. Poron, R. Gherardi, *Acta Neuropathol.* 91, 530 (1996).

[43] M. I. Alhatou, J. T. Sladky, O. Bagasra and J. D. Glass, *J. Mol. Histol.* 35, 615 (2004).

[44] P. Tacik, K. Traufeller, M. Deschauer, J. Weis and S. Zierz, *17th Meeting of the European Neurological Society Muscle Disorders*, 2007.

[45] M. Woo, S. J. Chung and I. Nonaka, *J. Neurol. Sci.* 88, 133 (1998).

[46] M. Molnar and J. M. Schröder, *Acta Neuropathol.* 96, 41 (1998).

[47] J. Arenas, M. R. Gonzalez-Crespo, Y. Campos, M. A. Martin, A. Cabello, J. J. Gomez-Reino, *Arthritis Rheum.* 39, 1869 (1996).

[48] E. A. Schon, R. Rizzuto, C. T. Moraes, H. Nakase, M. Zeviani and S. Di-Mauro, *Science* 244, 346 (1989).

[49] I. Nishino, A. Spinazzola and M. Hirano, *Science* 283, 689 (1999).

[50] J. Kaukonen, J. K. Juselius, V. Tiranti, A. Kyttälä, M. Zeviani, G. P. Comi, S. Keränen, L. Peltonen and A. Suomalainen, *Science* 289, 782 (2000).

[51] J. N. Spelbrink, F. Y. Li, V. Tiranti, K. Nikali, Q. P. Yuan, M. Tariq, S. Wanrooij, N. Garrido, G. Comi, L. Morandi, L. Santoro, A. Toscano, G. M. Fabrizi, H. Somer, R. Croxen, D. Beeson, J. Poulton, A. Suomalainen, H. T. Jacobs, M. Zeviani and C. Larsson, Nat. Genet. 28, 223 (2001).

[52] G. VanGoethem, B. Dermaut, A. Lofgren, J. J. Martin and C. VanBroeckhoven, *Nat. Genet.* 28, 211 (2001).

[53] A. Oldfors, A. R. Moslemi, L. Jonasson, M. Ohlsson, G. Kollberg and C. Lindberg, *Neurology* 66, S49 (2006).

[54] C. Richter, J. W. Park and B. N. Ames, *Proc. Natl. Acad. Sci U. S. A.* 85, 6465 (1988).

[55] R. A. Costa, C. D. Romagna, J. L. Pereira and N. C. Souza-Pinto, *J. Bioenerg. Biomembr.* 43, 25 (2011).

[56] A. P. West, I. E. Brodsky, C. Rahner, D. K. Woo, H. Erdjument-Bromage, P. Tempst, M. C. Walsh, Y. Choi, G. S. Shadel and S. Ghosh, *Nature* 472, 476 (2011).

[57] R. O. Vogel, R. J. Janssen, M. A. van den Brand, C. E. Dieteren, S. Verkaart, W. J. Koopman, P. H. Willems, W. Pluk, L. P. van den Heuvel, J. A. Smeitink and L. G. Nijtmans, *Genes Dev.* 21, 615 (2007).

[58] M. D. Brand and T. C. Esteves, *Cell Metab.* 2, 85 (2005).

[59] Y. Emre, C. Hurtaud, T. Nübel, F. Criscuolo, D. Ricquier and A. M. Cassard-Doulcier, *Biochem. J.* 402, 271 (2007).

[60] J. Tschopp and K. Schroder, *Nat. Rev. Immunol.* 10, 210 (2010).

[61] J. Sonoda, J. Laganière, I. R. Mehl, G. D. Barish, L. W. Chong, X. Li, I. E. Scheffler, D. C. Mock, A. R. Bataille, F. Robert, C. H. Lee, V. Giguère and R. M. Evans, *Genes Dev.* 21, 1909 (2007).

[62] J. J. Kim, S. B. Lee, J. K. Park and Y. D. Yoo, *Cell Death Differ.* 17, 1420 (2010).

[63] M. Mathy-Hartert, L. Hogge, C. Sanchez, G. Deby-Dupont, J. M. Crielaard and Y. Henrotin, *Osteoarthritis Cartilage* 16, 756 (2008).

[64] T. L. Horvath, S. Diano, S. Miyamoto, S. Barry, S. Gatti, D. Alberati, F. Livak, A. Lombardi, M. Moreno, F. Goglia, G. Mor, J. Hamilton, D. Kachinskas, B. Horwitz and C. H. Warden, *Int. J. Obes. Relat. Metab. Disord.* 27(4), 433 (2003).

[65] Y. Bai, H. Onuma, X. Bai, A. V. Medvedev, M. Misukonis, J. B. Weinberg, W. Cao, J. Robidoux, L. M. Floering, K. W. Daniel and S. Collins, *J. Biol. Chem.* 280, 19062 (2005).

[66] R. Zhou, A. S. Yazdi, P. Menu and J. Tschopp, *Nature* 469, 221 (2011).

[67] K. Nakahira, J. A. Haspel, V. A. Rathinam, S. J. Lee, T. Dolinay, H. C. Lam, J. A. Englert, M. Rabinovitch, M. Cernadas, H. P. Kim, K. A. Fitzgerald, S. W. Ryter and A. M. Choi, *Nat. Immunol.* 12(3), 222 (2011).

[68] M. C. Tal, M. Sasai, H. K. Lee, B. Yordy, G. S. Shadel and A Iwasaki, *Proc. Natl. Acad. Sci. U. S. A.* 106, 2770 (2009).

[69] X. Yu, L. Wester-Rosenlöf, U. Gimsa, S. A. Holzhueter, A. Marques, L. Jonas, K. Hagenow, M. Kunz, H. Nizze, M. Tiedge, R. Holmdahl and S. M. Ibrahim, *Hum. Mol. Genet.* 18, 4689 (2009).

[70] S. Vogler, R. Goedde, B. Miterski, R. Gold, A. Kroner, D. Koczan, U. K. Zettl, P. Rieckmann, J. T. Epplen and S. M. Ibrahim, *J. Mol. Med. (Berl).* 83(10), 806 (2005).

[71] M. Los, W. Dröge, K. Stricker, P. A. Baeuerle and K. Schulze-Osthoff. *Eur. J. Immunol.* 25, 159 (1995).

[72] B. T. Kurien and R. H. Scofield, *Autoimmun. Rev.* 7, 567 (2008).

[73] S. H. van Dooren, R. Raijmakers, H. Pluk, A. M. Lokate, T. S. Koemans, R. E. Spanjers, A. J. Heck, W. C. Boelens, W. J. van Venrooij and G. J. Pruijn, *Biochem. Cell Biol.* 89(6), 545 (2011).

[74] Q. Zhang, M. Raoof, Y. Chen, Y. Sumi, T. Sursal, W. Junger, K. Brohi, K. Itagaki and C. J. Hauser, *Nature* 464, 104 (2010).

[75] L. V. Collins, S. Hajizadeh, E. Holme, I. M. Jonsson and A. Tarkowski, *J. Leukoc. Biol.* 75, 995 (2004).

[76] Q. Zhang, K. Itagaki and C. J. Hauser, *Shock* 34, 55 (2010).

[77] H. Carp, *J. Exp. Med.* 155, 264 (1982).

[78] B. McDonald, K. Pittman, G. B. Menezes, S. A. Hirota, I. Slaba, C. C. Waterhouse, P. L. Beck, D. A. Muruve and P. Kubes, *Science* 330, 362 (2010).

[79] M. Raoof, Q. Zhang, K. Itagaki and C. J. Hauser, *J. Trauma* 68, 1328 (2010).

[80] E. D. Crouser, G. Shao, M. W. Julian, J. E. Macre, G. S. Shadel, S. Tridandapani, Q. Huang and M. D. Wewers. *Crit. Care Med.* 37, 2000 (2009).

[81] S. S. Iyer, W. P. Pulskens, J. J. Sadler, L. M. Butter, G. J. Teske, T. K. Ulland, S. C. Eisenbarth, S. Florquin, R. A. Flavell, J. C. Leemans and F. S. Sutterwala, *Proc. Natl. Acad. Sci U. S. A.* 106, 20388 (2009).

[82] S.A. Greenberg, *Neurology* 69, 2008 (2007).

[83] S. A. Nicholas, K. Coughlan, I. Yasinska, G. S. Lall, B. F. Gibbs, L. Calzolai and V. V. Sumbayev, *Int. J. Biochem. Cell Biol.* 43, 674 (2011).

[84] R. Marotta, K. Reardon, P. A. McKelvie, M. Chiotis, J. Chin, M. Cook and S. J. Collins, *J. Clin. Neurosci.* 16, 1223 (2009).

[85] M. C. Dalakas, I. Illa, G. H. Pezeshkpour, J. P. Laukaitis, B. Cohen, J. L. Griffin, *N. Engl. J. Med.* 322, 1098 (1990).

[86] R. Schreck, B. Meier, D. N. Männel, W. Dröge and P. A. Baeuerle, *J. Exp. Med.* 175, 1181 (1992).

[87] A. M. Sallum, M. H. Kiss, C. A. Silva, A. Wakamatsu, M. A. Vianna, S. Sachetti and S. K. Marie, *Autoimmun. Rev.* 5, 93 (2006).

[88] C. B. McMullen, E. Fleming, G. Clarke and M. A. Armstrong, *Mol. Cell Biol. Res. Commun.* 3, 231 (2000).

[89] S. Ayres Jr. and R. Mihan, *Cutis* 21, 321 (1978).

[90] D. C. Haas, *South Med. J.* 70, 1148 (1977).

[91] R. N. Killeen, S. Ayres Jr. and R. Mihan, *South Med. J.* 69, 1372 (1976).

[92] S. Ayres Jr., *Int. J. Dermatol.* 25, 668 (1986).

[93] T. Yamada, M. Minohara, Y. Imaiso, N. Sakae, H. Hara, K. Tanaka, T. Yamamoto, T. Taniwaki, H. Furuya, and J. Kira, *Fukuoka Igaku Zasshi* 92, 99 (2001).

Chapter 5

INVOLVEMENT OF INFLAMMATION ON SKELETAL MUSCLE ISCHEMIA-REPERFUSION DELETERIOUS EFFECTS

M. Guillot[a,b], J. Pottecher[a,c], J. Boisramé-Helms[a,d], A. Meyer[a,e], A. L. Charles[a], Z. Mansour[a,e], P. Diemunsch[c], J. Zoll[a,f] and B. Geny[*,a,f]*

[a]Equipe d'Accueil 3072, Institut de Physiologie, Faculté de Médecine,
Université de Strasbourg, Cedex, France
[b]Service de Réanimation Médicale, Pôle Urgences, Réanimations Médicales,
Centre Anti-Poison, Hôpital de Hautepierre, Hôpitaux Universitaires
de Strasbourg, Strasbourg, France
[c]Pôle Anesthésie Réanimation Chirurgicale, SAMU,
Hôpitaux Universitaires de Strasbourg, Strasbourg, France
[d]Service de Réanimation Médicale, Pôle Urgences, Réanimation Médicales,
Centre Anti-Poison, Nouvel Hôpital Civil, Hôpitaux Universitaires
de Strasbourg, Strasbourg, France
[e]Service de Chirurgie Cardio-Vasculaire, Pôle d'Activité Médico-Chirurgicale
Cardio-Vasculaire, Nouvel Hôpital Civil, Hôpitaux Universitaires
de Strasbourg, Strasbourg, France
[f]Service de Physiologie et d'Explorations Fonctionnelles, Pôle de Pathologie Thoracique,
Hôpitaux Universitaires, CHRU de Strasbourg, Cedex, France

ABSTRACT

Ischemia-reperfusion injury appears after a period of ischemia at the time of reperfusion. This is observed in many clinical situations such as surgery of the abdominal aorta, trauma or shock. Ischemia-reperfusion can induce muscle damage and triggers multiple organ failure and patient death. Several mechanisms are responsible for the

---

[*] Email address: bernard.geny@chru-strasbourg.fr

onset of ischemia-reperfusion injury. Impairment of mitochondrial respiratory chain and production of oxygen-derived free radicals during ischemia and reperfusion are the corner stones of skeletal muscle. However, inflammation is also one of the main mechanisms responsible for ischemia-reperfusion injury of skeletal muscle. Free radicals induce a production of pro-inflammatory mediators and trigger an activation of inflammation. The importance of inflammation in ischemia-reperfusion has been also described in heart muscle and other tissues. The activation of inflammation after ischemia-reperfusion has consequences at the local and also at the systemic level. Thus, infiltration of leukocytes during reperfusion is associated with an activation of complement and a production of immunoglobulin and pro-inflammatory cytokines like TNF-α, IL-1β, and IL-6. Several trials have demonstrated that inhibition of inflammation can protect skeletal muscle from ischemia-reperfusion. Future therapy would take care of inflammation in order to improve outcome of ischemia-reperfusion of skeletal muscle.

INTRODUCTION

It is common knowledge and current practice to perfuse an ischemic organ as soon as possible to prevent tissue infarction and permanent loss of function. However, if reperfusion is essential to maintain viability and actually restores blood flow to the ischemic organ, it also brings its own hazards, known as reperfusion injury [1].

Reperfusion injury can occur in almost every organ but takes on critical importance when it involves skeletal muscle. Although skeletal muscle is not considered a vital organ, it is particularly vulnerable to ischemia (in a limb, muscle is the primary affected tissue) and reperfusion of large muscular mass invariably leads to multi-organ failure and death. Skeletal muscle ischemia-reperfusion injury can occur following a wide range of circumstances, ranging from civilian or military casualties (e.g. traumatic vascular injury, crush syndrome, and compartment syndrome) to surgical procedures (e.g. vascular reconstruction, abdominal aneurysm repair, and orthopedic surgical operations) performed under vascular clamp or tourniquet. Indeed, skeletal muscles are occasionally subjected to prolonged or repeated episodes of ischemia as a result of unpredictable intraoperative or postoperative complications. This is all the more important as clinical conditions occurring with muscle ischemia-reperfusion are often associated with intravascular volume depletion which may worsen local and systemic outcome.

When limb reperfusion alleviates ischemic injury in skeletal muscle, it also drives an inflammatory response, initially confined in the muscle itself but rapidly spread to systemic circulation. Skeletal muscle endothelial cells are first exposed to the toxicity of molecular oxygen and generate significant amounts of reactive oxygen species that impair physiological vasodilatation and can attack the entire array of biomolecules found in tissues. Simultaneously, activated neutrophils get sequestrated in the disrupted muscular microvascular bed they encounter; transmigrate in the interstitium and increase tissue injury. Increased interstitial fluid pressure and a disrupted microvascular barrier trigger the failure of capillaries in postischemic tissues to reperfuse, known as no-reflow phenomenon. At that moment, circulating neutrophils and inflammatory mediators combine with complement membrane attack complexes to induce remote organ injury, involving lungs, kidneys, liver and heart. Fortunately, compelling support is provided by the observation that skeletal muscle ischemia-reperfusion may be amenable to therapeutic interventions that can alleviate both local and remote reperfusion injuries.

In the first part of this chapter, we describe the methods used to assess inflammation during ischemia inflammation. Then, we will describe the mechanisms responsible for the activation of inflammation and their effects on skeletal muscle. Finally, we will attempt to determine whether there are treatments to protect skeletal muscle from ischemia-reperfusion.

I. Measurements of Inflammation

A biomarker is 'a characteristic that is objectively measured and evaluated as an indicator of normal biological processes, pathogenic processes, or pharmacological responses to a therapeutic intervention' [2], a 'quantifiable measurement(s) of biological homeostasis that defines(s) what is 'normal', therefore providing a frame of reference for predicting or detecting what is 'abnormal'' [3]. Many novel biomarkers are regularly identified, thanks to research and developments in molecular biochemistry and proteomics. Some of them seem to be closely associated with a specific disease states or progression.

Measurement of circulating cytokines and acute-phase proteins as biomarkers of inflammation in pathological states may thus provide important prognostic information, by indicating poor disease control and high levels of systemic inflammation. It may also facilitate directing therapy or monitoring the response to a therapeutical intervention. The performances of a marker can be evaluated by its sensitivity and specificity and are referred to as endogen characteristics [4].

1. Role of the Endothelial Interface

Ischemia-reperfusion is associated with an intense activation of inflammation. The endothelium is one of the first targets of the inflammatory response, as it lies at the interface of circulating blood and the vessel wall and is precociously exposed to circulating signaling molecules and physical stresses [5]. Endothelial activation, defined by an up-regulation of adhesion molecules by pro-inflammatory cytokines, is related to many pathophysiological processes like inflammation and oxidative stress; it is characterized at varying degrees by procoagulant, proadhesive and vasoconstrictive changes [6] and is closely associated with markers of systemic inflammation [7]. Endothelial cells are able to adapt rapidly to chemical [e.g. pro-inflammatory cytokines, like tumor necrosis factor alpha (TNF-α), or interleukin 1 (IL-1)] and physical stimuli (e.g. shear stress or hypoxia).

Biomarkers of endothelial dysfunction (e.g. hemostatic cofactor von Willebrand factor; plasma metalloproteases; adhesion molecules like E-, P-, and L-selectins, VCAM-1, ICAM-1; endothelin; vascular endothelial growth factor (VEGF), soluble VEGF-receptor-1; platelet-derived growth factor; plasminogen activator inhibitor PAI-1; fibrinogen and fibrin monomers; ...) may thus help in the early detection of systemic inflammation and its complications. Soluble biomarkers have many advantages, as they are easy to collect and store and are generally rather cheap. High-sensitivity assays, however, are required for measurement of cytokines and inflammatory mediators. Moreover, the utility of cytokines track may remain limited as they are not stable in stored samples and thus not convenient for routine clinical evaluation in biochemistry laboratories. Finally, most of these biomarkers need further investigation and validation and are still not routinely measured [8].

2. Inflammation and Oxidative Stress

Oxidative stress is associated with inflammatory states and results from a decrease in antioxidant level and/or an increase in ROS level [9]. ROS usually persist for a very short period of time *in vivo* and can therefore not be measured directly. Serum level of H_2O_2 and $NO^\bullet$ seem however to be suitable for direct measurement. Different methods can be used to detect transient ROS, either by trapping them and dose trapped species, or by measuring oxidative damage correlated to the level of ROS. A major limitation of the estimation of plasma 'total antioxidant capacity' is that it doesn't take into account contributions from molecules like urate, ascorbate or SH-albumin. Electron spin resonance (ESR) is the only method that detects free radicals, by trapping the reactive species on body fluids and tissue samples with specific probes to form a stable radical. Because of potential toxicity, whole-body ESR techniques are not applicable to humans, although it has already been tested on animals [10]. Hydrogen peroxide is stable and has notably been proposed to be detectable in body fluids, like fresh urine or exhaled air and its level could increase during inflammation, although much interference can occur [10-12].

Another way to estimate the level of ROS is to measure their consequences, i.e. oxidative cellular damage. Indeed, interaction of ROS with nucleic acids, lipids and proteins can be responsible for mutagenic base and sugar formation, lipid peroxidation or may affect the function of receptors, enzymes, transport proteins; the molecules subsequently formed (chlorinated and nitrated lipids; isoprostanes which are specific end products of the peroxidation of polyunsaturated fatty acids; hydrocarbons, including ethane and pentane; 8-hydroxy-20-deoxyguanosine; reactive nitrogen and chlorine species; ...) can therefore be measured and interpreted as oxidative stress biomarkers [11].

3. Inflammation and Coagulation

Inflammation and coagulation are tightly linked, as pro-inflammatory cytokines like IL-6, TNF-α and IL-1 may take part both in coagulation activation and regulation [13], and conversely, coagulation factors like factor Xa, thrombin and TF/FVII complex are known to have pro-inflammatory properties. Endothelial production of pro-inflammatory cytokines and membrane expression of adhesion molecules (E-selectin, VCAM-1, ICAM-1) is thus enhanced by these coagulation factors and favor leukocytes adhesion and transmigration. Finally, protein C is responsible for the endothelium-mediated activation of inflammation *via* G-protein-coupled PARs, which increase the production of cytokines and subsequent inflammatory tissue damage [14]. Therefore, measurement of coagulation factors is also of special interest to study an inflammatory process. In addition to theses markers, detection of microparticles, which generation is associated with inflammatory and procoagulant states may constitute a new marker of great interest in clinical research and maybe one day in clinical daily practice.

4. Inflammation and Microparticles

Cellular activation notably leads to the production of microparticles (MPs), which are vectors of intercellular exchange of biologic information in health and several diseases [15]. MPs are submicron vesicles released in the extracellular environment after a membrane reorganization following cell activation or apoptosis. They express several cell surface markers, dependant on their cellular origin and on the process leading to their formation [16].

Circulating MPs may activate coagulation, inflammation and cells apoptosis, favouring multiple organic failures like in sepsis [17-19]. Although no standardized procedures in microparticle measurement are yet available, a consensus is arising on blood sampling and microparticle isolation by successive centrifugations that avoid exosomes contamination of MPs samples [20].

Microparticles can be analyzed through capture techniques using the high affinity of annexin V for PhtdSer for quantitative assessment or insolubilized antibodies for phenotyping combined to prothrombinase assay. Another method to measure MPs is flow cytometry, which also allows determination of MPs' cellular origin, but detects MPs of higher size ranges [21].

In summary, inflammation gathers numerous pathways, which implications vary according to the triggering stimulus, and closely interacts with oxidative stress and coagulation as well. Several biomarkers reflecting inflammation, coagulation or oxidative stress can therefore be correlated to specific pathogenic processes and mirror the severity of the pathology or the effects of therapeutic interventions. Measurement of inflammation depends on the underlying pathology we are interested in, and has to be considered in each type of disease according to the organ or system subjected to the inflammatory process. Many of these markers, however, cannot be measured in daily clinical practice and still falls within the research domain, either because measurement methods are too expensive or because it is too tedious in routine. Finally, two main concerns must be born in mind: first, as far as human pathology in clinical practice is concerned, inflammation is almost exclusively appreciated on circulating markers, which therefore only reflect a state at a precise moment, and second, the ideal sensitive and sensible biomarker of inflammation (quantitative or qualitative) doesn't exist yet and therefore inflammation should be appreciated on a body of evidence rather than on a single marker.

II. Inflammatory Mechanisms Involved in Local and Remote Injuries after Skeletal Muscle Ischemia-Reperfusion

Skeletal muscle injury after ischemia-reperfusion is characterised by endothelial injury, increase in vascular permeability and activation of inflammation and skeletal muscle cells lysis which may lead to a compartment syndrome, rhabdomyolysis and multiple organ failure (Figure 1) [22-27]. Ischemia-reperfusion induces activation, adhesion and migration of neutrophils in reperfused tissues. Activation of neutrophils is associated with cytokines release, reactive oxygen species production and the activation of complement. We will detail the mechanisms observed at the local level and those observed at the systemic level during ischemia-reperfusion of skeletal muscle.

1. Local Inflammation
Local activation of inflammation is seen in the first minutes of ischemia-reperfusion injury in skeletal muscle and production of reactive oxygen species (ROS), activation of complement and adhesion of neutrophils have a key role in the damage process [28].

Production of ROS

Ischemia-reperfusion of skeletal muscle induces several mechanisms of injury at local or systemic levels [23-25]. Production of reactive oxygen species (ROS) is probably the earliest stage and a keystone in the development of ischemia-reperfusion local injury [29]. Many studies have shown a protect effect of anti-oxidant therapies during ischemia-reperfusion [30-32]. The largest production of ROS is observed during reperfusion but there is also some evidence for a production during ischemia [33]. Indeed, although low, the oxygen pressure during ischemia is still above zero and can induce an increase in ROS production [34, 35]. The main sources of ROS during ischemia-reperfusion may be the complex III of the mitochondrial respiratory chain and the xanthine oxidase (XO) [33]. Xanthine oxidase usually exists in healthy skeletal muscle, predominantly as a dehydrogenase. During ischemia, xanthine dehydrogenase is converted to xanthine oxidase producing ROS. Moreover, ATP hydrolysis during ischemia promotes the accumulation of hypoxanthine and xanthine, substrates of XO and increases production of ROS. However, XO is not present in all animal species and muscles and can't always explain ROS production during ischemia-reperfusion.

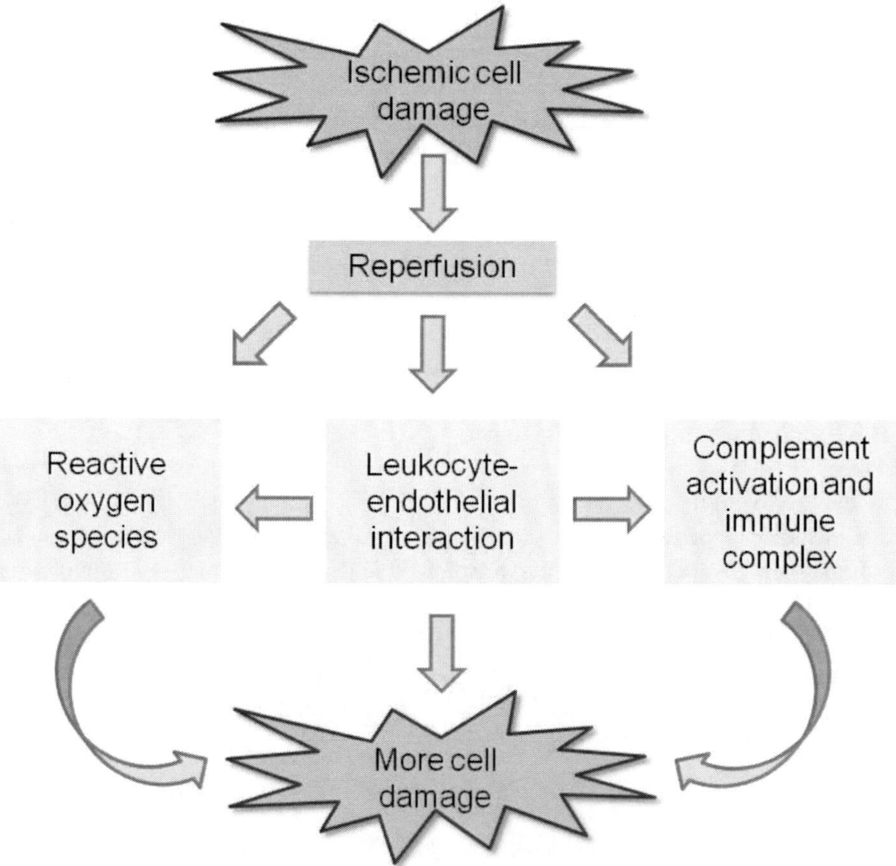

Figure 1. Mechanisms involved in reperfusion injury (adapted from Khalil et al. [28]).

The second main source of ROS during ischemia-reperfusion is the mitochondrial respiratory chain [36]. Several studies have shown a hypoxia-induced production of ROS in the mitochondrial respiratory chain [33, 37].

The production of ROS during ischemia leads to different injuries at local level. Anti-oxidant therapies decrease ROS concentration and can protect skeletal muscle from ischemia-reperfusion. Super-oxide dismutase, a superoxide anion scavenging enzyme can decrease tissues injuries after ischemia-reperfusion. Allopurinol, an inhibitor of xanthine oxidase can also protect skeletal muscle from ischemia-reperfusion. Many other anti-oxidant therapies have been shown to protect skeletal muscle and thus demonstrate the leading role of ROS in generation of injury during ischemia-reperfusion [32].

Whatever the ROS production during ischemia, the main production of ROS is seen during reperfusion when a large amount of oxygen is reintroduced in tissues [38]. Sources of ROS during reperfusion are almost the same during ischemia and reperfusion. In addition to XO and mitochondrial respiratory chain in endothelial and muscle cells, nicotinamide-adenine-dinucleotide phosphate (NADPH) seems also to produce ROS during reperfusion [39]. ROS production during ischemia-reperfusion leads to the activation of inflammation and induces several tissue injuries: alteration of microcirculation, increase in vascular permeability and formation of oedema.

Activation of Complement

Complement system is one the mediators of innate immune system and can contribute to various inflammatory diseases [40]. Three different pathways define this system: classical, alternate and mannose binding lectin (MBL) pathways. Theses pathways are involved in ischemia-reperfusion of skeletal muscle and seem to be related to ROS production [41-45]. However, two of them seem to have a critical role in ischemia-reperfusion of skeletal muscle: the classical pathway and the MBL pathways [46]. Indeed, ischemia-reperfusion of skeletal muscle induces an increase of C3a and C5a concentrations [42]. Activation of complement induces lesions in local and remote organs [47]. Complement can increase vascular permeability, oedema generation, and infiltration of leukocytes. Complement deletion models of ischemia- reperfusion have less local and systemic complications [43]. Moreover, many inhibitors of complement were developed in order to protect skeletal muscle from ischemia-reperfusion and showed a protective effect too [48-51]. However, there is no clinical evidence for use of such inhibitors. There are few trials in myocardial ischemia-reperfusion showing an improvement in troponin I level [52, 53].

Activation of Coagulation

ROS production during ischemia-reperfusion induces expression of tissue factor procoagulant activity in endothelial cells [33, 54, 55]. Tissue factor induces activation of the coagulation cascade with fibrin production [56]. Part of physiological regulation is that it induces fibrin own degradation. Fibrin degradation by plasmin products fragments like E-fragments and D-fragments. Fibrin and its derived fragments induce several effects like production of pro-inflammatory cytokines, interaction with cytokines (CD11c) and infiltration of leukocytes in reperfused tissues [56]. In fact, fibrin can induce production of IL-1ß and IL-8 and enhance chemotactism between leukocytes and endothelial cells [57, 58]. In order to prove these effects, several studies found that fibrin depletion can reduce infiltration of leukocytes across the endothelium [59, 60]. These mechanisms are seen during the first minutes of reperfusion [61, 62]. Many mechanisms have been observed. One of them is the interaction between the N-terminal of β-chain of E-fragments with VE-cadherins and CD11c. In order to demonstrate the role of fibrin-derived fragments in ischemia-reperfusion, Petzelbauer and al. used an E-fragment derived able to link VE-cadherines but not neutrophils CD11c [63]. They found this peptide decreases leukocytes infiltration in heart ischemia-reperfusion and prevents injuries.

Production of Cytokines

Ischemia-reperfusion of skeletal muscle induces production of cytokines. Cytokines, a large group of peptides involved in inflammation are one of the effectors responsible for causing tissue damage during ischemia-reperfusion [64] (Figure 2). Leukocytes and endothelial cells seem to be the major sources of cytokines production [65]. Activation of NF-κB increases the transcription of cytokines in response to ischemia-reperfusion of skeletal muscle [66-69]. Many pro-inflammatory cytokines are expressed during ischemia-reperfusion: TNF-α, IL-6, IL-1β [69-73]. Expression of these cytokines may have a dual role in skeletal muscle, leading to pro-apoptotic and protective effects in the same time [71]. However, the main effect of cytokines production during ischemia-reperfusion seems to be the alteration of the endothelium integrity and the activation of leukocytes.

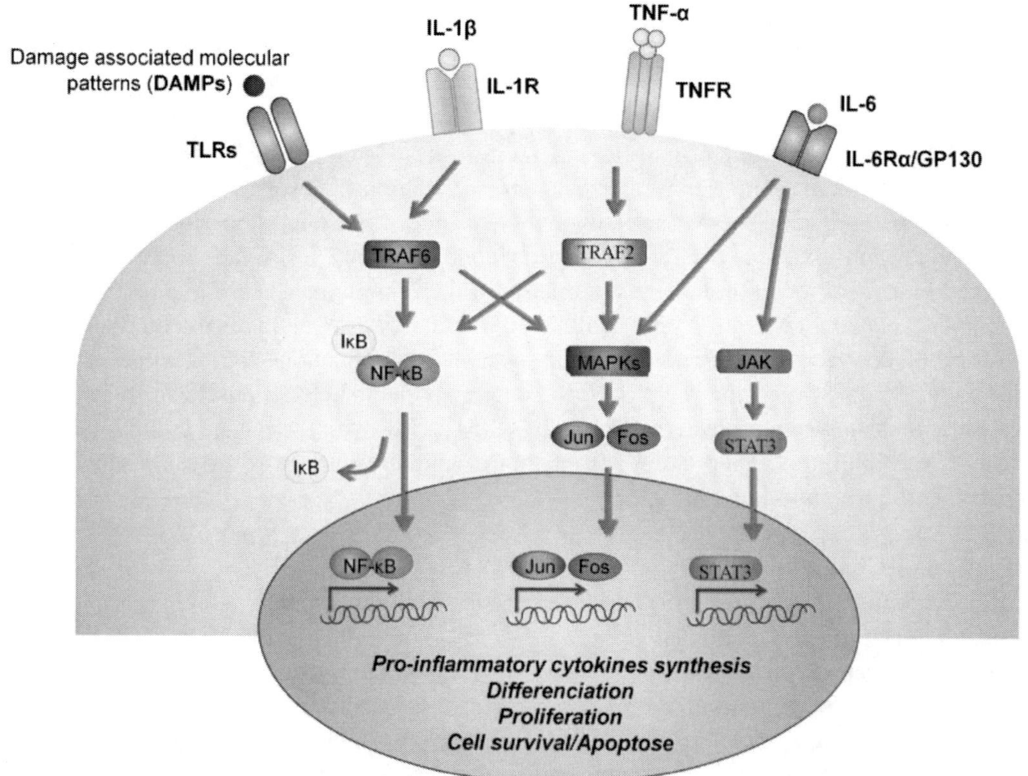

Figure 2. Cytokines receptors and TLRs signaling pathways (adapted from Verstrepen et al.[64]).

During reperfusion, damage associated molecular patterns (DAMPs), IL-1β, TNF-α and IL-6 are released which contribute to local cellular changes that lead to inflammation. This is driven by i) the transcription factors nuclear factor kappa-light-chain-enhancer of activated B cells (NF-κB), ii) activator protein 1 (AP-1) that consist of the dimerisation of the protein Fos Jun and iii) signal transducer and activator of transcription 3 (STAT3).

Both DAMPs/TLRs and IL-1 β/IL-1 receptor ligation trigger the dissociation of NF-κB from its inhibitor IκB through TNF receptor associated factor 6 (TRAF6) activation that allowed NF-κB to translocate in the nucleus. TRAF6 activation also leads to mitogen-activated protein kinases (MAPKs) activation that activate AP-1 nuclear translocation through Fos and Jun phosphorylation. NF-κB and AP-1 activations are also induced by TNF-α/TNF receptor (TNFR) ligation through TRAF2 signaling. Activation of the complex IL-6 receptor α (IL-6R α)/Glycoprotein 130 (GP130) lead to MAPKs-dependent phosphorylation of AP-1 but also to Janus kinase (JAK)-dependent nuclear translocation of STAT3.

Activation of Neutrophils

Ischemia-reperfusion is characterised by leukocytes accumulation in skeletal muscles. This has been described in other tissues like heart and kidney. This has been highlighted by the fact that leukopenic animals are protected from development of ischemia-reperfusion [74]. These results suggest that inflammation is the cause not the consequence of ischemia-reperfusion injuries. Mechanisms for infiltration of leukocytes are complex and are only beginning to be understood.

In summary, leukocyte infiltration seems to result from interactions between the endothelium, the inflammatory systems in the vascular bed and the coagulation cascade. Neutrophils are able to induce skeletal muscle cells lysis through a superoxide-dependent mechanism [38]. These observations were observed *in vitro* and *in vivo*, SOD treatment protects skeletal muscle cell from activated neutrophils [75, 76]. Superoxide anion production induces an increase of hydrogen superoxide which can damage cell membranes. However, theses oxidants have a short half time and can rapidly be converted by activated neutrophils myeloperoxidase (MPO) in lytic non-radical oxidants like hypochlorous acid [38, 76]. MPO can be released in serum and is used routinely as a neutrophils activation marker in skeletal muscle [26]. Ischemia-reperfusion induce production of pro-inflammatory cytokines like TNF-α, IL-6, and IL-8 [41, 77]

2. Remote Effects: Implication of Inflammation

Ischemia-reperfusion of skeletal muscle can induce multiple organ failure with dysfunctions of the lungs, the kidneys, the intestinal system or the liver [24, 72, 78, 79]. These injuries are induced by the local activation inflammation after ischemia-reperfusion of skeletal muscle [77]. Many factors are involved in the remote effects of ischemia-reperfusion of skeletal muscle. Indeed, aortic clamping induces an increase in circulating activated neutrophils and a rise in the cytokines circulation (IL-1, IL-8) [77, 80, 81]. Systemic activation of inflammation after ischemia-reperfusion of skeletal muscle induces an increase in vascular permeability and activation of neutrophils [80].

III. Therapeutic Approaches Aiming to Reduce Inflammation furing Ischemia-Reperfusion of Skeletal Muscle

1. Ischemic Conditioning

Originally described in the dog myocardium in 1986 by Murry *et al.* [82], ischemic preconditioning (IPreC), consisting of multiple brief ischemic episodes preceding a 40 min circumflex coronary artery occlusion, was shown to protect the heart from a subsequent sustained ischemic insult, reducing ATP depletion in ischemic cardiomyocytes and finally decreasing infarct size by 75%. Similar results were obtained by other teams in multiple animal models of myocardial ischemia, providing evidence that IPreC was a powerful cytoprotective measure against myocardial infarction. However, the conditioning stimulus varied in intensity and/or duration depending on the animal model under study [83].

Subsequent studies reported that IPreC could provide intra-organ protection (extending beyond the preconditioned myocardium [84]) and even interorgan protection with brief ischemic episodes in a distant organ (kidney [85], gut [86]) providing remote cardio-protection. This was called remote IPreC and set the stage for less invasive strategies, obviating the need of vascular clamping. Indeed, new therapeutic strategies (noninvasive remote IPreC) emerged in which remote cardioprotection was afforded non-invasively by intermittently inflating a tourniquet in the hind limb [87]. However, the therapeutic relevance of IPreC, remote IPreC and noninvasive remote IPreC was limited by the fact that the index sustained ischemia (e.g. myocardial infarction) is most of the time unpredictable and IPreC cannot be applied in the clinical situation.

Intense experimental research then focused on controlled reperfusion and ischemic post-conditioning (IPostC), characterized by repeated cycles of vessel occlusion and unclamping performed at the onset of reperfusion. This new paradigm of myocardial reperfusion also proved to be effective with an infarction sparing effect similar to that provided by IPreC in experimental models [88] and preliminary encouraging results in human [89].

More recent experimental and clinical evidence indicates that skeletal muscle is also amenable to ischemic conditioning and that IPreC, remote IPreC, noninvasive remote IPreC and IPostC can provide protection both locally (skeletal muscle) and in remote organs (heart, lung, kidney) that are involved in reperfusion injury. The mechanisms at work for these local and remote protections are not fully explored but imply a modulation of inflammatory pathways (decreased ROS generation, reduced cytokine concentration and attenuated neutrophilic infiltration) in a large proportion and converge to the mitochondrion as a central player.

Local Ischemic Preconditioning in Skeletal Muscle: Animal Studies

In 1995, Pang *et al.* [90] were the first to demonstrate that IPreC was efficient in reducing skeletal muscle infarction in swine (pig latissimus dorsi and gracilis muscle flaps subjected to 4 h ischemia and 48 h reperfusion) but the required conditioning stimulus (three cycles of 10 min occlusion and reperfusion) was stronger than in the heart (one single cycle). Muscle necrosis was ascertained by nitroblue tetrazolium staining technique and the authors were able to show that IPreC reduced skeletal muscle infarction by more than 40%. IPreC also provided an energy sparing effect ascertained by higher muscle concentrations of high-energy phosphates and lower lactate accumulation. It was then suggested that IPreC could reduce cellular energy demand through a reduction in mitochondrial ATPase activity [91] and slower destruction of adenine nucleotides. Similarly, Thaveau *et al.* demonstrated that ischemic preconditioning specifically restored complexes I and II activities of the mitochondrial respiratory chain in ischemic skeletal muscle [92].

Following experimental studies demonstrated that IPreC was able to modulate inflammatory pathways to provide both local and remote protection [93]. Indeed, in a rodent model of global hind limb ischemia induced by tourniquet application, Papanastasiou *et al.* [94] demonstrated that IPreC (3 cycles of 10 min ischemia and reperfusion) reduced myeloperoxidase muscle infiltration and retained muscle reduced glutathione content and systemic TNF-α concentrations to a level similar to that of the control group. It also raised the systemic nitric oxide content. In rat cremaster muscle, IPreC reduced leukocyte rolling and transmigration in postcapillary venules and restored functional capillary density (assessed through intravital microscopy) [95]. This effect was mediated through endothelial nitric oxide synthase (eNOS) and protein kinase C. Taken together, the aforementioned results suggest that IPreC can locally reduce skeletal muscle inflammation following ischemia and reperfusion.

Remote Ischemic Preconditioning In Skeletal Muscle: Animal Studies

Recent findings from our group and others also indicate that IPreC in a muscle group can also reduce inflammation and provide protection in another muscle group and also in more distant organs. Indeed, in rats submitted to three hours of hind limb ischemia by aortic cross clamping and two hours of reperfusion, IPreC restored mitochondrial oxidative capacity, normalized the Bax/Bcl2 ratio and tended to normalize reduced glutathione levels both in the preconditioned limb (local preconditioning) and in the contralateral limb (remote preconditioning) [96]. Harkin *et al.* [78] also demonstrated that IPreC before lower limb ischemia protected from acute lung injury in a swine model maintained normotensive. As a matter of fact, IPreC reduced systemic Il-6 concentrations, attenuated postischemic leukocyte adhesion and emigration, decreased lung tissue myeloperoxidase, decreased lung tissue wet to dry ratio and, *in fine*, decreased alveolar-arterial oxygen gradient, reflecting functional improve-ment and enhanced oxygenation. Finally, Eberlin *et al.* [97] showed that IPreC significantly reduced pulmonary neutrophil infiltration, decreased remote intestinal injury and reduced mortality in mice subjected to 2 hours ischemia and 24h reperfusion.

Remote noninvasive IPreC goes a step further using concise regional ischemia in non-vital and accessible organs and might provide protection in distant, vital or unreachable organs. Remote noninvasive IPreC can be induced by application of a tourniquet on the hind limb. In a swine model of multiple muscle groups ischemia-reperfusion (latissimus dorsi, gracilis and rectus abdominis muscle flaps) remote noninvasive IPreC was instigated by intermittent tourniquet application in a hind limb (three cycles of 10-min occlusion and reperfusion) [91]. Remote noninvasive IPreC reduced infarct size in all muscle groups and reduced muscle neutrophilic myeloperoxidase activity after 1.5 h of reperfusion. This protection was not mediated by neuronal pathway since ganglionic receptor antagonist didn't block its protective effects. Adenosine was also unlikely to be involved in the afforded protection. However, remote noninvasive IPreC was blocked by intravenous injection of opioids receptors antagonists, suggesting that remote IpreC releases endogenous opioid agonists, triggering an infarct-protective effect in all skeletal muscle in the body.

In another complementary study from the same group, Moses *et al.* [98] demonstrated that remote noninvasive IPreC specifically implicated the mitochondrial ATP-sensitive potassium channel (mK_{ATP}) since the infarct protective effect of remote noninvasive IPreC was totally blocked by the selective mK_{ATP} inhibitor 5-hydroxydecanoate but remained unaltered after intravenous injection of the sarcolemmal K_{ATP} channel inhibitor HMR 10-98. These authors also pointed out that mK_{ATP} channels had to remain open for more than 10 min to provide protection against skeletal muscle infarction. Furthermore, intravenous injection of the mK_{ATP} channel opener BMS-191095 mimicked the infarction sparing effect of remote non-invasive IPC with reduced rate of ATP depletion and reduced neutrophilic myelo-peroxidase activity in skeletal muscles of treated animals. It is suggested that mK_{ATP} activation mediates a reduction in mitochondrial ATPase activity, thereby reducing the rate of ATP depletion during hind limb ischemia. The mechanism mediating the reduced neutrophilic infiltration remains unknown.

Potential clinical applications for remote noninvasive IPreC are numerous. Indeed, it may provide prophylactic global protection for skeletal muscle that would undergo prolonged or repeated ischemic insults during vascular surgery, complex orthopedic procedures, plastic reconstructive surgery and every condition in which the predicted ischemia duration could be prolonged by unpredictable complications like vasospasm and thrombosis.

Ischemic Preconditioning in Skeletal Muscle: Human Studies

In humans, brief periods of forearm ischemia that preceded prolonged forearm ischemia prevented endothelial dysfunction (acetylcholine-dependent vasodilatation in the brachial artery) and activation of neutrophils (expression of CD11b and number of platelet-neutrophil complexes) in the circulating blood [99]. It is conceivable that ROS generated by activated neutrophils might inactivate nitric oxide synthesis and inhibit endothelium dependent vasodilatation in the absence of IPreC. As was demonstrated in animals, remote noninvasive IPreC is also able to provide protection in remote organs in man, including skeletal muscles, heart and kidney. For instance, induction of transient ischemia in one arm (3 cycles of 5min 200mmHg cuff inflation and deflation) maintained endothelial-dependent vasodilatation, protected the microcirculation during reperfusion and inhibited the no-reflow phenomenon in the contralateral arm submitted to prolonged ischemia (20 min 200mmHg cuff inflation) [100]. In a randomized, double blind human study, Ali *et al.* challenged the hypothesis that remote IPreC (iliac artery clamped 10 min twice) would decrease remote organ injury (myocardial injury and acute kidney failure) in patients undergoing abdominal aortic aneurysm repair [25]. Remote IPreC reduced the absolute risk of myocardial injury by 27% [95%CI: 8 to 45%] and the absolute risk of renal impairment by 23% [95%CI: 6.4 to 39%]. Remote ischemic preconditioning offers the opportunity to provide a widespread and systemic benefit, protecting remote organs from the deleterious effects of lower limb ischemia-reperfusion.

Ischemic Post-Conditioning in Skeletal Muscle

By alternating brief periods of vessel opening and re-occlusion at the very beginning of reperfusion, IPostC profoundly alters the hydrodynamics of early reperfusion. In myocardium, IPostC was shown to stimulate endogenous mechanisms that attenuate manifestations of reperfusion injury [101]. Indeed, IPostC was shown to decrease the surface expression of P-selectin, to improve vasodilatory response, to inhibit opening of the mitochondrial permeability transition pore (mPTP) and to increase the endogenous anti-oxidant defense by preserving reduced glutathione levels (the most abundant redox buffer in myocytes). Data are much scarcer in skeletal muscle.

In pig latissimus dorsi muscle flaps subjected to 4h ischemia and prolonged reperfusion (up to 72h), McAllister *et al.* demonstrated that IPostC (induced by 4 cycles of 30-s reperfusion/reocclusion) reduced ischemia/reperfusion-induced skeletal muscle injury [102]. Indeed, compared to time-matched controls, postconditioned muscles displayed reduced infarct size and decreased skeletal muscle myeloperoxidase activity indicating attenuated neutrophil accumulation. These protective effects involved inhibition of mPTP opening since they were mimicked by intravenous injection of cyclosporine A (mPTP opening inhibitor) and were abolished by intravenous injection of mPTP opener atractyloside.

In a rat extensor digitorum longus (EDL) muscle ischemia-reperfusion model, Park *et al.* demonstrated that IPostC reduced skeletal muscle inflammatory reactions induced by ischemia and reperfusion [102]. Indeed, IPostC reduced skeletal muscle edema (wet/dry ratio), tended to decrease myeloperoxidase activity and decreased polymorphonuclear infiltration in skeletal muscle. In the aforementioned model of aortic cross clamping in rats, Charles *et al.* demonstrated that IPostC performed through three short intervals of ischemia-reperfusion at the onset of reperfusion restored mitochondrial oxidative capacity, decreased superoxide anion generation and increased antioxidant defense (total glutathione levels) in gastrocnemius muscle [103].

Incremental reperfusion strategy may allow compensatory mechanisms to handle the most toxic early metabolic by-products of reperfusion. This was also observed by Gyurkovics *et al.* who demonstrated that postconditioned animals developed a lower increase in the level of TNF-α after 180min of lower limb ischemia followed by 240 min reperfusion [104]. They also exhibited a reduction in no reflow phenomenon in reperfused limbs, as assessed by laser Doppler flowmetry, and a decrease in remote organ dysfunction (lung and kidney).

Remote IPostC, performed through brief ischemia-reperfusion of the left limb in mice is also able to reduce reperfusion injury in the right limb and decreases wet/dry ratio, myelo-peroxidase activity and muscle necrosis in gastrocnemius. The molecular mediators that provide cell signaling during IPostC are not fully discovered but may include protein kinase C, mitochondrial K_{ATP} channels and survival kinases, like in IPreC. To our knowledge, no example of skeletal muscle IPostC was reported in the clinical area although experimental rationale is rather strong. Yet, performing IPostC through repetitive vessel cross clamping in atherosclerotic patients is neither always feasible, nor devoid of side effects [105, 106]. Recently and unexpectedly, local and remote ischemic post-conditioning further increased ischemia-reperfusion related injury by enhancing muscle mitochondrial dysfunction, oxidative stress production and inflammation [96]. Caution should thus be applied when considering ischemic post-conditioning in vascular surgery and there is obviously a need for alternative pharmacological approaches.

2. Pharmacological Approaches

Different strategies have been studied in order to protect skeletal muscle from ischemia-reperfusion injury. Most of these strategies were studied first in heart. Several pharma-cological agents were evaluated with the objective to improve muscle recovery [107].

Several anti-oxidants strategies have been studied during ischemia-reperfusion and seem to be useful [32, 108]. However, there are still conflicting results [109-111]. The differences between the molecules, protocols and models studied could explain these results. In addition, at present, there are no clinical studies that have demonstrated an interest of these antioxidant strategies. In order to directly modulate inflammation, different molecules have been developed to protect skeletal muscle from leukocytes extravasation during ischemia-reperfusion [56]. But, to our knowledge, these molecules have never been studied in skeletal muscle.

In the same idea, inhibitors of complement and cytokines have been studied during ischemia-reperfusion of skeletal muscle and may have a protective effect [112, 113]. Some of these inhibitors of complement are useful in hereditary angioedema [114]. But, they have not been studied in human ischemia-reperfusion of skeletal muscle.

CONCLUSION

Ischemia-reperfusion of skeletal muscle is observed in many clinical situations such as trauma, vascular surgery and shock. It is associated with high morbidity and mortality. Activation of inflammation is one the keystone of ischemia-reperfusion in skeletal muscle. Inflammation is mediated by the activation of leukocytes and production of various inflammatory molecules such as cytokines and complement. It results in the development of tissues injuries both locally and remotely and can induce a multiple organ failure.

Many strategies have been developed in order to reduce ischemia-reperfusion. However, their usefulness in clinical practice remains to be proved. A better understanding of the mechanisms involved in ischemia-reperfusion deleterious effects might help to improve the therapeutic and prognosis of patients suffering from skeletal muscle ischemia-reperfusion.

REFERENCES

[1]   F. W. Blaisdell, *Cardiovasc. Surg.* 10(6), 620 (2002).
[2]   E. Rivers, B. Nguyen, S. Havstad, J. Ressler, A. Muzzin, B. Knoblich, E. Peterson and M. Tomlanovich, *N. Engl. J. Med.* 345(19), 1368 (2001).
[3]   W. S. Dalton and S. H. Friend, *Science* 312(5777), 1165 (2006).
[4]   J. C. Marshall and K. Reinhart, *Crit. Care Med.* 37(7), 2290 (2009).
[5]   S. Danese, E. Dejana and C. Fiocchi, *J. Immunol.* 178(10), 6017 (2007).
[6]   W. C. Aird, *Pharmacol. Rep.* 60(1), 139 (2008).
[7]   F. Khan, B. Galarraga and J. J. Belch, *Nat. Rev. Rheumatol.* 6(5), 253 (2010).
[8]   P. Paulus, C. Jennewein and K. Zacharowski, *Biomarkers*, 16 Suppl 1, S11 (2011).
[9]   M .Valko, D. Leibfritz, J. Moncol, M. T. Cronin, M. Mazur and J. Telser, *Int. J. Biochem. Cell Biol.* 39(1), 44 (2007).
[10] B. Halliwell and M Whiteman, *Br. J. Pharmacol.* 142(2), 231 (2004).
[11] B. Halliwell, L. H. Long, T. P. Yee, S. Lim and R. Kelly, *Curr. Med. Chem.* 11(9), 1085 (2004).
[12] R. Q. Jobsis, S. L. Schellekens, A. Fakkel-Kroesbergen, R. H. Raatgeep and J. C. de Jongste, *Mediators Inflamm.* 10(6), 351 (2001).

[13] S. J. van Deventer, H. R. Buller, J. W. ten Cate, L. A. Aarden, C. E. Hack and A. Sturk, *Blood*, 76(12), 2520 (1990).

[14] M. Levi, *Hamostaseologie* 30(1), 10 (2010).

[15] F. Meziani, A. Tesse and R. Andriantsitohaina, *Pharmacol. Rep.* 60(1), 75 (2008).

[16] F. Meziani, X. Delabranche, P. Asfar and F. Toti, *Crit. Care* 14(5), 236 (2010).

[17] S. Martin, A. Tesse, B. Hugel, M. C. Martinez, O. Morel, J. M. Freyssinet and R. Andriantsitohaina, *Circulation* 109(13), 1653 (2004).

[18] G. N. Chironi, A. Simon, C. M. Boulanger, F. Dignat-George, B. Hugel, J. L. Megnien, M. Lefort, J. M. Freyssinet and A. Tedgui, *J. Hypertens.* 28(4), 789 (2010).

[19] S. Mortaza, M. C. Martinez, C. Baron-Menguy, M. Burban, M. de la Bourdonnaye, L. Fizanne, M. Pierrot, P. Cales, D. Henrion, R. Andriantsitohaina, A. Mercat, P Asfar and F. Meziani, *Care Med.* 37(6), 2045 (2009).

[20] C. Thery, L. Zitvogel and S. Amigorena, *Nat. Rev. Immunol.* 2(8), 569 (2002).

[21] W. Jy, L. L. Horstman, J. J. Jimenez, Y. S. Ahn, E. Biro, R. Nieuwland, A. Sturk, F. Dignat-George, F. Sabatier, J. Camoin-Jau, J. Sampol, B. Hugel, F. Zobairi, J. M. Freyssinet, S. Nomura, A. S. Shet, N. S. Key and R. P. Hebbel, *J. Thromb. Haemost.* 2(10), 1842 (2004).

[22] X. Bosch, E. Poch and J. M. Grau, *N. Engl. J. Med.* 361(1), 62 (2009).

[23] M. M. Yassin, D. W. Harkin, A. A. Barros D'Sa, M. I. Halliday and B. J. Rowlands, *World J. Surg.* 26(1), 115 (2002).

[24] C. Adembri, E. Kastamoniti, I. Bertolozzi, S. Vanni, W. Dorigo, M. Coppo, C. Pratesi, A. R. De Gaudio, G. F. Gensini and P. A. Modesti, *Crit. Care Med.* 32(5), 1170 (2004).

[25] Z. A. Ali, C. J. Callaghan, E. Lim, A. A. Ali, S. A. Nouraei, A. M. Akthar, J. R. Boyle, K. Varty, R. K. Kharbanda, D. P. Dutka and M. E. Gaunt, *Circulation* 116(11 Suppl), I98 (2007).

[26] J. Kang, H. Albadawi, V. I. Patel, T. A. Abbruzzese, J. H. Yoo, W. G. Austen Jr. and M. T. Watkins, *J. Vasc. Surg.* 48(3), 701 (2008).

[27] A. A. Khalil, F. A. Aziz and J. C. Hall, *Plast. Reconstr. Surg.* 117(3), 1024 (2006).

[28] S. H. Merchant, D. M. Gurule and R. S. Larson, *Am. J. Physiol. Heart Circ. Physiol.* 284(4), H1260 (2003).

[29] D. C. Gute, T. Ishida, K. Yarimizu and R. J. Korthuis, *Mol. Cell Biochem.* 179(1-2), 169 (1998).

[30] F. Ozkan, Y. Senayli, H. Ozyurt, U. Erkorkmaz and B. Bostan, *J. Surg. Res.* in press.

[31] J. W. Park, W. N. Qi, J. Q. Liu, J. R. Urbaniak, R. J. Folz and L. E. Chen, *Microsurgery* 25(8), 606 (2005).

[32] S. Cuzzocrea, D. P. Riley, A. P. Caputi and D. Salvemini, *Pharmacol. Rev.* 53(1), 135 (2001).

[33] N. Baudry, E. Laemmel and E. Vicaut, *Am. J. Physiol. Heart Circ. Physiol.* 294(2), H821 (2008).

[34] D. Troitzsch, S. Vogt, H. Abdul-Khaliq and R. Moosdorf, *J. Surg. Res.* 128(1), 9 (2005).

[35] M. L. Hess and N. H. Manson, *J. Mol. Cell Cardiol.* 16(11), 969 (1984).

[36] Q. Chen, E. J. Vazquez, S. Moghaddas, C. L. Hoppel and E. J. Lesnefsky, *J. Biol. Chem.* 278(38), 36027 (2003).

[37] L. B. Becker, T. L. vanden Hoek, Z. H. Shao, C. Q. Li and P. T. Schumacker, *Am. J. Physiol.* 277(6 Pt 2), H2240 (1999).

[38] J. G. Tidball, *Am. J. Physiol. Regul. Integr. Comp. Physiol.* 288(2), R345 (2005).

[39] L. Formigli, L. D. Lombardo, C. Adembri, S. Brunelleschi, E. Ferrari and G. P. Novelli, *Hum. Pathol.* 23(6), 627 (1992).

[40] D. Ricklin, G. Hajishengallis, K. Yanga and J. D. Lambris, *Nat. Immunol.* 11(9), 785 (2010).

[41] A. Brueggemann, A. Noltze, T. Lange, M. Kaun, J. Gliemroth, S. Goerg, L. Bahlmann, S. Klaus, F. Siemers, P. Mailaender and H. G. Machens, *J. Surg. Res.* 150(1), 125 (2008).

[42] T. V. Arumugam, I. A. Shiels, T. M. Woodruff, D. N. Granger and S. M Taylor, *Shock* 21(5), 401 (2004).

[43] D. Inderbitzin, G. Beldi, I. Avital, G. Vinci and D. Candinas, *Eur. Surg. Res.* 36(3), 142 (2004).

[44] C. D. Collard, A. Vakeva, C. Bukusoglu, G. Zund, C. J. Sperati, S. P. Colgan and G. L. Stahl, *Circulation* 96(1), 326 (1997).

[45] C. Mold and C. A. Morris, *Immunology* 102(3), 359 (2001).

[46] R. K. Chan, S. I. Ibrahim, K. Takahashi, E. Kwon, M. McCormack, A. Ezekowitz, M. C. Carroll, F. D. Moore Jr. and W. G. Austen Jr., *J. Immunol.* 177(11), 8080 (2006).

[47] T. F. Lindsay, J. Hill, F. Ortiz, A. Rudolph, C. R. Valeri, H. B. Hechtman and F. D. Moore Jr., *Ann. Surg.* 216(6), 677 (1992).

[48] A. E. Davis 3rd, P. Mejia and F. Lu, *Mol. Immunol.* 45(16), 4057 (2008).

[49] E. W. Nielsen, T. E. Mollnes, J. M. Harlan and R. K. Winn, *Scand. J. Immunol.* 56(6), 588 (2002).

[50] G. A. Toomayan, L. E. Chen, H. X. Jiang, W. N. Qi, A. V. Seaber, M. M. Frank and J. R. Urbaniak, *Microsurgery* 23(6), 561 (2003).

[51] C. Kyriakides, Y. Wang, W. G. Austen Jr., J. Favuzza, L. Kobzik, F. D. Moore Jr. and H. B. Hechtman, *Am. J. Physiol. Cell Physiol.* 281(1), C224 (2001).

[52] K. Fattouch, G. Bianco, G. Speziale, R. Sampognaro, C. Lavalle, F. Guccione, P. Dioguardi and G. Ruvolo, *Eur. J. Cardiothorac. Surg.* 32(2), 326 (2007).

[53] M. Thielmann, G. Marggraf, M. Neuhauser, J. Forkel, U. Herold, M. Kamler, P. Massoudy and H. Jakob, *Eur. J. Cardiothorac. Surg.* 30(2), 285 (2006).

[54] J. L. Zweier, P. Kuppusamy and G. A. Lutty, *Proc. Natl. Acad. Sci. U. S. A.* 85(11), 4046 (1988).

[55] P. Golino, M. Ragni, P. Cirillo, V. E. Avvedimento, A. Feliciello, N. Esposito, A. Scognamiglio, B. Trimarco, G. Iaccarino, M. Condorelli, M. Chiariello and G. Ambrosio, *Nat. Med.* 2(1), 35 (1996).

[56] K. Zacharowski, P. Zacharowski, S. Reingruber and P. Petzelbauer, *J. Mol. Med. (Berl)* 84(6), 469 (2006).

[57] R. L. Perez and J. Roman, *J. Immunol.* 154(4), 1879 (1995).

[58] J. Qi and D. L. Kreutzer, *J. Immunol.* 155(2), 867 (1995).

[59] A. F. Drew, H. Liu, J. M. Davidson, C. C. Daugherty and J. L. Degen, *Blood* 97(12), 3691 (2001).

[60] J. A. Wilberding, V. A. Ploplis, L. McLennan, Z. Liang, I. Cornelissen, M. Feldman, M. E. Deford, E. D. Rosen and F. J. Castellino, *Ann. N. Y. Acad. Sci.* 936, 542 (2001).

[61] A. G. Harris, R. Leiderer, F. Peer and K. Messmer, *Am. J. Physiol.* 271(6 Pt 2), H2388 (1996).

[62]   V. J. Milazzo, R. J. Ferrante, F. Sabido, M. B. Silva Jr., R. W. Hobson 2nd and W. N. Duran, *J. Surg. Res.* 61(1), 139 (1996).

[63]   P. Petzelbauer, P. A. Zacharowski, Y. Miyazaki, P. Friedl, G. Wickenhauser, F. J. Castellino, M. Groger, K. Wolff and K. Zacharowski, *Nat. Med.* 11(3), 298 (2005).

[64]   L. Verstrepen, T. Bekaert, T. L. Chau, J. Tavernier, A. Chariot and R. Beyaert, *Cell Mol. Life Sci.* 65(19), 2964 (2008).

[65]   J. l. Eliason and T. W. Wakefield, *Semin. Vasc. Surg.* 22(1), 29 (2009).

[66]   W. N. Qi, P. Chaiyakit, Y. Cai, D. M. Allen, L. E. Chen, A. V. Seaber and J. R. Urbaniak, *Microsurgery* 24(4), 316 (2004).

[67]   G. D. Rushing and L. D. Britt, *J. Am. Coll. Surg.* 204(5), 964 (2007).

[68]   S. T. Lille, S. R. Lefler, A. Mowlavi, H. Suchy, E. M. Boyle Jr., A. L. Farr, C. Y. Su, N. Frank and D. C. Mulligan, *Muscle Nerve* 24(4), 534 (2001).

[69]   D. P. Pearlstein, M. H. Ali, P. T. Mungai, K. L. Hynes, B. L. Gewertz and P. T. Schumacker, *Arterioscler. Thromb. Vasc. Biol.* 22(4), 566 (2002).

[70]   R. S. Crawford, H. Albadawi, M. D. Atkins, J. J. Jones, M. F. Conrad, W. G. Austen Jr., M. P. Fink and M. T. Watkins, *J. Trauma* 70(1), 103 (2011).

[71]   J. Jiang, J. Wang, C. Li, S. P. Yu and L. Wei L, *Vasc. Med.* 14(1), 37 (2009).

[72]   J. R. Scott, M. A. Cukiernik, M. C. Ott, A. Bihari, A. Badhwar, D. K. Gray, K. A. Harris, N. G. Parry and R. F. Potter, *Am. J. Physiol. Gastrointest. Liver Physiol.* 296(1), G9 (2009).

[73]   M. F. Conrad, D. H. Stone, H. Albadawi, H. T. Hua, F. Entabi, M. C. Stoner and M. T. Watkins, *Surgery* 138(2), 375 (2005).

[74]   E. G. Sheu, S. M. Oakes, C. Ahmadi-Yazdi, J. Afnan, M. C. Carroll and F. D. Moore Jr., *Surgery* 146(2), 340 (2009).

[75]   H. X. Nguyen, A. J. Lusis and J. G. Tidball, *J. Physiol.* 565(Pt 2), 403 (2005).

[76]   S. E. McAllister, M. A. Moses, K. Jindal, H. Ashrafpour, N. J. Cahoon, N. Huang, P. C. Neligan, C. R. Forrest, J. E. Lipa and C. Y. Pang, *J. Appl. Physiol.* 106(1), 20 (2009).

[77]   A. B. Groeneveld, P. G. Raijmakers, J. A. Rauwerda and C. E. Hack, *Eur. J. Vasc. Endovasc. Surg.* 14(5), 351 (1997).

[78]   D. W. Harkin, A. A. Barros D'Sa, K. McCallion, M. Hoper and F. C. Campbell, *J. Vasc. Surg.* 35(6), 1264 (2002).

[79]   S. Yagihashi, H. Mizukami, S. Ogasawara, S. Yamagishi, H. Nukada, N. Kato, C Hibi, S. Chung and S. Chung, *J. Pathol.* 220(5), 530 (2010).

[80]   P. G. Raijmakers, A. B. Groeneveld, J. A. Rauwerda, A. J. Schneider, G. J. Teule, C. E. Hack and L. G. Thijs, *Am. J. Respir. Crit. Care Med.* 151(3 Pt 1), 698 (1995).

[81]   E. Ascer, C. Mohan, M. Gennaro and S. Cupo, *Ann. Vasc. Surg.* 6(1), 69 (1992).

[82]   C. E. Murry, R. B. Jennings and K. A. Reimer, *Circulation* 74(5), 1124 (1986).

[83]   A. Nakano, G. Heusch, M. V. Cohen and J. M. Downey, *Basic Res. Cardiol.* 97(1), 35 (2002).

[84]   K. Przyklenk, B. Bauer, M. Ovize, R. A. Kloner and P. Whittaker, *Circulation* 87(3), 893 (1993).

[85]   T. J. Pell, G. F. Baxter, D. M. Yellon and G. M. Drew, *Am. J. Physiol.* 275(5 Pt 2), H1542 (1998).

[86]   Y. P. Wang, H. Maeta, K. Mizoguchi, T Suzuki, Y. Yamashita and M Oe, *Cardiovasc. Res.* 55(3), 576 (2002).

[87] T. Oxman, M. Arad, R. Klein, N. Avazov and B. Rabinowitz, *Am. J. Physiol.* 273(4 Pt 2), H1707 (1997).

[88] M. Cour, L. Gomez, N. Mewton, M. Ovize and L. Argaud, *J. Cardiovasc. Pharmacol. Ther.* 16(2), 117 (2011).

[89] Q. Ji, Y. Mei, X. Wang, J. Feng, D. Wusha, J. Cai and Y. Zhou, *Int. Heart J.* 52(5), 312 (2011).

[90] C. Y. Pang, R. Z. Yang, A Zhong, N. Xu, B. Boyd and C. R. Forrest, *Cardiovasc. Res.* 29(6), 782 (1995).

[91] P. D. Addison, P. C. Neligan, H. Ashrafpour, A. Khan, A. Zhong, M. Moses, C. R. Forrest and C. Y. Pang, *Am. J. Physiol. Heart Circ. Physiol.* 285(4), H1435 (2003).

[92] F. Thaveau, J. Zoll, O. Rouyer, N. Chafke, J. G. Kretz, F. Piquard, and B Geny, *J. Vasc. Surg.* 46(3), 541 (2007).

[93] S. Pasupathy and S. Homer-Vanniasinkam, *Eur. J. Vasc. Endovasc. Surg.* 29(2), 106 (2005).

[94] S. Papanastasiou, S. E. Estdale, S. Homer-Vanniasinkam and R. T. Mathie, *Br. J. Surg.* 86(7), 916 (1999).

[95] S. S. Huang, F. C. Wei, L. M. Hung, *Circ. J.* 70(8), 1070 (2006).

[96] Z. Mansour, J. Bouitbir, A. L. Charles, S. Talha, M. Kindo, J. Pottecher, J. Zoll and B. Geny, *J. Vasc. Surg.* 55(2), 497 (2012).

[97] K. R. Eberlin, M. C. McCormack, J. T. Nguyen, H. S. Tatlidede, M. A. Randolph and W. G. Austen Jr., *J. Surg. Res.* 148(1), 24 (2008).

[98] M. A. Moses, P. D. Addison, P. C. Neligan, H. Ashrafpour, N. Huang, M. Zair, A. Rassuli, C. R. Forrest, G. J. Grover and C. Y Pang, *Am. J. Physiol. Heart Circ. Physiol.* 288(2), H559 (2005).

[99] R. K. Kharbanda, M. Peters, B. Walton, M. Kattenhorn, M. Mullen, N. Klein, P. Vallance, J. Deanfield and R. MacAllister, *Circulation* 103(12), 1624 (2001).

[100] R. K. Kharbanda, U. M. Mortensen, P. A. White, S. B. Kristiansen, M. R. Schmidt, J. A. Hoschtitzky, M. Vogel, K. Sorensen, A. N. Redington and R. MacAllister, *Circulation* 106(23), 2881 (2002).

[101] Z. Q. Zhao and J. Vinten-Johansen, *Cardiovasc. Res.* 70(2), 200 (2006).

[102] S. E. McAllister, H. Ashrafpour, N. Cahoon, N. Huang, M. A. Moses, P. C. Neligan, C. R. Forrest, J. E. Lipa and C. Y. Pang, *Am. J. Physiol. Regul. Integr. Comp. Physiol.* 295(2), R681 (2008).

[103] A. L. Charles, A. S. Guilbert, J. Bouitbir, P. Goette-Di Marco, I Enache, J. Zoll J, F. Piquard and B. Geny, *Br. J. Surg.* 98(4), 511 (2011).

[104] E. Gyurkovics, P. Aranyi, R. Stangl, P. Onody, G. Ferreira, G. Lotz, P. Kupcsulik and A. Szijarto, *J. Surg. Res.* 169(1), 139 (2011).

[105] H. Y. Chen, J. A. Navia, S. Shafique and G. S. Kassab, *J. Biomech.* 43(2), 221 (2010).

[106] M. I. Dar, T. Gillott, F. Ciulli and G. J. Cooper, *Ann. Thorac. Surg.* 71(3), 794 (2001).

[107] Z. Mansour, A. L. Charles, J. Bouitbir, J. Pottecher, M. Kindo, J. P. Mazzucotelli, J. Zoll and B. Geny, *J. Vasc. Surg.* 55(2), 497 (2012).

[108] F. Thaveau, J. Zoll, J. Bouitbir, B. N'Guessan, P. Plobner, N. Chakfe, J. G. Kretz, R. Richard, F. Piquard and B. Geny, *Fundam. Clin. Pharmacol.* 24(3), 333 (2010).

[109] P. M. Walker, T. F. Lindsay, R. Labbe, D. A. Mickle and A. D. Romaschin, *J. Vasc. Surg.* 5(1), 68 (1987).

[110] L. Zhang, C. G. Looney, W. N. Qi, L. E. Chen, A. V. Seaber, J. S. Stamler and J. R. Urbaniak, *J. Appl. Physiol.* 94(4), 1473 (2003).

[111] K. A. Reimer and R. B. Jennings, *Circulation* 71(5), 1069 (1985).

[112] M. J. Hickey, K. R. Knight, D. A. Lepore, J. V. Hurley and W. A. Morrison, *Microsurgery* 17(9), 517 (1996).

[113] C. Kyriakides, W. G. Austen Jr., Y. Wang, J. Favuzza, F. D. Moore Jr. and H. B. Hechtman, *J. Trauma* 48(1), 32 (2000).

[114] G. C. Gaines, M. B. Welborn 3rd, L. L. Moldawer, T. S. Huber, T. R. Harward and J. M. Seeger, *J. Vasc. Surg.* 29(2), 370 (1999).

[115] J. E. Gadek, S. W. Hosea, J. A. Gelfand, M. Santaella, M. Wickerhauser, D. C. Triantaphyllopoulos and M. M. Frank, *N. Engl. J. Med.* 302(10), 542 (1980).

In: Skeletal Muscle
Editor: Mark Willems

ISBN: 978-1-62417-271-7
© 2013 Nova Science Publishers, Inc.

Chapter 6

EXERCISE-INDUCED MITOCHONDRIAL BIOGENESIS IN SKELETAL MUSCLE: MECHANISMS OF ACTIVATION THROUGH PEROXISOME PROLIFERATOR-ACTIVATED RECEPTOR GAMMA CO-ACTIVATOR 1 ALPHA (PGC-1A)

Jonathan P. Little[1] and Brendon Gurd[2]*
[1]School of Health and Exercise Science,
University of British Columbia Okanagan, Kelowna,
British Columbia, Canada
[2]School of Kinesiology, Queen's University,
Kingston, Ontario, Canada

ABSTRACT

Exercise training has long been known to increase mitochondrial content in skeletal muscle but only recently has research begun to characterize the molecular mechanisms underlying this adaptive response. The transcriptional co-activator peroxisome proliferator-activated receptor γ co-activator 1-α (PGC-1α) has emerged as a 'master regulator' of mitochondrial biogenesis in several tissues and accumulating evidence indicates that this protein plays a critical role in exercise-induced mitochondrial biogenesis in skeletal muscle. The regulation of PGC-1α is clearly complex with several post-translational modifications as well as sub-cellular localization influencing its activity. In this chapter, we will highlight the role of PGC-1α in the mitochondrial adaptive response to exercise with particular focus on the molecular mechanisms of PGC-1α activation. Further understanding of how PGC-1α is regulated by exercise may help with the basic comprehension of mitochondrial biogenesis and could also be relevant to aging and numerous chronic disease states where skeletal muscle mitochondrial impairments, and reduced PGC-1α activation, have been implicated.

---

* Email address: jonathan.little@ubc.ca

INTRODUCTION

In 1967, John Holloszy elegantly demonstrated that several weeks of endurance-based exercise training resulted in an increase in mitochondrial protein content, enzyme activity, and oxygen consumption in rat skeletal muscle [1]. These findings of exercise-induced mitochondrial biogenesis were subsequently confirmed in humans following cross-sectional studies comparing athletes to non-active controls [2] as well as longitudinal training studies using the percutaneous needle muscle biopsy technique [3]. Despite over 40 years of research examining the influence of exercise on skeletal muscle mitochondria, it is only recently that we have begun to unravel the cellular and molecular mechanisms that mediate the adaptive response to training. The application of modern molecular techniques, along with some findings in transgenic animals, has greatly improved our understanding of exercise-induced mitochondrial biogenesis. These studies have provided evidence that various signals (e.g., Ca^{2+}, adenosine monophosphate [AMP], reactive oxygen species [ROS]) produced within working skeletal muscle during acute exercise activate selected intracellular signaling pathways (e.g., calcium/calmodulin-dependent protein kinase [CaMK], AMP-activated protein kinase [AMPK], p38 mitogen activated protein kinase [MAPK]) that culminate in increased transcription of distinct target genes [4, 5]. Throughout the course of training, the cumulative effect of these acute pulses in gene trancsription results in the synthesis of new proteins that constitute the muscle adaptive response. In the case of mitochondrial biogenesis, incorporation of newly synthesized proteins into existing mitochondria results in an increase in mitochondrial density, elevated oxidative capacity, and improved muscle metabolic function. The discovery of peroxisome proliferator-activated receptor gamma co-activator (PGC)-1α represents a seminal moment in our understanding of the mechanisms underlying mitochondrial biogenesis [6]. PGC-1α is a transcriptional co-activator that functions as a critical regulator of mitochondrial and metabolic gene expression, particularly in skeletal muscle [7-9]. Recent findings suggest that PGC-1α plays a key role in integrating intramyocellular exercise-related signals to the downstream output of mitochondrial gene transcription [10-12]. In this chapter, we will discuss how exercise is thought to induce mitochondrial biogenesis, with a specific focus on the molecular mechanisms of PGC-1α activation in response to acute exercise. Elucidating how PGC-1α is activated has not only improved our basic understanding of muscle biology and adaptation to exercise, but may also shed important insight on therapeutic targets for aging and chronic diseases (e.g., type 2 diabetes) where mitochondrial biogenesis may be impaired.

PGC-1ALPHA: A BRIEF HISTORICAL OVERVIEW FOCUSING ON SKELETAL MUSCLE

In 1998, PGC-1α was discovered in brown adipocytes using a yeast two-hybrid screen for peroxisome proliferator-activated receptor γ (PPARγ) interacting proteins [6]. Characterization of this newly discovered protein indicated that it was upregulated in thermogenic brown fat and skeletal muscle of mice upon exposure to cold and its overexpression led to increases in several mitochondrial respiratory chain enzymes [6]. Soon thereafter, studies began exploring the molecular mechanisms involved in PGC-1α regulation.

As a transcriptional co-activator, PGC-1 interacts with numerous transcription factors that bind to the promoters of mitochondrial genes [8, 13-15]. In this fashion, PGC-1α controls the expression of entire gene networks to co-ordinate mitochondrial and metabolic gene adaptation. Specific to skeletal muscle, Wu et al. [13] used C2C12 myotubes to demonstrate that PGC-1α increased mitochondrial gene transcription through co-activation of nuclear respiratory factor (NRF)-1, a transcription factor that promotes expression of numerous nuclear-encoded subunits of the electron transport chain (ETC). This PGC-1α:NRF-1 interaction also induced the expression of mitochondrial transcription factor A (Tfam), a transcription factor required for transcription of 13 essential ETC subunits encoded within mitochondrial DNA. The description of this axis (PGC-1α:NRF-1:Tfam) demonstrated that activation of a single protein, PGC-1α, could co-ordinately activate nuclear- and mitochondrial DNA-encoded mitochondrial proteins [13]. This co-ordinated activation of both genomes is likely to be essential for effective activation of mitochondrial biogenesis.

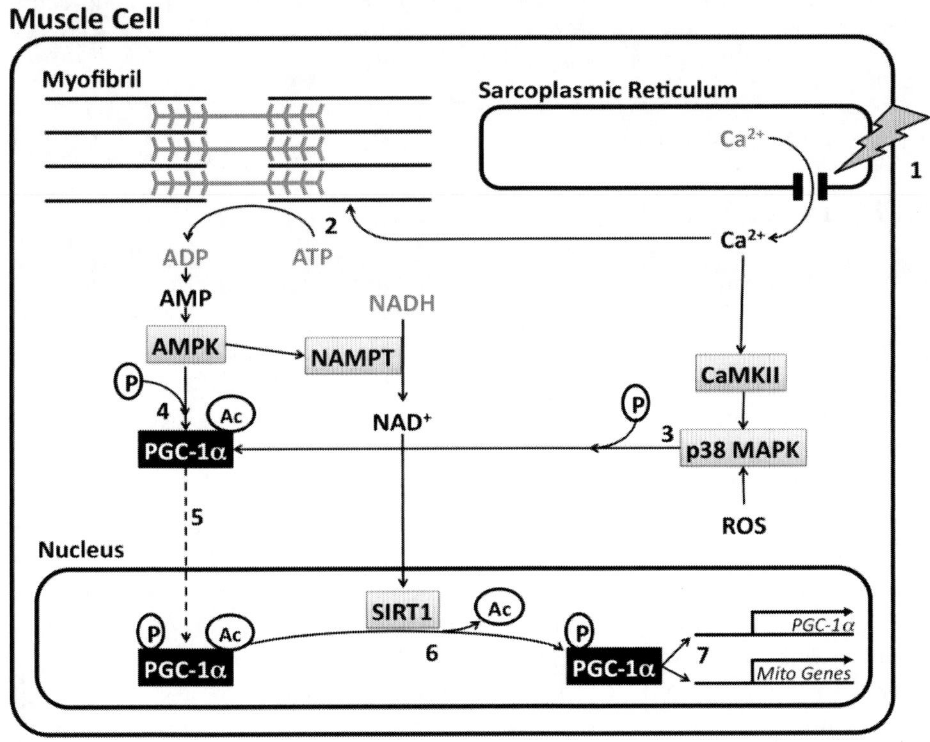

Figure 1. Putative mechanisms underlying the activation of PGC-1α following muscle contraction.Excitation of the muscle membrane results in release of calcium (Ca^{2+}) from the sarcoplasmic reticulum (**1**) and subsequent contraction and breakdown of ATP at the myofibril (**2**). The increase in intramuscular Ca^{2+} also leads to activation of the CaMK signalling pathway. Both CaMK and reactive oxygen species (ROS) are postulated to activate p38 MAPK which phosphorylates PGC-1α (**3**). Concomitantly, accumulation of AMP activates AMPK triggering both phosphorylation of PGC-1α (**4**) and the activation of NAMPT. While we have depicted phosphorylation of PGC-1α inducing its translocation from the cytosol into the nucleus (**5**), it is also possible that phosphorylation events occur within the nucleus. In the nucleus, PGC-1α is believed to be activated by SIRT1 through deacetylation (**6**).

PGC-1ALPHA AND EXERCISE

Early research examining the importance of PGC-1α in skeletal muscle demonstrated concurrent increases in PGC-1α expression at both the mRNA [16] and protein [17] levels following 3-7 days of swim training in rats. While these findings indicated that increased PGC-1α expression might be involved in mediating an increase in mitochondrial gene transcription with exercise, they could not explain the rapid increase in mitochondrial gene transcription that is evident almost immediately following a single, acute bout of exercise [18,19]. Thus, it was subsequently argued that if PGC-1α is involved in the skeletal muscle adaptive response to exercise, a mechanism of acutely activating this protein must exist [18]. Indeed, accumulating evidence indicates that PGC-1α activity is acutely induced by exercise and that this activation may be regulated by a myriad of intramuscular events including sub-cellular localization [18-21], post-translational modifications [10], and interaction with activator and/or inhibitor proteins [22] (See Figure 1). The following sections will present the evidence highlighting how acute activation of PGC-1α may be involved in the upregulation of mitochondrial gene transcription seen following exercise in skeletal muscle.

MECHANISMS OF EXERCISE-INDUCED PGC-1ALPHA ACTIVATION

Sub-Cellular Localization

For a transcriptional co-activator to be active, it must be present in the nucleus of the cell. There are numerous examples of transcriptional regulatory proteins that are regulated through changes in sub-cellular localization (e.g., shuttling of NFκB from the cytosol to the nucleus in response to an inflammatory trigger [23]). Initial reports using immunocytochemistry indicated that ectopically expressed PGC-1α protein was expressed exclusively in the nucleus using COS-7 cells (a fibroblast-like cell line derived from monkeys) [6]. However, several subsequent reports, across a range of tissues, indicate that PGC-1α is also present in the cytosol [24-27]. In skeletal muscle from rodents, Wright and colleagues [18] demonstrated that the majority of PGC-1α protein, detected by Western blotting following sub-cellular fractionation, was detected in cytosolic compared to nuclear fractions. Further, an increase in nuclear PGC-1α protein was detected following acute exercise (2 h swimming) that corresponded with increased NRF-2 activation and mitochondrial gene transcription [18]. These results suggested that PGC-1α shuttled from the cytosol to the nucleus following exercise and implicated this change in cellular localization in the acute control of PGC-1α activity. These findings were subsequently confirmed in human skeletal muscle, where acute endurance [21] and interval-based [20] exercise resulted in an increase in the nuclear abundance of PGC-1α that coincided with increases in mRNA expression of several mitochondrial enzymes. These results suggest that an increase in total PGC-1α protein may not be required for PGC-1α-mediated mitochondrial biogenesis; exercise may cause a redistribution of PGC-1α from the cytosol to the nucleus where it can function to activate mitochondrial gene transcription. It must be noted that direct evidence of cytosolic-nuclear shuttling in skeletal muscle has remained elusive. It would be of interest to employ immunofluorescence techniques of skeletal muscle cross-sections to more definitively demonstrate cytosolic-nuclear shuttling, although antibody specificity issues appear to limit use of measuring native PGC-1α with immunostaining techniques.

With respect to PGC-1α sub-cellular localization, several immunofluorescence-based studies in cell lines indicate that PGC-1α, and several of its interacting proteins, can be detected in punctate sub-nuclear domains [28-30]. The potential significance of these findings in skeletal muscle mitochondrial biogenesis is currently unknown, but there is evidence of dynamic regulation of PGC-1α between nuclear sub-domains [24, 30]. The sub-nuclear localization of PGC-1α in skeletal muscle remains an intriguing potential mechanism of activation that requires further study in the context of exercise.

Post-Translational Modifications

The acute exercise-induced increase in nuclear PGC-1α, along with increased transcription of genes controlled by PGC-1α, raises several intriguing questions. What signals might promote nuclear redistribution of PGC-1α? Is intrinsic transactivational capacity of nuclear PGC-1α affected by exercise? Post-translational modifications to PGC-1α are emerging as key players that may help answer these questions.

Phosphorylation by 5'AMP-Activated Protein Kinase (AMPK)

AMPK is a highly-conserved energy sensing enzyme that is intimately involved in regulating metabolism in skeletal muscle [31]. AMPK is activated in response to increases in intramuscular AMP:ATP or ADP:ATP ratios, both of which change in proportion to exercise intensity. When active, AMPK increases phosphorylation of several downstream targets generally promoting an increase in cellular processes that generate energy (e.g., fatty acid oxidation, glucose uptake) while inhibiting processes that consume ATP (e.g., protein synthesis) [31]. AMPK is also intimately involved in regulating gene transcription [32]. A role for AMPK as a direct activator of PGC-1α was provided when Jäger et al. [11] discovered two AMPK phosphorylation sites (Thr-177, Ser-538) on PGC-1α. Mutation of these phosphorylation sites reduced PGC-1α activation and mitochondrial gene expression following pharmacological activation of AMPK in muscle cells [11]. Moreover, pharmacological activation of AMPK in mice lacking PGC-1α in skeletal muscle (muscle-specific PGC-1α knock-out mice) failed to show increases in mitochondrial biogenesis [11]. These findings suggested that the well-established links between AMPK activation and mitochondrial biogenesis [32, 33] are mediated by PGC-1α.

The degree of AMPK activation is proportional to exercise intensity [34, 35], as well as the extent of glycogen depletion [36]. It could therefore be surmised that maximal activation of PGC-1α would occur following high-intensity, glycogen-depleting exercise. If this is the case, high-intensity interval-type exercise, which results in substantial increases in intramuscular AMP/ADP/ATP ratio [37] and rapidly depletes muscle glycogen [38], may be a potent strategy for inducing mitochondrial biogenesis. Preliminary evidence is in agreement that exercise intensity is positively related to PGC-1α activation [39, 40] and that high-intensity intermittent exercise, even when total volume of work is low, robustly activates PGC-1α and increases mitochondrial gene expression [20]. These findings suggest that activation of the AMPK-PGC-1α pathway may be involved in regulating mitochondrial adaptations seen following high-intensity interval training [41-44].

Phosphorylation by p38 Mitogen Activated Protein Kinase (MAPK)

p38 MAPK is also implicated as a potential activator of PGC-1α. Primary support for a relationship between p38 MAPK and PGC-1α activation following exercise comes from data that indicate p38 is itself activated following exercise in skeletal muscle of rodents [18, 45] and humans [20, 21, 46, 47]. Work in isolated skeletal muscle indicates that this activation may occur in a calcium dependent fashion following a pathway that leads from CaMK to p38 MAPK [48]. Transient increases in ROS during exercise may also be involved in p38 MAPK activation [4, 5]. Pharmacological inhibition of p38 MAPK suppresses PGC-1α activation and blunts exercise-induced mitochondrial gene transcription in mice [49]. Evidence from transgenic mouse models also supports a role of p38 MAPK in the control of PGC-1α transcriptional activity. Specifically, mice with constitutively active p38 MAPK have increased PGC-1α-mediated mitochondrial biogenesis [45]; while muscle-specific p38 MAPK knock-out mice display reduced activation of PGC-1α and an absence of mitochondrial biogenesis in response to exercise [50].

While the exact mechanisms linking p38 MAPK phosphorylation to PGC-1α-mediated transcriptional activity remains undefined there are several compelling possibilities. First, early mechanistic studies demonstrated that p38 MAPK could directly phosphorylate PGC-1αat three amino acid residues (Thr-262, Ser-265, and Thr-298) leading to increased PGC-1α protein stability and activation of mitochondrial gene transcription [51]. Alternatively, or perhaps additionally, p38 MAPK also increased PGC-1α activation by relieving inhibition from a repressive interacting protein called p160 myb binding protein [22, 52]. Although interaction of PGC-1α with p38 MAPK in human skeletal muscle has been shown via co-immunoprecipitation[47], direct phosphorylation by p38 MAPK and interaction with p160 myb binding protein requires further exploration, particularly with respect to exercise.

In our studies in humans, exercise appears to cause a predominant activation of p38 MAPK in the cytosol [20, 21], followed temporally by increased nuclear abundance of PGC-1α and activation of mitochondrial gene transcription [20]. It is therefore tempting to speculate that p38 MAPK may be involved in regulating the sub-cellular distribution of PGC-1α, although this speculative hypothesis requires confirmation.

Deacetylation by SIRT1 and Acetylation by GCN5

In addition to modification by phosphorylation, PGC-1α activity is also altered by reciprocal acetylation:deacetylation on 13 identified lysine residues [29]. The deacetylase SIRT1 (silent mating type information regulation 2 homolog 1) activates, and the acetyl-transferase GCN5 (general control non-repressed protein 5) inhibits PGC-1αtranscriptional activity in skeletal muscle cells [53]. While the role of SIRT1 in skeletal muscle remains controversial [54], increases in SIRT1 activity (measured by commercially available assay) have been observed in humans [55] and rodents [19,56] following exercise and/or chronic contractile activity. In addition, and consistent with SIRT1 acutely activating PGC-1α, SIRT1 is activated [19] and PGC-1αdeacetylated[10,57] following an acute bout of exercise in rodents. The activation of SIRT1 by exercise is linked to changes in the metabolic state of the cell as SIRT1 activity has long been known to be sensitive to changes in NAD+ [28,29]. More recently SIRT1 activity has been shown to be impacted by AMPK activation [10, 58, 59], providing a mechanism whereby exercise-mediated metabolic disturbances could activate SIRT1 leading to deacetylation of PGC-1α.

Interestingly, a recent report has demonstrated normal PGC-1αdeacetylation and activation following exercise in mice expressing an inactive form of SIRT1 in skeletal muscle [57]. This illustrates the controversy currently surrounding SIRT1 in skeletal muscle, but also highlights the potential importance of GCN5 in exercise-mediated activation of PGC-1α. GCN5 acetylates and inhibits PGC-1α activity in C2C12 myotubes [53]. Further, GCN5 activity is regulated by changes in nutrient availability and cellular energy status [60]. Interestingly, it has been shown that there is a reduction in nuclear GCN5 and reduced interaction between PGC-1α and GCN5 following exercise in mouse skeletal muscle [57], suggesting that GCN5 may play a predominant role in regulating PGC-1α acetylation status in response to exercise. While this evidence suggests a role for GCN5 in controlling PGC-1α activity, there is little information available regarding changes in GCN5 activity following either exercise training or an acute bout of exercise.

Other Post-Translational Modifications

Several other post-translational modifications to PGC-1α have been described in the literature, although it must be acknowledged that not all of these have been demonstrated to occur in skeletal muscle. Nonetheless, some speculations as to the role of these modifications in exercise-induced mitochondrial biogenesis can be drawn and may enlighten future research.

Phosphorylation by Glycogen Synthase Kinase (GSK)3β

Anderson et al. [24] demonstrated that glycogen synthase kinase (GSK) 3β phosphorylated PGC-1α at Thr-295 in cells and skeletal muscle from mice. GSK3β phosphorylation appeared to tag PGC-1α for ubiquitination and subsequent degradation in the nuclear proteosome. Since the activation of GSK3β was induced by both oxidative stress and caloric restriction in this study, this mechanism may be relevant to exercise, where transient increases in reactive oxygen species (ROS) and changes in nutrient availability are thought to act as signaling molecules involved in exercise adaptation [10,61,62]. Olson and colleagues [63] extended this work by showing that prior p38 MAPK phosphorylation was required for GSK3β to act upon PGC-1α, followed by ubiquitin-mediated degradation. This suggests that activation of PGC-1α by p38 MAPK may also precipitate its destruction via GSK3β and the ubiquitin-proteosome system. Because PGC-1α can activate its own promoter, through an autoregulatory loop involving co-activation of the myocyte enhancer factor (MEF) 2 transcription factor [64](see below for more details), this sequence of events could help explain how PGC-1α is dynamically activated, transcribed, synthesized, and degraded in the cell. If this model is correct, the relative degree of p38 MAPK to GSK3β activation may be critical in dictating PGC-1α protein content and the potential for mitochondrial biogenesis. While GSK3β remains an intriguing potential controller of PGC-1α activity there is a need for more work examining this pathway in skeletal muscle following exercise.

Phosphorylation by Akt

In liver, increased insulin inhibits the ability of PGC-1α to activate gene transcription through direct inhibitory phosphorylation of the Ser-570 site by Akt/protein kinase b (PKB) [65]. A recent study suggests that this pathway may also operate in skeletal muscle, as hyperinsulinemia resulting from short-term high-fat feeding in rodents led to elevated basal

Aktphoshorylation, reduced PGC-1α protein, and impaired mitochondrial biogenesis (potentially as a result of reduced PGC-1α transcriptional activity) [66]. Interestingly, acute endurance and interval exercise in humans results in a robust decrease in skeletal muscle Akt phosphorylation (Little, JP, unpublished observations). Short-term exercise training in overweight adults also leads to reduced Akt phosphorylation with a concomitant increase in total PGC-1α protein as well as markers of mitochondrial content [67]. These results raise the interesting possibility that relief of Akt inhibition may contribute to exercise-induced activation of PGC-1α in human skeletal muscle.

Phosphorylation by Cdc-Like Kinase 2 (Clk2)

Also in liver, the Akt associated protein Clk2 has been shown to directly phosphorylate and inhibit PGC-1α [68]. Because PGC-1α is intimately involved in hepatic gluconeogenesis, these effects appear central in insulin-mediated reduction of liver glucose production. We are unaware of any data on interactions between exercise, Clk2, and PGC-1α in skeletal muscle.

Other post-translational modifications affecting PGC-1α activity include inhibition by SUMOylation [69] and activation by methylation via protein arginine methyl transferase (PRMT) 1 [70]. O-linked β-N-acetyl glucosamination has also been shown but functional effects of this modification are unknown [71]. At present, the possibility that SUMOylation, methylation and O-linked β-N-acetyl glucosamination may be involved in exercise-induced PGC-1α activation in skeletal muscle has not been assessed.

A Final Word on Post-Translational Modifications

The influence of exercise on post-translational modification of PGC-1α appears to be an attractive line of inquiry for delineating the role of PGC-1α in skeletal muscle biology. However, it appears that techniques for measuring post-translational modifications to PGC-1α limit research in this area, particularly for *in vivo* studies involving human skeletal muscle biopsy specimens. Commercial antibodies specific to PGC-1α post-translational modifications are currently unavailable and attempts to immunoprecipitate PGC-1α with subsequent assessment of modified amino acid residues have proven largely unsuccessful, except in mouse models examining changes in acetylation status [10, 57, 59]. Advancements in techniques to measure PGC-1α post-translational modifications could allow for enhanced insight into PGC-1α-mediated mitochondrial biogenesis.

Activating and Inhibiting Proteins

In addition to subcellular local and post-translational modifications, there is also evidence that PGC-1α can form both activating and inhibiting complexes with other proteins. Indeed, there are several proteins that appear to interact with PGC-1α and either increase, or decrease its transactivational capacity. While there is a paucity of data on the influence of exercise on the relative abundance, and protein-protein interactions, of PGC-1αwith activators or inhibitors we feel a brief discussion of these potential modifiers is of interest.

Early work exploring how PGC-1α regulates gene expression indicated that upon docking to a transcription factor, a conformational change in PGC-1α promotes its binding to the histone acetyl transferase (HAT) proteins steroid receptor co-activator (SRC)-1 and CREB binding protein (CBP)/p300 [72]. Acetylation of histones leads to chromatin remodeling that allows for transcription factor binding to DNA and promoter activation [73]. In this manner, an additional function of PGC-1α as a molecular scaffold was described, with a role in orchestrating the molecular machinery for gene transcription [72]. Presumably interactions between HAT proteins and PGC-1α are involved in exercise-induced mitochondrial gene transcription, but research examining the role of histone modifications in the adaptive response to exercise is in its infancy [73, 74]. It is therefore unknown if exercise influences PGC-1α co-activator function by altering its interaction with HAT proteins. As mentioned above, cell culture studies have demonstrated that p160 myb binding protein can repress PGC-1α activity and reduce mitochondrial biogenesis [22, 52]. These inhibitory effects are reduced following p38 MAPK phosphorylation of PGC-1α [22, 52]. A relatively rapid means to increase PGC-1α activity may therefore occur though p38 MAPK derepression of p160. To our knowledge, the influence of exercise on skeletal muscle p160 myb binding protein, and its interaction with PGC-1α, have not been studied.

Another protein that appears to inhibit PGC-1α through direct interaction is RIP140 [75]. RIP140 is a transcriptional co-repressor that appears to act in opposition to PGC-1α by reducing mitochondrial gene transcription through inhibition of transcription factors [76]. In addition to these transcriptional regulatory functions, studies in adipocytes found that RIP140 can bind to PGC-1α and repress its activity [75]. Whether this also occurs in skeletal muscle remains to be determined. A few recent studies have examined the influence of exercise on RIP140 in skeletal muscle. Acute or chronic exercise in rodents or humans has no effect on total or nuclear RIP140 protein content in skeletal muscle [67, 77]. Thus, it appears that reductions in RIP140 are not involved in PGC-1α activation with exercise, although alterations in protein-protein interactions have not been studied. Further work examining the relative abundance of activating and inhibiting proteins, and regulation of protein-protein interactions, is needed to determine how exercise might influence mitochondrial biogenesis through alterations in PGC-1α protein complexes.

PGC-1α Autoregulation: mRNA as an Indicator of Activation

The most common method employed in the literature to assess "PGC-1α activation" has been to measure PGC-1α mRNA expression. The validity of this approach stems from findings in C2C12 myotubes demonstrating that PGC-1α protein regulates its own gene transcription through coactivation of the MEF2 transcription factor [64]. Since calcineurin and CaMKs activate MEF2 in response to changes in intracellular Ca^{2+} [64], this "autoregulatory loop" links muscle contraction with activation of existing PGC-1α protein to enhance PGC-1α gene transcription. Indeed, many studies have demonstrated a robust increase in skeletal muscle PGC-1α mRNA expression following various forms of exercise in rodents [10, 16-18, 45, 49, 78] and humans [20, 39, 40, 43, 46, 47, 79-83]. Although the approach of using PGC-1α mRNA expression as an indicator of PGC-1α activation is certainly valid and appealing, it should be remembered that this output is indicative of PGC-1α:MEF2 co-activation may not necessarily reflect the transactivational activity of PGC-1α on other transcription factors. For this reason, comprehensive analyses of PGC-1α regulation in skeletal muscle should include multiple outputs, including sub-cellular localization, activation of upstream signaling pathways, (ideally) post-translational modifications, abundance or interaction with activator/inhibitor proteins, and perhaps most importantly, markers of downstream mitochondrial gene expression.

CONCLUSION

It is widely accepted that skeletal muscle PGC-1α plays an integral role in the mitochondrial biogenic response to exercise. In this chapter, we have highlighted several possible mechanisms involved in activating PGC-1α protein in response to acute exercise. It appears that an increase in nuclear PGC-1α in the immediate post-exercise recovery period may represent a rapid means to activate mitochondrial gene transcription. The dynamic regulation of PGC-1α protein, including sub-nuclear localization and pathways involved in its degradation, represent emerging lines of research that could further our understanding of PGC-1α-directed mitochondrial biogenesis. Several post-translational modifications appear to influence exercise-induced PGC-1α activation, including phosphorylation by AMPK and p38 MAPK and deacetylation by SIRT1. These modifications have all been shown to occur in muscle cells and circumstantial evidence implicates these pathways in PGC-1α-mediated exercise adaptation *in vivo*. Other post-translational modifications, including phosphorylation by GSK3β and Akt, may be involved in skeletal muscle PGC-1α regulation but these require further investigation, especially in the context of exercise. Finally, interactions of PGC-1α with inhibiting and activating proteins may be an area of study that could improve our understanding of mitochondrial biogenesis. Further research exploring how PGC-1α is activated by exercise holds promise for enhancing the basic understanding of muscle mitochondrial adaptation to exercise and could provide potential therapeutic targets for increasing mitochondrial biogenesis to benefit the many chronic diseases where muscle mitochondrial content is reduced.

REFERENCES

[1] J.O. Holloszy, *J. Biol. Chem.*242, 2278 (1967).

[2] H. Hoppeler, P. Luthi, H. Claassen, E. R. Weibeland H. Howald,*Pflugers Arch.*344, 217(1973).

[3] P. D. Gollnick, R. B. Armstrong, B. Saltin, C. W. Saubert, W. L. Sembrowichand R. E. Shepherd, *J. Appl Physiol.* 34, 107 (1973).

[4] D. A. Hood, I. Irrcher, V. Ljubicicand A. M. Joseph, *J. Exp. Biol.*209, 2265 (2006).

[5] J. A. Hawley, M. Hargreaves and J. R. Zierath, *Essays Biochem.* 42, 1 (2006)

[6] P. Puigserver, Z. Wu, C. W. Park, R. Graves, M. Wrightand B. M. Spiegelman, *Cell*92, 829 (1998).

[7] V. A. Lira, C. R. Benton, Z. Yanand A. Bonen, *Am. J. Physiol. Endocrinol. Metab.* 299, E145 (2010).

[8] B. N. Finck and D. P. Kelly, *J. Clin. Invest.*116, 615 (2006).

[9] B. M. Spiegelman and R. Heinrich, *Cell*119, 157 (2004).

[10] C. Canto, L. Q. Jiang, A. S. Deshmukh, C. Mataki, A. Coste, M. Lagouge, J. R. Zierath and J. Auwerx, *Cell Metab.* 11, 213 (2010).

[11] S. Jäger, C. Handschin, J. St-Pierre and B. M. Spiegelman, *Proc. Natl. Acad. Sci. U. S. A.* 104, 12017 (2007).

[12] T. Geng, P. Li, M. Okutsu, X. Yin, J. Kwek, M. Zhang and Z. Yan, *Am. J. Physiol. Cell. Physiol.* 298, C572 (2010).

[13] Z. Wu, P. Puigserver, U. Andersson, C. Zhang, G. Adelmant, V. Mootha, A. Troy, S. Cinti, B. Lowell, R. C. Scarpulla and B. M. Spiegelman, *Cell* 98, 115 (1999).

[14] R. Ventura-Clapier, A. Garnierand V. Veksler, *Cardiovasc. Res.* 79, 208 (2008).

[15] R. C. Scarpulla, *Physiol. Rev.* 88, 611 (2008).

[16] M. Goto, S. Terada, M. Kato, M. Katoh, T. Yokozeki, I. Tabata and T. Shimokawa, *Biochem. Biophys. Res. Commun.* 274, 350 (2000).

[17] K. Baar, A. R. Wende, T. E. Jones, M. Marison, L. A. Nolte, M Chen, D. P. Kelly and J. O. Holloszy, *FASEB J* 16, 1879 (2002).

[18] D. C. Wright, D. H. Han, P. M. Garcia-Roves, P. C. Geiger, T. E. Jonesand J. O. Holloszy, *J. Biol. Chem.* 282, 194 (2007).

[19] B. J. Gurd, Y. Yoshida, J. T. McFarlan, G. P. Holloway, C. D. Moyes, G. J. Heigenhauser, L. Spriet and A. Bonen, *Am. J. Physiol. Regul. Integr. Comp. Physiol.* 301, R67 (2011).

[20] J. P. Little, A. Safdar, D. Bishop, M. A. Tarnopolsky and M. J. Gibala, *Am. J. Physiol. Regul. Integr. Comp. Physiol.* 300, R1303 (2011).

[21] J. P. Little, A. Safdar, N. Cermak, M. A. Tarnopolskyand M. J. Gibala, *Am. J. Physiol. Regul. Integr. Comp. Physiol.* 298, R912 (2010).

[22] M. Fan, J. Rhee, J. St-Pierre, C. Handschin, P. Puigserver, J. Lin, S. Jäeger, H. Erdjument-Bromage, P. Tempst and B. M. Spiegelman, *Genes Dev.* 18, 278 (2004).

[23] Q. Li and I. M. Verma, *Nat. Rev. Immunol.* 2, 725 (2002).

[24] R. M. Anderson, J. L. Barger, M. G. Edwards, K. H. Braun, C. E. O'Connor, T. A. Prolla and R. Weindruch, *Aging Cell* 7, 101 (2008).

[25] R. M. Cowell, K. R. Blakeand J. W. Russell, *J. Comp. Neurol.* 502, 1 (2007).

[26] J. S. Rim, B. Xue, B. Gawronska-Kozak and L. P. Kozak, *J. Biol. Chem.* 279, 25916 (2004).

[27] M. A. Lomax, F. Sadiq, G. Karamanlidis, A. Karamitri, P. Trayhurn and D. G. Hazlerigg, *Endocrinology* 148, 461 (2007).

[28] J. E. Dominy Jr., Y. Lee, Z. Gerhart-Hinesand P. Puigserver, *Biochimica Biophysica Acta* 1804, 1676 (2010).

[29] J. T. Rodgers, C. Lerin, Z. Gerhart-Hinesand P. Puigserver, *FEBS Letters* 582, 46 (2008).

[30] M. Sano, S. Tokudome, N. Shimizu, N. Yoshikawa, C. Ogawa, K. Shirakawa, J. Endo, T. Katayama, S. Yuasa, M. Ieda, S. Makino, F. Hattori, H. Tanaka, K. Fukuda, *J. Biol. Chem.* 282, 25970 (2007).

[31] D. G. Hardie and K. Sakamoto, *Physiology* 21, 48 (2006).

[32] S. L. McGee and M. Hargreaves, *Clin. Sci. (Lond).* 118, 507 (2010).

[33] C. T. Putman, M. Kiricsi, J. Pearcey, I. M. MacLean, J. A. Bamford, G. K. Murdoch, W. T. Dixon and D. Pette, *J. Physiol.* 551, 169 (2003).

[34] G. D. Wadley, R. S. Lee-Young, B. J. Canny, C. Wasuntarawat, Z. P. Chen, M. Hargreaves, B. E. Kemp and G. K. McConell, *Am. J. Physiol. Endocrinol. Metab.* 290, E694 (2006).

[35] Z. P. Chen, T. J. Stephens, S. Murthy, B. J. Canny, M. Hargreaves, L. A. Witters, B. E. Kemp and G. K. McConell, *Diabetes*52, 2205 (2003).

[36] J. F. P. Wojtaszewski, C. MacDonald, J. N. Nielsen, Y. Hellsten, D. G. Hardie, B. E. Kemp, B. Kiens and E. A. Richter, *Am. J. Physiol. Endocrinol. Metab.* 284, E813 (2003).

[37] M. L. Parolin, A. Chesley, M. P. Matsos, L. L. Spriet, N. L. Jonesand G. J. F. Heigenhauser *Am. J. Physiol. Endocrinol. Metab.*277, E890 (1999).

[38] J. D. MacDougall, G. R. Wardand J. R. Sutton, *J. Appl. Physiol.* 42, 129 (1977).

[39] B. Egan, B. P. Carson, P. M. Garcia-Roves, A. V. Chibalin, F. M. Sarsfield, N. Barron, N. McCaffrey, N. M. Moyna, J. R. Zierath and D. J. O'Gorman, *J. Physiol.*588, 1779 (2010).

[40] N. B. Nordsborg, C. Lundby, L. Leick and H. Pilegaard, *Med. Sci. Sports Exerc.* 42, 1477(2010).

[41] K. A. Burgomaster, K. R. Howarth, S. M. Phillips, M. Rakobowchuk, M. J. Macdonald, S. L. McGee, M. J. Gibala,*J. Physiol.* 586, 151 (2008).

[42] J. P. Little, A. Safdar, G. P. Wilkin, M. A. Tarnopolsky and M. J. Gibala, *J. Physiol.*588, 1011 (2010).

[43] C. G. Perry, J. Lally, G. P. Holloway, G. J. Heigenhauser, A. Bonen and L. L. Spriet, *J. Physiol.*588, 4795 (2010).

[44] J. L. Talanian, S. D. Galloway, G. J. Heigenhauser, A. Bonen and L. L. Spriet, *J. Appl. Physiol.*102, 1439 (2007).

[45] T. Akimoto, S. C. Pohnert, P. Li, M. Zhang, C. Gumbs, P. B. Rosenberg, R. S. Williams and Z. Yan,*J. Biol. Chem.*280, 19587 (2005).

[46] M. J. Gibala, S. L. McGee, A. P. Garnham, K. F. Howlett, R. J. Snowand M. Hargreaves, *J. Appl. Physiol.*106, 929 (2009).

[47] S. L. McGee and M. Hargreaves, *Diabetes*53, 1208 (2004).

[48] D. C. Wright, P. C. Geiger, D. H. Han, T. E. Jonesand J. O. Holloszy, *J. Biol. Chem.*282, 18793 (2007).

[49] S. Ikeda, T. Kizaki, S. Haga, H. Ohnoand T. Takemasa, *Biochem. Biophys. Res. Commun.*368, 323 (2008).

[50] A. R. Pogozelski, T. Geng, P. Li, X. Yin, V. A. Lira, M. Zhang, J. T. Chi and Z. Yan, *PLoS One*4, e7934 (2009).

[51] P. Puigserver, J. Rhee, J. Lin, Z. Wu, J. C. Yoon, C. Y. Zhang, S. Krauss, V. K. Mootha, B. B. Lowell and B. M. Spiegelman, *Mol. Cell.*8, 971 (2001).

[52] D. Knutti, D. Kressler and A. Kralli, *Proc. Natl. Acad. Sci. U. S. A.*98, 9713 (2001).

[53] Z. Gerhart-Hines, J. T. Rodgers, O. Bare, C. Lerin, S. H. Kim, R. Mostoslavsky, F. W. Alt, Z. Wu and P. Puigserver, *EMBO J.* 26, 1913 (2007).

[54] B. J. Gurd, J. P. Little, C. G. R. Perry, *J. Appl. Physiol.*112(5), 926 (2012).

[55] B. J. Gurd, C. G. R Perry, G. J. F. Heigenhauser, L. L. Sprietand A. Bonen, *Appl Physiol. Nutr. Metab.*35, 350 (2010).

[56] B. J. Gurd, Y. Yoshida, J. Lally, G. P. Holloway and A. Bonen, *J Physiol*587, 1817 (2009).

[57] A. Philp, A. Chen, D. Lan, G. A. Meyer, A. N Murphy, A. E. Knapp, I. M. Olfert, C. E. McCurdy, G. R. Marcotte, M. C. Hogan, K. Baar and S. Schenk, *J. Biol. Chem.*286, 30561 (2011).

[58] C. Canto and J. Auwerx, *Curr. Opin. Lipidol.*20, 98 (2009).

[59] C. Canto, Z. Gerhart-Hines, J. N. Feige, M. Lagouge, L. Noriega, J. C. Milne, P. J. Elliott, P. Puigserver and J. Auwerx, *Nature*458, 1056 (2009).

[60] E. H. Jeninga, K. Schoonjans and J. Auwerx, *Oncogene*29, 4617 (2010).

[61] C. Kang, K. M. O'Moore, J. R. Dickman and L. L. Ji, *Free Radic. Biol. Med.*47, 1394 (2009).

[62] M. Ristow, K. Zarse, A. Oberbach, N. Klöting, M. Birringer, M. Kiehntopf, M. Stumvoll, C. R. Kahn and M. Blüher, *Proc. Natl. Acad. Sci. U. S. A.*106, 8665 (2009).

[63] B. L. Olson, M. B. Hock, S. Ekholm-Reed, J. A. Wohlschlegel, K. K. Dev, A. Kralli and S. I. Reed, *Genes Dev.*22, 252 (2008).

[64] C. Handschin, J. Rhee, J. Lin, P. T. Tarr and B. M. Spiegelman, *Proc. Natl. Acad. Sci. U. S. A.* 100, 7111 (2003).

[65] X. Li, B. Monks, Q. Geand M. J. Birnbaum, *Nature*447, 1012 (2007).

[66] H. Y. Liu, T. Hong, G. B. Wen, J. Han, D. Zuo, Z. Liu Z and W. Cao, *Am. J. Physiol. Endocrinol. Metab.*297, E898 (2009).

[67] M. S. Hood, J. P. Little, M. A. Tarnopolsky, F. Myslik and M. J. Gibala, *Med. Sci. Sports Exerc.* 43, 1849 (2011).

[68] J. T. Rodgers, W. Haas, S. P. Gygi and P. Puigserver, *Cell Metab*11, 23 (2010).

[69] M. M. Rytinki and J. J. Palvimo, *J. Biol. Chem.*284, 26184 (2009).

[70] C. Teyssier, H. Ma, R. Emter, A. Kralli and M. R. Stallcup, *Genes Dev.* 19, 1466 (2005).

[71] M. P. Housley, N. D. Udeshi, J. T. Rodgers, J. Shabanowitz, P. Puigserver, D. F. Hunt and G. W. Hart, *J. Biol. Chem.*284, 5148 (2009).

[72] P. Puigserver, G. Adelmant, Z. Wu, M. Fan, J. Xu, B. O'Malley and B. M. Spiegelman, *Science*286, 1368 (1999).

[73] S. L. McGee and M. Hargreaves, *J. Appl. Physiol.* 110, 258 (2011).

[74] S. L. McGee, E. Fairlie, A. P. Garnham and M. Hargreaves, *J. Physiol.*587, 5951 (2009).

[75] M. Hallberg, D. L. Morganstein, E. Kiskinis, K. Shah, A. Kralli, S. M. Dilworth, R. White, M. G. Parker and M. Christian, *Mol. Cell. Biol.*28, 6785 (2008).

[76] A. M. Powelka, A. Seth, J. V. Virbasius, E. Kiskinis, S. M. Nicoloro, A. Guilherme, X. Tang, J. Straubhaar, A. D. Cherniack, M. G. Parker and M. P. Czech, *J. Clin. Invest.*116, 125 (2006).

[77] B. C. Frier, C. R. Hancock, J. P. Little, N. Fillmore, T. A. Bliss, D. M. Thomson, Z. Wan and D. C. Wright, *J. Appl. Physiol.*111, 688 (2011).

[78] W. Aoi, Y. Naito, K. Mizushima, Y. Takanami, Y. Kawai, H. Ichikawa and T. Yoshikawa, *Am. J. Physiol. Endocrinol. Metab.*298, E799 (2010)

[79] A. J. Cochran, J. P. Little, M. A. Tarnopolsky and M. J. Gibala, *J. Appl. Physiol.*108, 628 (2010).

[80] E. De Filippis, G. Alvarez, R. Berria, K. Cusi, S. Everman, C. Meyer and L. J. Mandarino, *Am. J. Physiol. Endocrinol. Metab.* 294, E607 (2008).

[81] A. S. Mathai, A. Bonen, C. R. Benton, D. L. Robinson and T. E. Graham, *J. Appl. Physiol.*105, 1098 (2008).

[82] J. Norrbom, C. J. Sundberg, H. Ameln, W. E. Kraus, E. Jansson and T. Gustafsson *J. Appl. Physiol.*96, 189 (2004).

[83] H. Pilegaard, B. Saltin and P. D. Neufer,*J. Physiol.*546, 851(2003).

In: Skeletal Muscle
Editor: Mark Willems

ISBN: 978-1-62417-271-7
© 2013 Nova Science Publishers, Inc.

Chapter 7

LIMB-GIRDLE MUSCULAR DYSTROPHIES: DIFFERENT TYPES AND DIAGNOSIS

Marija Meznaric[*] *and Karin Writzl*[**]
University of Ljubljana, Medical Faculty, Institute of Anatomy,
Ljubljana, Slovenia
University Medical Center Ljubljana, Institute of Medical Genetics,
Ljubljana, Slovenia

ABSTRACT

Modern classification of limb-girdle muscular dystrophies (LGMD) is based on the mode of inheritance and molecular-genetic abnormalities. It is continuously complementing: at present, the literature distinguishes 8 subtypes of autosomal dominant and 17 subtypes of autosomal recessive LGMD (http://www.musclegenetable.fr/, http://www.kegg.jp/kegg/disease). The absence or reduction of structural proteins localized to sarcolemma, nuclear envelope and sarcomeric proteins or functional abnormalities of enzymes could be detected in subtypes of LGMD. Genetic diagnosis of LGMD is at present possible in about 60% of cases, if we exclude recently described mutations of the ANO5 gene, which seem to cause a very frequent form of LGMD, at least in Central and Northern Europe. As the clinical presentation is usually not specific (proximal muscle weakness starting in lower limbs first and later spreading to upper limbs), immune analysis (immunohistochemistry and Western blot) on muscle biopsy is necessary to direct molecular-genetic testing. Unfortunately, the results of immune analysis on muscle biopsy could not be specific and are not very sensitive. From a practical point of view, it is important to exclude X-linked disorder (DMD in a child and BMD in adolescent or young adult), which could be done by molecular-genetic analysis of the dystrophin gene. Moderately elevated serum CK and neurogenic EMG should direct molecular genetic testing of the SMN gene in a patient with limb-girdle distribution of muscle weakness. FSH muscular dystrophy has a typical clinical presentation and genetic testing is possible, but some FSH cases may be difficult to diagnose; diagnostic process is further complicated since some subtypes of LGMD may

---

[*] E-mail address: marija.meznaric@mf.uni-lj.si
[**] E-mail address: karinwritzl@gmail.com

have affected facial muscles (mutations of TRIM32 gene). Further heterogeneity of LGMD is observed in phenotypic diversity of the same mutation (for instance mutations of DYSF gene can cause LGMD2B, Miyoshi myopathy, tibial muscular dystrophy and several "atypical" phenotypes). Defects in α-dystroglycan glycosylation can present as congenital forms, characteristically affecting at once brain and skeletal muscle or later in life as LGMD, with a more favorable prognosis. It is well known that geographic differences exist in the frequency of different subtypes of LGMD, which should also be applied in the diagnostic process (mutations of FKRP gene are frequent in Northern part of Europe, while mutations in CAPN3 gene are more frequent in Mediterranean countries). Genetic diagnosis of LGMD could be quite a demanding process; cooperation between expert centers/networks is necessary. Research in developing a neuromuscular disease chip (NMD-chip) (www.nmd-chip.eu) to allow the assessment of all known genes implicated in a group of diseases (as LGMD), will provide results quickly, be patient friendly, and cheaper than the present approach.

INTRODUCTION

Limb-girdle muscular dystrophies (LGMD) are named according to the characteristic distribution of muscle weakness: pelvic girdle, shoulder girdle and proximal limb muscles are affected. The disease usually presents with walking difficulties. With disease progression, muscle weakness might spread to upper limbs and eventually also to distal regions predominantly of lower limbs. LGMD are clinically heterogeneous - according to the age of onset, rate of progression, degree of elevated serum creatine kinase (sCK), associated signs such as cardiomyopathy, neuropathy, myoglobinuria, myalgia, muscle rippling movements and other signs. It is impossible to reach a precise diagnosis on clinical grounds alone, since phenotype-genotype correlation is generally weak [1]. Autosomal dominant forms (LGMD1) and autosomal recessive forms (LGMD2) are distinguished based on the mode of inheritance. Genetic heterogeneity of LGMD means that one phenotype has different genes, but also one gene might have different phenotypes [1]. As new genes are being discovered, the list of LGMD1 and LGMD2 is continuously enlarging. In October 2012, the literature distinguished 8 LGMD1 and 17 LGMD2 [2]. The recently described autosomal recessive LGMD due to a mutation of the α-dystroglycan gene (DAG1) [3] is in the Kegg disease database [4] named LGMD2P and in the Gene table of Neuromuscular Disorders [2] as recessive LGMD with primary α-dystroglycan defect. However, about one third of LGMD patients do not have a genetic diagnosis [5]. This data is valid for expert neuromuscular centers; in regions where molecular genetic tests are not easy available, the percentage is much higher. If no molecular diagnosis can be established, the clinical diagnosis of LGMD relies on familial occurrence, clinical examination, laboratory, EMG, muscle imaging and muscle biopsy findings [6].

LGMD are usually labeled with an alphabetical letter indicating the order of gene mapping [5]. As the list of LGMD forms is continuously enlarging, this might be unpractical. HUGO (human genome organization) Gene Nomenclature Committee – HGNC [7], the world wide authority that assigns standardized nomenclature to human genes has nearly withdrawn "alphabetical labeling" and approved symbols for many LGMD genes according to the protein coded; for instance LGMD2A gene is withdrawn and replaced by CAPN3 gene for gene coding calpain-3 (p94).

Limb-girdle distribution of muscle weakness is not specific for LGMD; other muscle diseases may have similar presentation [8, 9]. It is important to exclude X-linked dystrophinopathy; Pompe disease, myotonic dystrophy type 2, spinal muscular atrophy type III and congenital myastenic syndrome also present with limb-girdle muscle weakness. If contractures are outstanding, Bethlem myopathy has to be considered and muscle MRI is important for demonstrating a characteristic pattern of muscle involvement in Bethlem myopathy [10].

Mutations in individual LGMD genes can cause different phenotypes (beyond LGMD), which is particularly characteristic for some LGMD1 [11, 12], but also known in LGMD2 [13, 14]. Members of the same family may have different phenotypes [13] and intrafamilar variability of the disease severity is well documented [15]. Age of onset varies between childhood to late adult onset. Defects in α-dystroglycan glycosylation present with congenital muscular dystrophy and different degrees of central nervous abnormalities [16], in other LGMD congenital onset is exceptional [17].

Known protein abnormalities in LGMD are: deficiency of structural proteins – sarcolemmal, sarcomeric and nuclear envelope proteins and deficiency or functional abnormalities of enzymes. Genetic and phenotypic heterogeneity make molecular-genetic diagnosis of LGMD difficult [8]. Muscle biopsy is important for excluding other diseases with limb-girdle presentation and for revealing protein abnormalities by immunohistochemistry and Western blotting [9], which should help to select specific genetic test(s) to establish the precise diagnosis with DNA analysis. Interpretation of protein abnormalities might be difficult, since the abnormalities could be neither specific nor very sensitive.

Recent advances in technologies have dramatically increased the speed and efficiency and reduced the cost of DNA testing. The mutation detection was improved first with array-based technologies using comparative genome hybridization (CGH), and later with massively parallel sequencing.

The goal of CGH is to detect copy number variations (deletions and duplications). In brief, CGH uses two genomes, a test and a control, which are differentially labeled and competitively hybridized to slides arrayed with small segments of DNA as the targets for analysis. Whole-genome or target sequence (for example LGMD genes) arrays can be designed. The latter are used for detection of exonic deletions and duplications [18]. The level of resolution is determined by probe size and the genomic distance between DNA probes. The fluorescence ratio of the test, and control hybridization signals provide information on the relative copy number of sequences in the test genome as compared to the normal genome.

The development of massively parallel sequencing, also termed next-generation sequencing (NGS), has been introduced as a method of choice to analyze large numbers of genes in a comprehensive and cost-effective way [19]. In brief, genomic DNA is hybridized to beads containing probes designed to capture all of the exons, deep intronic mutation spots, splice sites and the immediately adjacent intron sequences of (for example known LGMD) genes. The DNA captured on the beads is then subjected to next-generation sequencing. After NGS reads have been generated, they are aligned to a known reference sequence. At present, it is advised that apparently positive tests are validated by standard Sanger sequencing of the patient's DNA, before results are reported to the patient. This method was first used for detection of small mutations including point mutations and short insertions/deletions (indels) spanning several base pairs, but recently has also been shown to capture large deletions and duplications [20].

The development of NGS revolutionized our ability to detect new human disease genes [21]. Currently, the most commonly used way is either targeted sequencing of the regions identified by linkage analyses or whole-exome sequencing. The goal of exome sequencing is to selectively sequence 1% to 2% of the genome, containing coding genes, microRNAs, and other noncoding RNAs. Current knowledge of the genome reveals that the large majority of DNA changes causing human genetic diseases are within the exome.

The EU-funded (FP7) neuromuscular diseases NMD-Chip project [22], which formally finished in September 2011, aimed to design chips for known and candidate NMD genes, using CGH arrays for insertion or deletion detection and a combination of enrichment of targeted genes and NGS for point mutation detection. The CGH NMD-chips, including all the known genes, were said to be validated and proposed to become part of diagnostics techniques. Lately the NGS panel, including 14 known genes for autosomal recessive LGMD, has been designed and offered by genetic laboratories as a research service.

The development of new genetic technologies offers the potential for fast, inexpensive and accurate genetic diagnostics and for discovery of still unknown LGMD genes.

AUTOSOMAL DOMINANT FORMS

At present eight loci have been identified and mutated genes and their protein products only in first four LGMD1 (myotilin, lamin A/C, caveolin-3, DNAJB6 or HSP-40 homologue, subfamily B, number 6) [2]. Phenotypes beyond LGMD are characteristic, particularly for LGMD1B and 1C [11, 12]. Progression of disease is in general considered slow [1]. De novo mutations are particularly frequent in LGMD1B [11].

LGMD1A (HGNC: MYOT)

Myotilin, mutated in LGMD1A [23], is localized to the Z line and plays a significant role in sarcomeric assembly by acting together with α-actinin and filamin C to cross-link actin into tightly packed bundles [24]. Either initial proximal or distal muscle weakness in legs with variable spreading to other muscle groups of lower and/or upper limbs characterizes the disease [25]. Limb-girdle presentation seems to be rare, while distal presentation may be more frequent – myofibrillar myopathy myotilin related. Onset is in late adulthood (around or after 40 years). Nasal speech was reported in some family members. Later onset could be associated with respiratory insufficiency and loss of ambulation. Cardiomyopathy and neuropathy may be associated and sCK is 3- to 4-times normal level. In muscle biopsy intracytoplasmic deposits, strongly immunoreactive to myotilin (Figure 1), multiple rimmed and non-rimmed vacuoles and streaming of Z-lines are detected [25]. Myotilin mutations also need to be considered as a rare cause of adult-onset, dominant LGMD without clear-cut myofibrillar pathology [26].

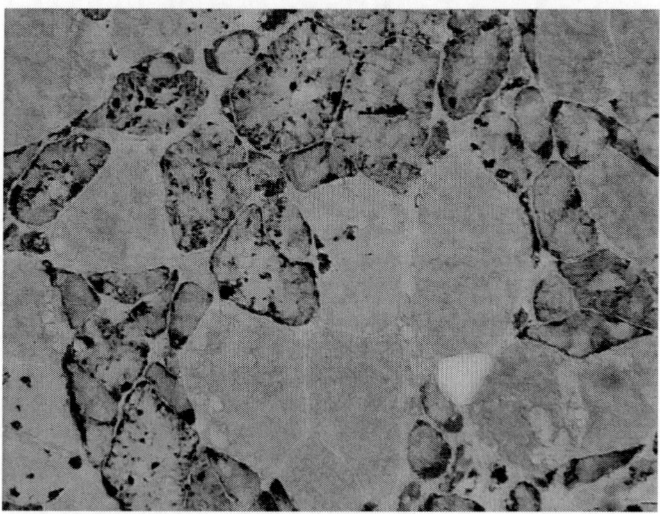

Figure 1. Myotilin accumulation within muscle fibers in a patient with LGMD1A.

LGMD1B (HGNC: LMNA)

Lamin A/C is a structural protein of nuclear envelope. De novo mutations are frequent. LGMD with contractures, arrhythmogenic and dilated cardyomyopathy is not a frequent phenotype of LMNA gene mutations. Onset is variable (4 to 38 years) with sCK up to 6-times normal level [11]. Typical phenotype is Emery-Dreifuss muscular dystrophy with childhood onset, humeroperoneal muscle weakness, rigid spine and other contractures. Other phenotypes are: isolated dilated cardiomyopathy, familial partial lipodystrophy, Hutchinson-Gilford progeria syndrome, mandibuloacral dysplasia, restrictive dermopathy, axonal Charcot-Marie-Tooth disease type 2B1 [11] and hand-heart syndrome of Slovenian type [27]. Immunohistochemical demonstration of lamin A/C on muscle biopsy is not helpful, since it is normal.

LGMD1C (HGNC: CAV3)

Caveolin-3 is a muscle specific isoform of caveolin, a small sized (21-24 kDa) integral protein of caveolae (small invagination or subcompartment of plasmalemma) [28]. Caveolae appear to mediate the selective uptake and transport of several molecules via different processes (transcytosis, endocytosis, and potocytosis). Mutations of the CAV3 gene have been associated with various phenotypes, including LGMD 1C, distal myopathy with caveolin defect, long QT syndrome 9, idiopathic hyperCKemia, familial hypertrophic cardiomyopathy and rippling muscle disease [2]. Typically onset is in childhood. Originally described family members with LGMD1C [29] had initial normal motor milestones, onset at about age 5 years, all had calf hypertrophy and mild to moderate proximal muscle weakness, each adult patient showed a positive Gowers sign and sCK levels were elevated 4- to 25-fold. In patients with caveolinopathy from the UK [12] myalgia was the most prominent symptom and was in half of the patients the reason for referral. In their study, rippling muscle movement was present in 80% of the patients and was more frequent than muscle weakness, present in 50% of the patients. Recently, a patient with asymmetric scapular winging, asymmetric proximal muscle weakness in lower limbs and asymmetric posterior thigh muscles involvement on MRI was reported [30]. On muscle biopsy caveolin-3 immunolabeling is typically lost and on blots caveolin-3 level ranges from absent to 20% of control [31], however, normal immunolabeling of caveolin-3 on muscle was reported in 2 patients with mutations of CAV3 gene [32]. In these patients, CAV3 is properly localized, but may not function normally. It seems that immunolabeling is not completely sensitive for screening of CAV3 deficiency. Mutations of CAV3 gene are infrequent cause of LGMD [33].

LGMD1D (HGNC: DNAJB6)

LGMD1 locus was mapped to chromosome 7q36 in two American families more than ten years ago [34]. The nomenclature of the locus LGMD1D has been incoherent: The locus has been termed also LGMD1E in the European Neuromuscular Council report [35] and in later reviews [5, 36]. Further studies in Finnish families [37, 38] confirmed linkage to 7q36, refined the linked area and described the phenotype more precisely, i.e. onset of muscle weakness in the pelvic girdle muscles between the fourth and sixth decade, later involvement of the shoulder girdle, mildly elevated sCK and marked walking difficulties in the eighth decade. Muscle biopsies showed myopathic and/or dystrophic features, rimmed vacuoles, desmin positive myofibrillar aggregates and Z line streaming. Mutations in DNAJB6 gene have been recently identified as the molecular cause of LGMD1D [39]. DNAJB6 or HSP-40 homologue, subfamily B, number 6 is ubiquitously expressed co-chaperone of the J-domain protein family and has been shown to suppress the aggregation of polyglutamine-containing proteins [40]. LGMD1D mutations significantly impair the anti-aggregation activity of the cytoplasmatic DNAJB6 isoform [39]. In addition, DNAJB6 interacts with HSPA8, one of the key components of chaperone assisted selective autophagy, which is important for selective maintenance of Z-line [41]. As Z-line disintegration is marked in LGMD1D and interactions of DNAJB6 with most components of chaperone assisted selective autophagy have been demonstrated [41], DNAJB6 is apparently of great importance for correct Z-line maintenance, possibly via chaperone assisted selective autophagy.

LGMD1E, LGMD1F, LGMD1G, LGMD1H

Only locus has been identified in these LGMD1 (6q23, 7q32, 4q21, 3p23-p25) [2] and the range of phenotypic variation is not well known [37]. HGNC classifies LGMD1E as synonym for CMD1F (cardyomyopathy dilated 1F) [7]. In fact, cardiomyopathy is usually observed in LGMD1E, but not in others. The progression is reported to be slow, except for LGMD1F, the age of onset is in adulthood or is more variable (LGMD1F and LGMD1H), sCK is up 10 times the normal value [5].

AUTOSOMAL RECESSIVE FORMS

At present 17 gene defects have been identified [2, 4]. LGMD2 generally have a more rapid progression than LGMD1.

LGMD2A (HGNC: CAPN3)

LGMD2A is caused by mutations in the CAPN3 gene and is the most frequent subtype of LGMD2 (20%-40%) in several countries/regions: Netherlands [42], Bulgaria [43], Northern Italy [33, 44] and Northern England [45]. However, compared to other European countries, five- to six-fold lower prevalence of LGMD2A patients was reported in Denmark [46]. In the USA, LGMD2A represents 12% of LGMD cases [47] and 8% in Australia [48]. Calpain-3 or p94, skeletal-muscle-specific member of the Ca^{2+}-activated intracellular "modulator proteases" undergoes Na^+-dependent autolysis and is Na^+-dependent intracellular protease [49]. In the sarcomeres, calpain-3 binds to connectin/titin and changes its localization from M-lines to the N2A regions as the sarcomeres extend. Stretch-induced dynamic redistribution of calpain-3 is dependent on its protease activity and is essential to protect muscle following physical stress [50]. Besides myofibrillar and cytosolic localization, calpain-3 has been shown to localize to other subcellular compartments: myonuclei [51], membrane, during myogenesis [52], triads, in mature skeletal muscle [53]. It is reasonable to suggest that in different locations, calpain-3 may have different substrate/binding partners and may serve different roles [53]. Indeed, calpain-3 is able to cleave multiple substrates [54]. Like other ubiquitous calpains, calpain-3 does not have a consensus cleavage site and calpain-3 has been implicated in such diverse processes as: apoptosis [51, 55], myogenesis when calpain-3 is necessary for regulation of the membrane-associated M-cadherin–β-catenin complex [52], sarcomeric or cytoskeletal remodelling [56, 57, 58, 59], structural integrity of the triad-associated protein complex with regulation of calcium release in skeletal muscle [53]. Molecular mechanisms of the LGMD2A are complex and remain unclear. Main phenotypes associated with LGMD2A are: pelvifemoral (Leyden-Möbius) phenotype with initial muscle weakness in the pelvic girdle, posterior and medial thigh compartment, scapulohumeral (Erb) phenotype with initial weakness in the shoulder girdle with scapular winging and later lower limb weakness, and isolated hyperCKemia during presymptomatic period [60]. The characteristic pattern of muscle involvement can be presented by muscle MRI: affection of adductor magnus and semimembranosus muscle, with preserved quadriceps (except vastus intermedius) and preserved sartorius and gracilis muscles [61]. Other variable phenotypes, including Becker muscular dystrophy-like phenotypes have been demonstrated by extensive scanning of CAPN3 gene [62]. The initial sign may be toe walking due to Achilles tendon contractures. Onset of the disease is between 2 to 40 years with sCK levels elevated 5- to 10-fold [60]. In general, affected muscles are atrophic, pseudohypertrophy is unusual, but is described in Brazilian patients [63]. Cardiomyopathy and respiratory insufficiency are usually not part of the clinical picture. Intrafamilial variability with respect to disease severity is known [15] with later disease onset in female patients.

Western blot analysis is at present the method of choice to identify LGMD2A on muscle biopsy: absence of calpain-3 (Figure 2) is characteristic for LGMD2A [64], but sensitivity is incomplete: at some LGMD2A patients calpain-3 is present but is non-functional – its autocatalytic activity is lost [65, 66]. Reduction of calpain-3 on Western blot is less specific: secondary calpain-3 reductions could be detected in dysferlinopathies [67] and titinopathies [68]. Nevertheless, the probability to detect genetic defect after established protein abnormality is rather high: 75% of patients with abnormal calpain-3 on Western blot (abnormal amount or abnormal autocatalytic activity) also have a mutation in the CAPN3 gene [69].

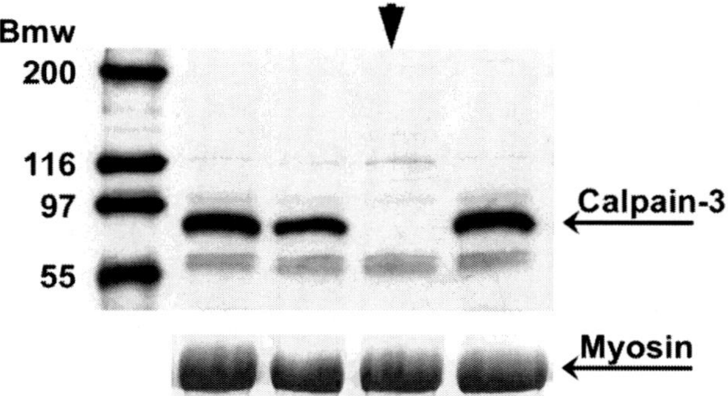

Figure 2. Western blot - absence of calpain-3 (arrow head) in a patient with LGMD2A.

Bmw - biotinylated molecular weight markers.

Abnormal autocatalytic activity of calpain-3 was always associated with mutations; a reduced amount of calpain-3 in 50% and absent calpain-3 in 86% of cases. Also, normal autocatalytic activity does not exclude LGMD2A, but the yield is low: 5% of patients with a normal amount and normal autocatalytic activity have mutations in the CAPN3 gene [69]. Another diagnostic test for LDMD2A [70] is using inactive calpain-3 as a substrate to assess proteolytical activity of calpain-3. The study showed that 32% of patients with normal proteolytical activity have mutations in the CAPN3 gene.

Histopathological alterations in muscle biopsy are not specific; a high frequency of lobulated fibers was reported [71]. Patients with eosinophilic infiltrates in their muscle biopsy were discovered to have CAPN3 mutations [72], but the finding is not specific [73].

Mutation detection is usually carried out by DNA analysis of all exons [5]. A high frequency of c.550delA mutations in Eastern Europe [43] may direct the DNA analysis in patients from this region. In patients with typical clinical presentation, some intronic splice mutations can be overlooked [74] and heterozygous deletions missed [75]. The sensitivity is improved by adding mRNA testing [46]. However, interpretation in these cases may be difficult, since many missense mutations await experimental proof of pathogenicity [5].

LGMD2B (HGNC: DYSF)

LGMD2B is caused by mutations in the DYSF gene [76, 77] and is in many geographical regions the second most frequent subtype of LGMD (15-25%) [33, 44]. In the USA, it is the most frequent LGMD, its frequency is a little higher than LGMD2A - 18% [47]. In Australia, it is the second most frequent LGMD with a frequency of 5% [48].

Dysferlin is a 237 kDa transmembrane protein localized to sarcolemma and is expressed in different tissues such as skeletal and heart muscle, liver and kidneys [77]. It is of interest that dysferlin is also expressed in immune cells- peripheral blood monocytes (PBM) and Ho et. al. [78] developed a blood-based diagnostic assay for dysferlin. In the sarcolemma, dysferlin is associated with AHNAK, caveolin-3, annexin A1 and A2 and is not a component of the dystrophin-glycoprotein complex [79]. Different studies have demonstrated the role of dysferlin in calcium-mediated sarcolemma repair [80, 81]. Dysferlin has also been shown to be involved in muscle differentiation *in vitro* [82] and *in vivo* [83].

Main phenotypes associated with mutations in DYSF gene are LGMD 2B [76, 77], Miyoshi myopathy [77] and tibial muscular dystrophy [84]. A recent study demonstrated that different phenotypes share a similar magnetic resonance pattern [85]. In addition, it is well known that in some families the same mutation is associated with LGMD2B, Miyoshi myopathy or tibial muscular dystrophy which suggests a role for modifier gene(s) [86]. Other "atypical" phenotypes widen the clinical spectrum of dysferlinopathies: mixed proximal-distal phenotype with simultaneous proximal weakness of lower limbs and calf atrophy at onset, pseudo-metabolic phenotype, asymptomatic increased CK level [13], late onset dysferlin-opathy with exercise induced muscle stiffness [87] and congenital onset of proximal lower limbs and neck flexor weakness [17]. Early adult onset and very high levels of sCK (up to 100 times normal) are characteristic [88]. In congenital forms, sCK is normal [17]. During the presymptomatic period, the patients are often active in sport, which is characteristic also for LGMD2L, but is not observed in other LGMD. Cardiomyopathy and respiratory insufficiency, as with LGMD2A, are usually not observed. The Western blot is the method of choice for identifying dysferlinopathy on muscle biopsies: Dysferlin reduction to 20% or less (Figure 3) is always caused by a mutation in the DYSF gene and is specific for dysferlinopathy [89].

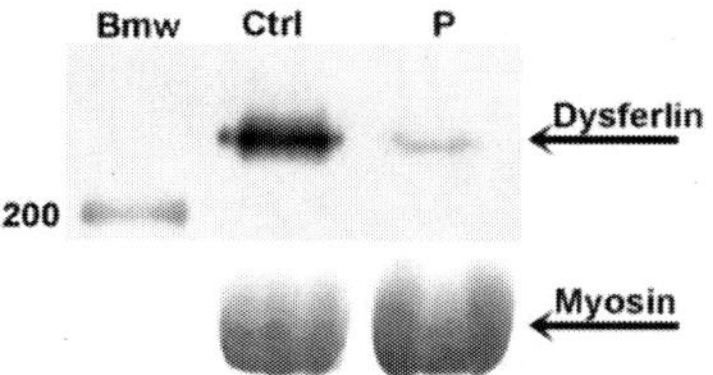

Figure 3. Western blot - reduction of dysferlin to 15% of control (Ctrl) in a patient (P) with LGMD2B.

Bmw - biotinylated molecular weight marker.

Dysferlin reduction of milder degree has also been reported in symptomatic dysferlin mutation carriers [90] and definite heterozygotes of LGMD 2B [91]. Interestingly, when dysferlin mRNA levels were analyzed in symptomatic heterozygotes, they showed normal expression [90].

Secondary dysferlin reduction (of milder degree and in some studies detected by immunofluorescence only) has been described in patients with mutations of the CAPN3 gene [92, 93], CAV3 gene [94] and in other muscular dystrophies [48, 95]. Therefore, it seems that diminished dysferlin immunofluorescence is more unspecific than severe reduction (less than 20%) of dysferlin on Western blot.

On the other hand, 3 patients with mutations in the DYSF gene with increased levels of dysferlin on WB and reduced or absent immunofluorescence of sarcolemma have recently been described [96], but the pathogenicity of discovered mutations is less certain and in two mutations were detected in one allele only. It was speculated that increased dysferlin on WB might at least in one case be related to cytoplasmatic accumulation of the mutant dysferlin that fails to insert into the sarcolemma. In addition, amyloid deposited in the smooth muscle layer of skeletal muscle arterioles and in perymisial connective tissue co-localized with dysferlin immunoreactivity. Amyloid deposits in muscle are unique to LGMD2B and sarcolemmal and interstitial amyloid deposits in LGMD2B have been reported before [97]. These patients also had unusual clinical features: atypical enlargements of deltoid and calf muscles [96].

Western blot of dysferlin on PBM [78] has been established as an alternative to Western blot analysis of dysferlin on muscle biopsy. WB of dysferlin on PBM is as informative as WB of dysferlin on skeletal muscle tissue [98]; since it is less invasive than muscle biopsy, it could in fact be a preferred method for screening dysferlinopathies for mutational studies.

The presence of inflammatory infiltrates, mainly constituted of macrophages, is a hallmark of dysferlinopathies [99, 100]. The absence of dysferlin has also been reported to cause abnormalities in immune cells- enhance phagocytosis in murine monocytes [101]. In experimental models of dysferlinopathy, it was demonstrated that restoring of dysferlin in skeletal muscle abolishes signs of muscular dystrophy and inflammation [102]. Genetic ablation of complement [103] and anti TNF-therapy [104] attenuates muscle pathology and inflammation in murine models of dysferlinopathy. Inflammation is therefore not the main pathogenetic factor, but worsens the dystrophic process.

Inflammation might be such an outstanding feature in muscle biopsy, that a dysferlinopathy patient is erroneously diagnosed with polymyositis [105]. It is therefore recommended that the diagnosis of "polymyositis" unresponsive to therapy should be checked by dysferlin testing.

Genetic testing is laborious for the huge number of exons to be screened and lack of mutational hot spots [5]. Dysferlin mRNA from PBM is also a reliable source for mutational analysis [98]. In 12% [98] and 32% [13] of patients with dysferlin abnormalities in skeletal muscle, mutational analysis did not detect mutation in both alleles. When mutational analysis detects a mutation in one allele only, dysferlin [106] and/or dysferlin mRNA [107] abnormalities are diagnostic for dysferlinopathy. Analysis of dysferlin mRNA can also be helpful for distinguishing symptomatic heterozygotes, in whom dysferlin mRNA is normal [90], from patients with mutations in one allele only in whom reduction of dysferlin mRNA is present [107].

LGMD 2C, LGMD2D, LGMD2E, LGMD2F (HGNC: SGCG, SGCA, SGCB, SGCD)

Sarcoglycans (α- β- γ- and δ-) are N-glycosylated transmembrane proteins that form a subcomplex in skeletal and cardiac muscle within the major dystrophin-associated protein complex, which is essential for membrane integrity during muscle contraction and provides a scaffold for important signalling molecules [108]. Proper assembly, trafficking and targeting of the sarcoglycan complex is of vital importance for structural integrity of sarcolemma. Gene defects in one sarcoglycan cause the absence or reduced concentration of the other subunits [109].

Mutations in sarcoglycan genes are a frequent cause (36%) of childhood-onset LGMD [33], of these mutations of SGCA gene are most frequent (19%). In childhood-onset, LGMD mutations in sarcoglycan genes are even more frequent than mutations in the CAPN3 gene (30%), but in adult-onset LGMD they constitute only about 4% of LGMD [33]. The rate of progression is usually severe and sarcoglycanopathies are the most frequent cause of severe LGMD [110].

The typical phenotype resembles Duchenne muscular dystrophy (DMD like phenotype), mild affection and asymptomatic hyperCKemia are also possible [111]. For the clinical distinction towards DMD, it is important that intelligence is normal, winging of scapula is probably more frequent in sarcoglycanopathies than in DMD and sCK is high but lower than in DMD [8, 9]. With disease progression, cardiomyopathy and respiratory insufficiency develop [112]. Different subtypes of sarcoglycanopathies differ with respect to occurrence of cardiomyopathy: arrhythmogenic and dilatative cardiomyopathy are most frequent in δ- and β-sarcoglycanopathy [112] and rarely described in α-sarcoglycanopathy [113].

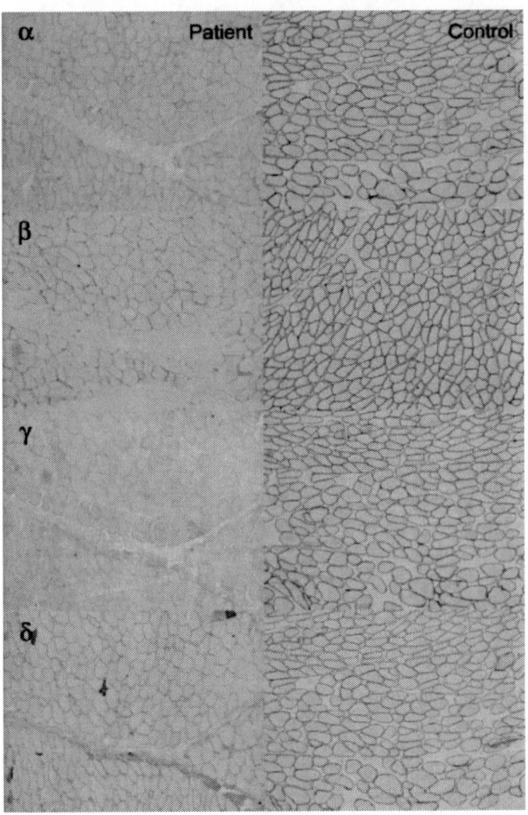

Figure 4. Reduction of all sarcoglycans in a patient with LGMD2D. Skeletal muscle sections were stained with sarcoglycans antibodies.

Reduction or absence of all four sarcoglycans (Figure 4) can be detected by immuno-fluorescence or Western blot in muscle biopsy [114, 115], since a primary defect in any of the sarcoglycans leads in most cases to disruption of the whole sarcoglycan subcomplex, and thus to a secondary deficiency of the other sarcoglycans [114]. An algorithm for mutation analysis cannot be established from abnormalities detected by immune analysis on muscle in most cases [115]. It is important to demonstrate normal immunoreactiviy of sarcolemma with dystrophin antibodies, since secondary reductions in sarcoglycans occur in dystrophinopathies [116]. The genetic analysis is oriented to genotype common mutations or to sequence the exonic regions of a specific sarcoglycan, but more than in other LGMD the sensitivity is lower without muscle mRNA testing [5].

LGMD2G (HGNC: TCAP)

Mutations in the TCAP (titin cap) gene, also known as telethonin gene, cause one of the rarest forms of LGMD originally described in Brazilian patients [117]. Telethonin is one of the titin interacting Z-disc proteins involved in the regulation and development of normal sarcomeric structure. A main novel function of telethonin in apoptosis was described recently and loss of telethonin mRNA and nuclear accumulation of this protein was demonstrated in human heart failure [118]. Mutations in the TCAP gene are also associated with cardiomyopathy [119] and common variants in TCAP may play a role in genetic susceptibility to dilated cardiomyopathy [120].

Onset of LGMD2G is in the first or second decade [117] or congenital [121]. Loss of ambulation around 40 years, frequent cardiomyopathy and sCK 3- to 30-fold normal level were described in Brazilian patients [117]. In the first European case, described in 2008 [122], proximal-distal muscle weakness was present with calf pseudohypertrophy.

Absence of telethonin can be demonstrated by immunofluorescence or Western blot [121, 122] on muscle biopsy. Full sequencing of TCAP gene may be cost-effective in all cases, since the gene is composed only of 2 exons [5].

LGMD2H (HGNC: TRIM32)

Tripartite motif-containing protein 32, coded by TRIM32 gene, is ubiquitous E3 ubiquitin ligase, member of the tripartite motif (TRIM) family. The TRIM motif includes three zinc-binding domains, a ring, a B box type 1 and 2, and a coiled-coil region [123]. TRIM32 is expressed in skeletal muscle, where it interacts with myosin and may ubiquitinate actin [124]. The second mechanism by which TRIM32 is believed to operate involves binding of proteins to the C-terminal NHL repeat, which has been shown to activate miRNAs [125]. A mutation in the B-box region of TRIM32 has been shown to result in a multisystem disorder, Bardet-Biedl Syndrome [126]. Although TRIM32 appears to be expressed ubiquitously, it is still not clear why certain mutations of TRIM32 would result in a phenotype relatively restricted to skeletal muscle [127]. It is likely that C-terminal mutations in TRIM32 affect the ability of muscle proteins to be degraded by the ubiquitin-proteasome pathway [127]. LGMD2H is a slowly progressive LGMD with onset in the second to third decade with mild to moderately raised levels of sCK. Loss of ambulation may occur after 60 years. Facial weakness, scapular winging, hypertrophied calves, and Achilles tendon contractions are not uncommon and the phenotype to some extend resembles facioscapulohumeral muscular dystrophy. LGMD2H was originally described in the Hutterite population that resides in central Canada and the Dakotas of the USA [128]. TRIM32 mutations were later described in the European population [129] and constitute about 3% of LGMD [5]. Mutations in TRIM32 also cause more severe sarcotubular myopathy, which has a histological picture of microvacuolar myopathy [130, 131]. LGMD2H cannot be diagnosed without genetic studies – DNA sequencing of the unique coding exon is routinely performed [5].

DYSTROGLYCANOPATHIES: LGMD2I, LGMD2K, LGMD2M, LGMD2N, LGMD2O, LGMD2P (HGNC: FKRP, POMT1, FKTN, POMT2, POMGNT1, DAG1)

Proteins encoded by the above listed genes (except DAG1 gene) are enzymes (known or putative glycosyltransferases) necessary for posttranslational modification (glycosylation) of α-dystroglycan [16, 132]. Alpha-dystroglycan (dystrophin-associated glycoprotein 1, dystroglycan 1), encoded by DAG1 gene, is a structural sarcolemmal protein, a component of dystrophin-glycoprotein complex, which interacts with proteins of the extracellular matrix (laminin α-2) [133]. POMT1 and POMT2 are located in the endoplasmic reticulum, FKRP, FKTN and POMGNT1 in Golgi apparatus. Details of individual enzymes can be found at the website of the National Centre for Biotechnology Information [134]. Mutations in these genes cause congenital muscular dystrophies with central nervous system abnormalities: Muscle-eye brain syndrome, Walker-Warburg syndrome and Fukuyama congenital muscular dystrophy [16, 132]. Variable reduction of α-dystroglycan expression is observed in skeletal muscle biopsy. In addition, several cases showed a deficiency of laminin α-2 either by immunofluorescence or Western blotting [132].

Milder allelic variants of dystroglycanopathies have phenotypes of LGMD [135], which are less frequent than congenital muscular dystrophies. In general, defects in glycolysation of α-dystroglycan are not a frequent cause of LGMD except mutations of FKRP gene [136].

The phenotype of LGMD2I is similar to Becker muscular dystrophy with onset in the beginning of the second decade, calf and tongue hypertrophy, high sCK and cardiomyopathy [137]. Respiratory impairment is frequent. Cardiomyopathy may be isolated [138] or exceeds skeletal muscle pathology [139]. Myoglobinuria and myalgia following exercise are common [140]. Mild cognitive impairment has also been reported [141]. The most consistent protein abnormality found on muscle biopsy was a reduction of laminin α-2 immunolabeling, either on muscle sections or Western blotting alone [137]. Patients homozygous for the common c.826C>A mutation had later disease onset and milder phenotype than patients who were compound heterozygotes [140].

Missense mutation in DAG1 gene was reported recently [3] in a patient with LGMD and mild cognitive impairment. The mutation interferes with LARGE-dependent glycosylation of α-dystroglycan affecting its binding to laminin α-2 [3].

LGMD2J (HGNC: TTN)

TTN gene codes the gigantic protein titin which spans from sarcomeric Z to M line, with C-terminus located at the M line [142]. Titin is implicated in many functions – for instance: guiding of myofibrillogenesis, providing elasticity to myofibrils, distribution of forces across sarcomers and maintenance of sarcomeric structure during contraction [143]. Mutations in the distal part of the C-terminus of titin, when present in both alleles, cause LGMD2J [144] and early onset proximal muscular dystrophy [145]. When present in one allele, the phenotype is tibial muscular dystrophy (TMD), late onset distal myopathy with predominant muscle weakness of foot dorsi-flexors [144, 145]. Further study has demonstrated large variability of phenotypic expression caused by the Finnish founder mutation [146]. Mutation presumably leads to domain misfolding and loss of protein interactions of the affected titin region [144]. M-band titin binds calpain-3 within the region affected by the TMD/LGMD2J mutations [147, 148]. Interactions with titin may protect calpain-3 from autolytic activation [147] and their disruption in titinopathies could result in secondary calpain deficiency in LGMD2J [149]. Calpain-3 deficiency as a primary abnormality is present in LGMD2A [64]. In addition, two recent studies are relevant for the molecular pathomechanisms of titinopathies. Sarparanta et al. [149] demonstrated interactions of titin with myospryn in the sarcomeric M-band and Charton et al. [150] showed that the disruption of calpain-3 regulation seems to be a major factor in the pathological mechanism in the experimental model of titinopathy, since removal of calpain-3 protease reverses the myopathology. In addition to the Finnish founder mutation [144], six other mutations were detected in the C-terminal of the titin gene in unrelated Belgian, Spanish, French, and Italian patients [151, 152, 153, 154].

LGMD2L (HGNC: ANO5)

ANO5 encodes a member of anoctamin family of transmembrane proteins, a putative calcium activated chloride channel [155]. Phenotypes associated with recessive mutations in ANO5 are: LGMD2L and Miyoshi myopathy (not linked to DYSF gene) [14, 156]. Gnathodiaphyseal dysplasia is caused by a dominant mutation in ANO5 [157]. In Northern England, a founder mutation – c.191 dupA in exon 5 of ANO5 represents a common cause of adult onset LGMD [158]. Onset is between 20 to 40 years with sCK several ten times the normal value. Before the onset of muscle weakness, persons are often active in sports, which is, unusual for muscular dystrophies, except for Miyoshi myopathy caused by DYS gene mutations. On lower limbs, the most affected muscles are quadriceps, knee flexors and the medial head of the gastrocnemius muscle. With disease progression, the biceps muscle on upper limbs and brachioradialis became affected. In most cases (80%) muscle weakness is asymmetrical. At present, antibodies to anoctamin are not available. It was suggested that due to the high frequency of anoctamin mutations in English and German patients, the screening of funder mutation should be included in early diagnostic evaluation of patients who fit the clinical description [158].

LGMD2Q (HGNC: PLEC1)

So far, three Turkish families are described with mutations [159]. Plectin1 is a member of an important family of plakins or cytolinkers that are capable of interlinking different elements of the cytoskeleton [160]. LGMD2Q is an early childhood onset LGMD2 with progression during the teenage years, without skin involvement. Walking ability could be lost around the age of 20 years. Motor milestones are delayed with sCK several ten times above normal [159]. Absent immunoreactivity of sarcolemma with antibodies to plectin 1f was demonstrated on fast muscle fibers [159].

CONCLUSION

LGMD are relatively infrequent muscular dystrophies with only a few families described in some subtypes. Clinical heterogeneity is evident with respect to onset of the disease, degree of muscle weakness and occurrence of individual signs as: pseudohypertrophy (frequent with mutations of SGCA and FKRP genes), scapular winging (frequent with mutations of CAPN3 and SGCA genes), degree of hyperCKemia (highest values with mutations of DYSF gene), scoliosis (often with sarcoglycanopathies), cardiac arrhythmia and cardiomyopathy (frequent with mutations of LMNA gene), cardiomyopathy and respiratory insufficiency (frequent with mutations of SGCD, SGCB and FKRP genes), myalgia and myoglobinuria (frequent with mutations of FKRP gene). The most frequent childhood-onset LGMD are sarcoglycanopathies which constitute the main part of severe LGMD in general. Regional differences are noted with respect to frequency of individual subtypes. The most frequent form of adult-onset LGMD in Mediterranean countries are mutations of the CAPN3 gene, in Northern Europe mutations of FKRP and ANO5 genes. Mutations in the same gene may cause different phenotypes (frequent with mutations of LMNA gene and also with mutations of CAV3, DYSF and ANO5 genes). Negative family history does not exclude LMNA mutations, since de novo mutations are frequent. For selection of genetic test(s), immune analysis on muscle is important (particularly accumulation of myotilin, absence of calpain-3 or absence of autocatalytic activity of calpain-3, dysferlin reduction to 20% or less, reduction of sarcoglycans). Moderate reductions of proteins may be secondary and sometimes protein tests are not very sensitive. The prospect is that new technologies (next generation sequencing) will reverse the order – first muscle biopsy than genetic tests- and will provide results quickly, will be patient friendly and cheaper than the present approach. Abnormal splicing is present in about 10% and this requires sequence analysis of mRNA usually from a muscle biopsy.

REFERENCES

[1]   M. Zatz, M. Vainzof and M. R. Passos-Bueno, *Curr. Opin. Neurol.* 13, 511 (2000).

[2]   Gene Table of Neuromuscular Disorders, J.C. Kaplan and D. Hamroun, http://www.musclegenetable.fr/

[3]   Y. Hara, B. Balci-Hayta, T. Yoshida-Moriguchi, M. Kanagawa, D. Beltrán-Valero de Bernabé, H. Gündeşli, T. Willer, J. S. Satz, R. W. Crawford, S. J. Burden, S. Kunz, M. B. Oldstone, A. Accardi, B. Talim, F. Muntoni, H. Topaloğlu, P. Dinçer and K. P. Campbell, *N. Engl. J. Med.* 364, 939 (2011).

[4]   Kanehisa Laboratories, M. Kanehisa, http://www.kegg.jp/kegg/disease/

[5]   V. Nigro, S. Aurino and G. Piluso, *Curr. Opin. Neurol.* 24, 429 (2011).

[6]   I. Dalkilic and L. M. Kunkel, *Curr. Opin. Genet. Dev.* 13, 231 (2003).

[7]   HGNC, HUGO Gene Nomenclature Committee, http://www.genenames.org/

[8]   M. Gulieri and K. Bushby, *Neurol. India,* 56, 271 (2008).

[9]   K. Bushby, *Pract. Neurol.* 9, 314 (2009).

[10]   E. Mercuri, A. Lampe, J. Allsop, R. Knight, M. Pane, M. Kinali, C. Bonnemann, K. Flanigan, I. Lapini, K. Bushby, G. Pepe and F. Muntoni, *Neuromuscul. Disord.* 5, 303 (2005).

[11]   H. J. Worman and G. Bonne, *Exp. Cell Res.* 313, 2121 (2007).

[12]   A. Aboumousa, J. Hoogendijk, R. Charlton, R. Barresi, R. Herrmann, T. Voit, J. Hudson, M. Roberts, D. Hilton-Jones, M. Eagle, K. Bushby and V. Straub, *Neuromuscul. Disord.* 18, 572 (2008).

[13]   K. Nguyen, G. Bassez, M. Krahn, R. Bernard, P. Laforêt, V. Labelle, J. A. Urtizberea, D. Figarella-Branger, N. Romero, S. Attarian, F. Leturcq, J. Pouget, N. Lévy and B. Eymard, *Arch. Neurol.* 64, 1176 (2007).

[14]   V. Bolduc, G. Marlow, K. M. Boycott, K. Saleki, H. Inoue, J. Kroon, M. Itakura, Y. Robitaille, L. Parent, F. Baas, K. Mizuta, N. Kamata, I. Richard, W. H. Linssen, I. Mahjneh, M. de Visser, R. Bashir and B. Brais, *Am. J. Hum. Genet.* 86, 213 (2010).

[15]   J. Schessl, M. C. Walter, G. Schreiber, U. Schara, C. R. Müller, H. Lochmüller, C. G. Bönnemann, R. Korinthenberg and J. Kirschner, *Acta Myol.* 27, 54 (2008).

[16]   C. Godfrey, E. Clement, R. Mein, M. Brockington, J. Smith, B. Talim, V. Straub, S. Robb, R. Quinlivan, L. Feng, C. Jimenez-Mallebrera, E. Mercuri, A. Y. Manzur, M. Kinali, S. Torelli, S. C. Brown, C. A. Sewry, K. Bushby, H. Topaloglu, K. North, S. Abbs and F. Muntoni, *Brain*, 130, 2725 (2007).

[17]   C. Paradas, L. González-Quereda, N. De Luna, E. Gallardo, I. García-Consuegra, H. Gómez, A. Cabello, I. Illa and P. Gallano, *Neuromuscul. Disord.* 19, 21 (2009).

[18]   Y. Saillour, M. Cossée, F. Leturcq, A. Vasson, C. Beugnet, K. Poirier, V. Commere, S. Sublemontier, M. Viel, F. Letourneur, J. C. Barbot, N. Deburgrave, J. Chelly and T. Bienvenu, *Hum. Mutat.* 29, 1083 (2008).

[19]   M. L. Metzker, *Nat. Rev. Genet.* 11, 31 (2010).

[20]   B. C. Lim, S. Lee, J. Y. Shin, J. I. Kim , H. Hwang, K. J. Kim, Y. S. Hwang, J. S. Seo and J. H. Chae, *J. Med. Genet.* 48, 731 (2011).

[21]   C. Gilissen, A. Hoischen, H. G. Brunner and J. A. Veltman, *Genome Biol.* 12, 228 (2011).

[22]   NMD chip, N. Lévy, http://www.nmd-chip.eu/

[23]   M. A. Hauser, S. K. Horrigan, P. Salmikangas, U. M. Torian, K. D. Viles, R. Dancel, R. W. Tim, A. Taivainen, L. Bartoloni, J. M. Gilchrist, J. M. Stajich, P. C. Gaskell, J. R. Gilbert, J. M. Vance, M. A. Pericak-Vance, O. Carpen, C. A. Westbrook and M. C. Speer, *Hum. Mol. Genet.* 9, 2141 (2000).

[24]   P. Salmikangas, P. van der Ven, M. Lalowski, A. Taivainen, F. Zhao, H. Suila, R. Schröder, P. Lappalainen, D. O. Fűrst and O. Carpén, *Hum. Mol. Genet.* 12, 189 (2003).

[25] M. Olivé, L. G. Goldfarb, A. Shatunov, D. Fischer and I. Ferrer, *Brain*, 128, 2315 (2005).

[26] P. Reilich , S. Krause , N. Schramm , U. Klutzny , S. Bulst , B. Zehetmayer , P. Schneiderat, M. C. Walter, B. Schoser and H. Lochmüller, *J. Neurol.* 258, 1437 (2011).

[27] L. Renou , S. Stora , R. B. Yaou , M. Volk , M. Sinkovec , L. Demay, P. Richard, B. Peterlin and G. Bonne, *J. Med. Genet.* 45, 666 (2008).

[28] B. Razani, S. E. Woodman and M. P. Lisanti, *Pharmacol. Rev.* 54, 431 (2002).

[29] C. Minetti, F. Sotgia, C. Bruno, P. Scartezzini, P. Broda, M. Bado, E. Masetti, M. Mazzocco, A. Egeo, M. A. Donati, D. Volonte, F. Galbiati, G. Cordone, F. D. Bricarelli, M. P. Lisanti and F. Zara, *Nat. Genet.* 18, 365 (1998).

[30] K. Hankiewicz, R. Rojas-García, J. Díaz-Manera, E. Gallardo, J. Llauger, P. Gallano and I. Illa, *Neuromuscul. Disord.* 21, 669 (2011).

[31] L. Fulizio, A. C. Nascimbeni, M. Fanin, G. Piluso, L. Politano, V. Nigro and C. Angelini, *Hum. Mutat.* 25, 82 (2005).

[32] J. C. Reijneveld, I. B. Ginjaar, W. S. Frankhuizen and N.C. Notermans, *Muscle Nerve*, 34, 656 (2006).

[33] M. Fanin, A. C. Nascimbeni, S. Aurino, E. Tasca, E. Pegoraro, V. Nigro and C. Angelini, *Neurology*, 72, 1432 (2009).

[34] M. C. Speer, J. M. Gilchrist, J. G. Chutkow, R. McMichael, C. A. Westbrook, J. M. Stajich, E. M. Jorgenson, P. C. Gaskell, B. L. Rosi, R. Ramesar, J. M. Vance, L. H. Yamaoka, A. D. Roses and M. A. Pericak-Vance, *Am. J. Hum. Genet.* 57, 1371 (1995).

[35] K. M. Bushby, *Neuromuscul. Disord.* 5, 71 (1995).

[36] F. Norwood, M. de Visser, B. Eymard, H. Lochmüller and K. Bushby, *Eur. J. Neurol.* 14, 1305 (2007).

[37] S. Sandell, S. Huovinen, J. Sarparanta, H. Luque, O. Raheem, H. Haapasalo, P. Hackman and B. Udd, *J. Neurol. Neurosurg. Psychiatry*, 81, 834 (2010).

[38] P. Hackman, S. Sandell, J. Sarparanta, H. Luque, S. Huovinen, J. Palmio, A. Paetau, H. Kalimo, I. Mahjneh and B. Udd, *Neuromuscul. Disord.* 21, 338 (2011).

[39] J. Sarparanta, P. H. Jonson, H. Luque and B. Udd, *Neuromuscul. Disord.* 21, 668 (2011).

[40] UniProt, R. Apweiler, C. Wu and I. Xenarios, http://www.uniprot.org/uniprot/O75190

[41] P. H. Jonson, J. Sarparanta, H. Luque, A. Vihola, and B. Udd, *Neuromuscul. Disord.* 21, 667 (2011).

[42] A. J. van der Kooi, W. S. Frankhuizen, P. G. Barth, C. J. Howeler, G. W. Padberg, F. Spaans, A. R. Wintzen, J. H. J. Wokke, G. J. B. van Ommen, M. de Visser, E. Bakker and H. B. Ginjaar, *Neurology*, 68, 2125 (2007).

[43] A. Todorova, B. Georgieva, I. Tournev, T. Todorov, N. Bogdanova, V. Mitev, C. R. Mueller, I. Kremensky and J. Horst, *Neurogenetics*, 8, 225 (2007).

[44] M. Guglieri, F. Magri, M. G. D'Angelo, A. Prelle, L. Morandi, C. Rodolico, R. Cagliani, M. Mora, F. Fortunato, A. Bordoni, R. Del Bo, S. Ghezzi, S. Pagliarani, S. Lucchiari, S. Salani, C. Zecca, C. Lamperti, D. Ronchi, M. Aguennouz, P. Ciscato, C. Di Blasi, A. Ruggieri, I. Moroni, A. Turconi, A. Toscano, M. Moggio, N. Bresolin and G. P. Comi, *Hum. Mutat.* 29, 258 (2008).

[45] F. L. M. Norwood, C. Harling, P. F. Chinnery, M. Eagle, K. Bushby and V. Straub, *Brain*, 132, 3175 (2009).

[46] M. Duno, M. L. Sveen, M. Schwartz and J. Vissing, Europ. *J. Hum. Genet.* 16, 935 (2008).

[47] S. A. Moore, C. J. Shilling, S. Westar, C. Wall, M. P. Wicklund, C. Stolle, C. A. Brown, D. E. Michele, F. Piccolo, T. L. Winder, A. Stence, R. Barresi, N. King, W. King, J. Florence, K. P. Campbell, G. M. Fenichel, H. H. Stedman, J. T. Kissel, R. C. Griggs, S. Pandya, K. D. Mathews, A. Pestronk, C. Serrano, D. Darvish and R. J. Mendell, *J. Neuropathol. Exp. Neurol.* 65, 995 (2006).

[48] H.P. Lo, S. T. Cooper, F. J. Evesson, J. T. Seto, M. Chiotis, V. Tay, A. G. Compton, A. G. Cairns, A. Corbett, D. G. MacArthur, N. Yang, K. Reardon and K. N. North, *Neuromuscul. Disord.* 18, 34 (2008).

[49] Y. Ono, K. Ojima, F. Torii, E. Takaya, N. Doi, K. Nakagawa, S. Hata, K. Abe and H. Sorimachi, *J. Biol. Chem.* 285, 22986 (2010).

[50] K. Ojima, Y. Kawabata, H. Nakao, K. Nakao, N. Doi, F. Kitamura, Y. Ono, S. Hata, H. Suzuki, H. Kawahara, J. Bogomolovas, C. Witt, C. Ottenheijm, S. Labeit, H. Granzier, N. Toyama-Sorimachi, M. Sorimachi, K. Suzuki, T. Maeda, K. Abe, A. Aiba and H. Sorimachi, *J. Clin. Invest.* 120, 2672 (2010).

[51] S. Baghdiguian , M. Martin , I. Richard, F. Pons, C. Astier, N. Bourg, R. T. Hay, R. Chemaly, G. Halaby, J. Loiselet, L. V. Anderson, A. Lopez de Munain, M. Fardeau, P. Mangeat, J. S. Beckmann and G. Lefranc, *Nat. Med.* 5, 503 (1999).

[52] I. Kramerova, E. Kudryashova, B. Wu, C. Ottenheijm, H. Granzier and M. J. Spencer, *Mol. Cell Biol.* 26, 8437 (2006).

[53] I. Kramerova, E. Kudryashova, B. Wu, C. Ottenheijm, H. Granzier and M. J. Spencer, *Hum. Mol. Genet.* 17, 3271 (2008).

[54] D.E. Goll, V. F. Thompson, H. Li, W. Wei and J. Cong. *Physiol. Rev.* 83, 731 (2003).

[55] B. Benayoun, S. Baghdiguian, A. Lajmanovich, M. Bartoli, N. Daniele, E. Gicquel, N. Bourg, F. Raynaud, M. A. Pasquier, L. Suel, H. Lochmuller, G. Lefranc and I. Richard, *FASEB J.* 22, 1521 (2008).

[56] M. Taveau, N. Bourg, G. Sillon, C. Roudaut, M. Bartoli and I. Richard, Mol. Cell Biol. 23, 9127 (2003).

[57] I. Kramerova, E. Kudryashova, G. Venkatraman and M. J. Spencer, *Hum. Mol. Genet.* 14, 2125 (2005).

[58] S. Duguez, M. Bartoli and I. Richard, *FEBS J.* 273, 3427 (2006).

[59] A. de Morrée, D. Lutje Hulsik, A. Impagliazzo, H. H. van Haagen, P. de Galan, A. van Remoortere, P. A. 't Hoen, G. B. van Ommen, R. R. Frants and S. M. van der Maarel, *PLoS One*, 5, 11940 (2010).

[60] C. Angelini, L. Nardetto, C. Borsato, R. Padoan, M. Fanin, A.C. Nascimbeni and E. Tasca, *Neurol Res.* 32, 41 (2010).

[61] E. Mercuri, K. Bushby, E. Ricci, D. Birchall, M. Pane, M. Kinali, J. Allsop, V. Nigro, A. Saenz, A. Nascimbeni, L. Fulizio, C. Angelini and F. Muntoni, *Neuromuscul. Disord.* 15, 164 (2005).

[62] G. Piluso, L. Politano, S. Aurino, M. Fanin, E. Ricci, V. M. Ventriglia, A. Belsito, A. Totaro, V. Saccone, H. Topaloglu, A. C. Nascimbeni, L. Fulizio, A. Broccolini, N. Canki-Klain, L. I. Comi, G. Nigro, C. Angelini and V. Nigro, *J. Med. Genet.* 42, 686 (2005).

[63] M. R. Passos-Bueno, M. Vainzof, E. S. Moreira and M. Zatz, *Am. J. Med. Genet.* 82, 392 (1999).

[64] L. V. Anderson, K. Davison, J. A. Moss, I. Richard, M. Fardeau, F. M. Tomé, C. Hübner, A. Lasa, J. Colomer and J. S. Beckmann, *Am. J. Pathol.* 153, 1169 (1998).

[65] M. Fanin, A. C. Nascimbeni, L. Fulizio, C. P. Trevisan, M. Meznaric-Petrusa and C. Angelini, *Am. J. Pathol.* 163, 1929 (2003).

[66] M. Fanin, A. C. Nascimbeni, E. Tasca and C. Angelini, *Eur. J. Hum. Genet.* 17, 598 (2009).

[67] L. V. Anderson, R. M. Harrison, R. Pogue, E. Vafiadaki, C. Pollitt, K. Davison, J. A. Moss, S. Keers, A. Pyle, P. J. Shaw, I. Mahjneh, Z. Argov, C. R. Greenberg, K. Wrogemann, T. Bertorini, H. H. Goebel, J. S. Beckmann, R. Bashir and K. M. Bushby, *Neuromuscul. Disord.* 10, 553 (2000).

[68] H. Haravuori, A. Vihola, V. Straub, M. Auranen, I. Richard, S. Marchand, T. Voit, S. Labeit, H. Somer, L. Peltonen, J. S. Beckmann and B. Udd, *Neurology*, 56, 869 (2001).

[69] M. Fanin, A. C. Nascimbeni, L. Fulizio and C. Angelini, *Neuromuscul. Disord.* 15, 218 (2005).

[70] A. Milic, N. Daniele, H. Lochmüller, M. Mora, G. P. Comi, M. Moggio, F. Noulet, M. C. Walter, L. Morandi, J. Poupiot, C. Roudaut, R. E. Bittner, M. Bartoli and I. Richard, *Neuromuscul. Disord.* 17, 148 (2007).

[71] M. Fanin, L. Nardetto, A. C. Nascimbeni, E. Tasca, M. Spinazzi, R. Padoan and C. Angelini, *J. Med. Genet.* 44, 609 (2007).

[72] M. Krahn, A. Lopez de Munain, N. Streichenberger, R. Bernard, C. Pécheux, H. Testard, J. L. Pena-Segura, E. Yoldi, A. Cabello, N. B. Romero, J. J. Poza, S. Bouillot-Eimer, X. Ferrer, M. Goicoechea, F. Garcia-Bragado, F. Leturcq, J. A. Urtizberea and N. Lévy, *Ann. Neurol.* 59, 905 (2006).

[73] S. K. Baumeister, S. Todorović, V. Milić-Rasić, G. Dekomien, H. Lochmüller and M. C. Walter, *Neuromuscul. Disord.* 19, 167 (2009).

[74] A. C. Nascimbeni, M. Fanin, E. Tasca and C. Angelini, *Hum. Mutat.* 31, E1658 (2010).

[75] M. Krahn, C. Pécheux, F. Chapon, C. Béroud, V. Drouin-Garraud, P. Laforet, N. B. Romero, I. Penisson-Besnier, R. Bernard, J. A. Urtizberea, F. Leturcq and N. Lévy, *Clin. Genet.* 2, 582 (2007).

[76] R. Bashir, S. Britton, T. Strachan, S. Keers, E. Vafiadaki, M. Lako, I. Richard, S. Marchand, N. Bourg, Z. Argov, M. Sadeh, I. Mahjneh, G. Marconi, M. R. Passos-Bueno, S. Moreira Ede, M. Zatz, J. S. Beckmann and K. Bushby, *Nat. Genet.* 20, 37 (1998).

[77] J. Liu, M. Aoki, I. Illa, C. Wu, M. Fardeau, C. Angelini, C. Serrano, J. A. Urtizberea, F. Hentati, M. B. Hamida, S. Bohlega, E. J. Culper, A. A. Amato, K. Bossie, J. Oeltjen, K. Bejaoui, D. McKenna-Yasek, B. A. Hosler, E. Schurr, K. Arahata, P. J. de Jong and R. H. Brown Jr., *Nat. Genet.* 20, 31 (1998).

[78] M. Ho, E. Gallardo, D. McKenna-Yasek, N. De Luna, I. Illa and R. H. Brown Jr., *Ann. Neurol.* 51, 129 (2002).

[79] Y. Huang, S. H. Laval, A. van Remoortere, J. Baudier, C. Benaud, L. V. B. Anderson, V. Straub, A. Deelder, R. R. Frants, J. T. den Dunnen, K. Bushby and S. M. van der Maarel, *FASEB J.* 21, 732 (2007).

[80] D. Bansal, K. Miyake, S. S. Vogel, S. Groh, C. C. Chen, R. Williamson, P. L. McNeil and K. P. Campbell, *Nature*, 423, 168 (2003).

[81] R. Han and K. P. Campbell, *Curr. Opin. Cell Biol.* 19, 409 (2007).

[82] N. de Luna N, E. Gallardo, M. Soriano, R. Dominguez-Perles, C. de la Torre, R. Rojas-García, J. M. García-Verdugo and I. Illa, *J. Biol. Chem.* 281, 17092 (2006).

[83] C. De la Torre, I. Illa, G. Faulkner, L. Soria, R. Robles-Cedeño, R. Dominguez-Perles, N. De Luna, E. Gallardo, *Proteomics Clin. Appl.* 3, 486 (2009).

[84] I. Illa I, C. Serrano-Munuera, E. Gallardo, A. Lasa, R. Rojas-García, J. Palmer, P. Gallano, M. Baiget, C. Matsuda and R. H. Brown, *Ann. Neurol.* 49, 130 (2001).

[85] C. Paradas, J. Llauger, J. Diaz-Manera, R. Rojas-García, N. De Luna, C. Iturriaga, C. Márquez, M. Usón, K. Hankiewicz, E. Gallardo and I. Illa, *Neurology*, 75, 316 (2010).

[86] T. Weiler, R. Bashir, L. V. Anderson , K. Davison, J. A. Moss, S. Britton, E. Nylen, S. Keers, E. Vafiadaki, C. R. Greenberg, C. R. Bushby and K. Wrogemann, *Hum. Mol. Genet.* 8, 871 (1999).

[87] L. Klinge, A.F. Dean, W. Kress, P. Dixon, R. Charlton, J. S. Müller, L. V. Anderson, V. Straub, R. Barresi, H. Lochmüller and K. Bushby, *Neuromuscul. Disord.* 18, 288 (2008).

[88] K. M. Bushby, *Acta Neurol. Belg.* 100, 142 (2000).

[89] M. Cacciottolo, G. Numitone, S. Aurino, I. R. Caserta, M. Fanin, L. Politano, C. Minetti, E. Ricci, G. Piluso, C. Angelini and V. Nigro, *Eur. J. Hum. Genet.* 19, 974 (2011).

[90] I. Illa , N. De Luna, R. Dominguez-Perles, R. Rojas-Garcia, C. Paradas, J. Palmer, C. Márquez, P. Gallano and E. Gallardo, *Neurol*ogy, 68, 1284 (2007).

[91] M. Fanin, A. C. Nascimbeni and C. Angelini. *Neuromuscul. Disord.* 16, 792 (2006).

[92] T. Chrobakova, M. Hermanova, I. Kroupova, P. Vondracek, T. Marikova, R. Mazanec, J. Zámecník, J. Stanek, M. Havlová, and L. Fajkusová, *Neuromuscul. Disord.* 14, 659 (2004).

[93] M. Hermanova, E. Zapletalova, J. Sedlackova, T. Chrobakova, O. Letocha, I. Kroupova, J. Zámecník, P. Vondrácek , R. Mazanec, T. Maríková , S. Vohánka and L. Fajkusová, *Muscle Nerve*, 33, 424 (2006).

[94] C. Matsuda, Y. K. Hayashi, M. Ogawa, M. Aoki, K. Murayama, I. Nishino, I. Nonaka, K. Arahata and R. H. Brown Jr., *Hum. Mol. Gen.* 10, 1761 (2001).

[95] F. Piccolo, S. A. Moore, G. C. Ford and K. P. Campbell. *Ann. Neurol.* 48, 902 (2000).

[96] X. Q. Rosales, J. M. Gastier-Foster, S. Lewis, M. Vinod, D. L. Thrush, C. Astbury, R. Pyatt, S. Reshmi, Z. Sahenk, and J. R. Mendell, *Muscle Nerve*, 42, 14 (2010).

[97] S. Spuler, M. Carl, J. Zabojszcza, V. Straub, K. Bushby, S. A. Moore, S. Bähring, K. Wenzel, U. Vinkemeier and C. Rocken, *Ann. Neurol.* 63, 323 (2008).

[98] N. De Luna, A. Freixas, P. Gallano, L. Caselles, R. Rojas-García, C. Paradas, G. Nogales, R. Dominguez-Perles, L. Gonzalez-Quereda, J. J. Vílchez, C. Márquez, J. Bautista, A. Guerrero, J. A. Salazar, A. Pou, I. Illa and E. Gallardo, *Neuromuscul. Disord.* 17, 69 (2007).

[99] E. Gallardo, R. Rojas-García, N. de Luna, A. Pou, R. H. Brown Jr. and I. Illa, *Neurology*, 57, 2136 (2001).

[100] M. Fanin and C. Angelini, *Neuropathol. Appl. Neurobiol.* 28, 461(2002).

[101] K. Nagaraju, R. Rawat, E. Veszelovszky, R. Thapliyal, A. Kesari, S. Sparks, N. Raben, P. Plotz and E. P. Hoffman, *Am. J. Pathol.* 172, 774 (2008).

[102] D. P. Millay, M. Maillet, J. A. Roche, M. A.Sargent, E. M. McNally, R. J. Bloch and J. D. Molkentin, *Am. J. Pathol.* 175, 1817 (2009).

[103] R. Han, E. M. Frett, J. R. Levy, E. P. Rader, J. D. Lueck, D. Bansal, S. A. Moore, R. Ng, D. de Bernabé Beltrán-Valero, J. A. Faulkner and K. P. Campbell, *J. Clin. Invest.* 120, 4366 (2010).

[104] H. Nemoto, S. Konno, H. Sugimoto, H. Nakazora, N. Nomoto, M. Murata, H. Kitazono and T. Fujioka, *Exp. Mol. Pathol.* 90, 264 (2011).

[105] A. A. Amato and R. H. Brown Jr., *Handb. Clin. Neurol.* 101, 11 (2011).

[106] A. Gal, E. Siska, Z. Nagy, G. Karpati and M. J. Molnar, *Clin. Neuropathol.* 27, 289 (2008).

[107] M. Meznaric, L. Gonzalez-Quereda, E. Gallardo, N. de Luna, P. Gallano, M. Fanin, C. Angelini, B. Peterlin and J. Zidar, *Eur. J. Neurol.* 18, 1021 (2011).

[108] E. Ozawa, Y. Mizuno, Y. Hagiwara, T. Sasaoka and M. Yoshida, Muscle and Nerve, 32, 563 (2005).

[109] D. Sandona and R. Betto, *Expert Rev. Mol. Med.* 11, e28, (2009).

[110] M. Vainzof, M. R. Passos-Bueno, R. C. Pavanello, S. K. Marie, A. S. Oliveira and M. J. Zatz, *Neurol. Sci.* 164, 44 (1999).

[111] C. Angelini, M. Fanin, M. P. Freda, D. J. Duggan, G. Siciliano and E. P. Hoffman, *Neurology*, 52, 176 (1999).

[112] L. Politano, V. Nigro, L. Passamano, V. Petretta, L. I. Comi, S. Papparella, G. Nigro, P. F. Rambaldi, P. Raia, A. Pini, M. Mora, M. A. Giugliano, M. G. Esposito and G. Nigro, *Neuromuscul. Disord.* 11, 178 (2001).

[113] M. Meznaric-Petrusa, E. Kralj, C. Angelini, M. Fanin and D. Trinkaus, *FSISUP.* 1, 58 (2009).

[114] V. Handa, A. Mital, M. Gupta and S. Goyle, *Neurol. India.* 49, 19 (2001).

[115] L. Klinge, G. Dekomien, A. Aboumousa, R. Charlton, J. T. Epplen, R. Barresi, K. Bushby and V. Straub, *Neuromuscul. Disord.* 18, 934 (2008).

[116] K. Matsumura, F. M. S. Tome, V. Ionasescu, J. M. Ervasti, R. D. Anderson, N. B. Romero, D. Simon, D. R6can, J. C. Kaplan, M. Fardeau and K. P. Campbell, *J. Clin. Invest.* 92, 866 (1993).

[117] E. S. Moreira, M. Vainzof, S. K. Marie, A. L. Sertié, M. Zatz and M. R. Passos-Bueno, *Am. J. Hum. Genet.* 61, 151 (1997).

[118] R. Knöll, W. A. Linke, P. Zou, S. Miocic, S. Kostin, B. Buyandelger, C. H. Ku, S. Neef, M. Bug, K. Schäfer, G. Knöll, L. E. Felkin, J. Wessels, K. Toischer, F. Hagn, H. Kessler, M. Didié, T. Quentin, L. S. Maier, N. Teucher, B. Unsöld, A. Schmidt, E. J. Birks, S. Gunkel, P. Lang, H. Granzier, W. H. Zimmermann, L. J. Field, G. Faulkner, M. Dobbelstein, P. J. Barton, M. Sattler, M. Wilmanns and K. R. Chien, *Circ. Res.* 109, 758 (2011).

[119] T Hayashi, T Arimura, M Itoh-Satoh, K Ueda, S Hohda, N Inagaki, M Takahashi, H Hori, M Yasunami, H Nishi, Y Koga, H Nakamura, M Matsuzaki, B. Y. Choi, S. W. Bae, C. W. You, K. H. Han, J. E. Park, R. Knöll, M. Hoshijima, K. R. Chien and A. Kimura, *J. Am. Coll. Cardiol.* 44, 2192 (2004).

[120] E. Rampersaud, D. D. Kinnamon, K. Hamilton, S. Khuri, R. E. Hershberger and E. R. Martin, *Ann. Hum. Genet.* 74, 110 (2010).

[121] A. Ferreiro, M. Mezmezian, M. Olivé, D. Herlicoviez, M. Fardeau, P. Richard and N. B. Romero, *Neuromuscul. Disord.* 21, 433 (2011).

[122] M. Olivé, A. Shatunov, L. Gonzalez, O. Carmona, D. Moreno, L. G. Quereda, J. A. Martinez-Matos, L. G. Goldfarb and I. Ferrer, *Neuromuscul. Disord.* 18, 929 (2008).

[123] M. Locke, C. L. Tinsley, M. A. Benson and D. J. Blake, *Hum. Mol. Genet.* 18, 2344 (2009).

[124] E. Kudryashova, D. Kudryashov, I. Kramerova and M. J. Spencer, *J. Mol. Biol.* 354, 413 (2005).

[125] J. Schwamborn, E. Berezikov and J. Knoblich, *Cell,* 136, 913 (2009).

[126] A. P. Chiang, J. S. Beck, H. J. Yen, M. K. Tayeh, T. E. Scheetz, R. E. Swiderski, D.Y. Nishimura, T. A. Braun, K. Y. Kim, J. Huang, K. Elbedour, R. Carmi, D. C. Slusarski, T. L. Casavant, E. M. Stone and V.C. Sheffield, *Proc. Natl. Acad. Sci. U.S.A.* 103, 6287 (2006).

[127] P. B. Shieh, E. Kudryashova and M. J. Spencer, *Handb. Clin. Neurol,* 101, 125 (2011).

[128] P. Frosk, T. Weiler, E. Nylen, T. Sudha, C. R. Greenberg, K. Morgan, T. M. Fujiwara and K. Wrogemann, *Am. J. Hum. Genet.* 70, 663 (2002).

[129] V. Saccone, M. Palmieri, L. Passamano, G. Piluso, G. Meroni, L. Politano and V. Nigro, *Hum. Mutat.* 29, 240 (2008).

[130] B. G. Schoser, P. Frosk, A. G. Engel, U. Klutzny, H. Lochmüller and K. Wrogemann, *Ann. Neurol.* 57, 591 (2005).

[131] K. Borg, R. Stucka, M. Locke, E. Melin, G. Ahlberg, U. Klutzny, M. Hagen, A. Huebner, H. Lochmüller, K. Wrogemann, L. E. Thornell, D. J. Blake and B. Schoser, *Hum. Mutat.* 30, E831 (2009).

[132] C. Godfrey, A. R. Foley, E. Clement and F. Muntoni, *Curr. Opin. Genet. Dev.* 21, 278 (2011).

[133] J. M. Ervasti and K. P. Campbell, *J. Cell. Biol.* 122, 809 (1993).

[134] NCBI, http://www.ncbi.nlm.nih.gov

[135] M. Brockington, Y. Yuva, P. Prandini, S. C. Brown, S. Torelli, M. A. Benson, R. Herrmann, L. V. Anderson, R. Bashir, J. M. Burgunder, S. Fallet, N. Romero, M. Fardeau, V. Straub, G. Storey, C. Pollitt, I. Richard, C. A. Sewry, K. Bushby, T. Voit, D. J. Blake and F. Muntoni, *Hum. Mol. Genet.* 10, 2851 (2001).

[136] E. Stensland, S. Lindal, C. Jonsrud, T. Torbergsen, L. A. Bindoff, M. Rasmussen, A. Dahl, F. Thyssen and Ø. Nilssen, *Neuromuscl. Disord.* 21, 41 (2011).

[137] M. Poppe, L. Cree, J. Bourke, M. Eagle, L. V. B. Anderson, D. Birchall, M. Brockington, M. Buddles, M. Busby, F. Muntoni, A. Wills and K. Bushby, *Neurology,* 60, 1246 (2003).

[138] A. D'Amico, S. Petrini, F. Parisi, A. Tessa, P. Francalanci, G. Grutter, F. M. Santorelli and E. Bertini, *Neuromuscul. Disord.* 18, 153 (2008).

[139] M. Margeta, A. M. Connolly, T. L. Winder, A. Pestronk and S. A. Moore, *Muscle Nerve,* 40, 883 (2009).

[140] K. D. Mathews, C. M. Stephan, K. Laubenthal, T. L. Winder, D. E. Michele, S. A. Moore and K. P. Campbell, *Neurology,* 76, 194 (2011).

[141] A. Palmieri, R. Manara, L. Bello, G. Mento, L. Lazzarini, C. Borsato, L. Bortolussi, C. Angelini and E. Pegoraro, *J. Neurol.* 258, 1312 (2011).

[142] D. O Fürst, M. Osborn, R. Nave and K. Weber, *J. Cell Biol.* 106, 1563 (1988).

[143] H. Granzier and S. Labeit, *Muscle Nerve,* 36, 740 (2007).

[144] P. Hackman, A. Vihola, H. Haravuori, S. Marchand, J. Sarparanta, J. De Seze, S. Labeit, C. Witt, L. Peltonen, I. Richard and B. Udd, *Am. J. Hum. Genet.* 71, 492 (2002).

[145] B. Udd, H. Kääriänen and H. Somer, *Muscle Nerve,* 14, 1050 (1991).

[146] B. Udd, A. Vihola, J. Sarparanta, I. Richard and P. Hackman, *Neurology*, 64, 636 (2005).

[147] H. Sorimachi, K. Kinbara, S. Kimura, M. Takahashi, S. Ishiura, N. Sasagawa, N. Sorimachi, H. Shimada, K. Tagawa, and K. Maruyama, *J. Biol. Chem.* 270, 31158 (1995).

[148] K. Kinbara, H. Sorimachi, S. Ishiura and K. Suzuki, *Arch. Biochem. Biophys.* 342, 99 (1997).

[149] J. Sarparanta, G. Blandin, K. Charton, A. Vihola, S. Marchand, A. Milic, P. Hackman, E. Ehler, I. Richard and B. Udd, *J. Biol. Chem.* 285, 30304 (2010).

[150] K. Charton, N. Danièle, A. Vihola, C. Roudaut, E. Gicquel, F. Monjaret, A. Tarrade, J. Sarparanta, B. Udd and I. Richard, *Hum. Mol. Genet.* 19, 4608 (2010).

[151] P.Y. Van den Bergh, O. Bouquiaux, C. Verellen, S. Marchand, I. Richard, P. Hackman and B. Udd, *Ann. Neurol.* 54, 248 (2003).

[152] P. Hackman, S. Marchand, J. Sarparanta, A. Vihola, I. Penisson-Besnier, B. Eymard, J. M. Pardal-Fernandez, el-H. Hammouda, I. Richard, I. Illa and B. Udd, *Neuromuscul. Disord.* 18, 922 (2008).

[153] I. Pénisson-Besnier, P. Hackman, T. Suominen, J. Sarparanta, S. Huovinen, I. Richard-Crémieux and B. Udd, J. Neurol. Neurosurg. *Psychiatry*, 11, 1200 (2010).

[154] M. Pollazzon, T. Suominen, S. Penttila, A. Malandrini, M. A. Carluccio, M. Mondelli, A. Marozza, A. Federico, A. Renieri, P. Hackman, M. T. Dotti and B. Udd, *J. Neurol.* 257, 575 (2010).

[155] H.C. Hartzell, K. Yu, Q. Xiao, L. T. Chien and Z. Qu, *J. Physiol.* 587, 2127 (2009).

[156] I. Mahjneh, J. Jaiswal, A. Lamminen, M. Somer, G. Marlow, S. Kiuru-Enari and R. Bashir, *Neuromuscul. Disord.* 12, 791 (2010).

[157] S. Tsutsumi, N. Kamata, T. J. Vokes, Y. Maruoka, K. Nakakuki, S. Enomoto, K. Omura, T. Amagasa, M. Nagayama, F. Saito-Ohara, J. Inazawa, M. Moritani, T. Yamaoka, H. Inoue and M. Itakura, *Am. J. Hum. Genet.* 74, 1255 (2004).

[158] D. Hicks, A. Sarkozy, N. Muelas, K. Koehler, A. Huebner, G. Hudson, P. F. Chinnery, R. Barresi, M. Eagle, T. Polvikoski, G. Bailey, J. Miller, A. Radunovic, P. J. Hughes, R. Roberts, S. Krause, M. C. Walter, S. H. Laval, V. Straub, H. Lochmüller and K. Bushby, *Brain*, 134, 171 (2011).

[159] H. Gundesli, B. Talim, P. Korkusuz, B. Balci-Hayta, S. Cirak, N. A. Akarsu, H. Topaloglu and P. Dincer, *Am. J. Hum. Genet.* 87, 834 (2010).

[160] J. Uitto, L. Pulkkinen, F. J. Smith and W. H. McLean, *Exp. Dermatol.* 5, 237 (1996).

In: Skeletal Muscle
Editor: Mark Willems

ISBN: 978-1-62417-271-7
© 2013 Nova Science Publishers, Inc.

Chapter 8

STRETCH-ACTIVATED CHANNELS IN MUSCULAR DYSTROPHY

Ella W. Yeung[*]

Muscle Physiology Laboratory, Department of Rehabilitation Sciences,
The Hong Kong Polytechnic University, Hong Kong

ABSTRACT

Duchenne muscular dystrophy is a devastating genetic disease caused by the absence of the protein dystrophin. The pathogenic mechanisms are complex and there is currently no cure for the disease. Dystrophic muscles are susceptible to damage induced by muscle contractions, and excessive calcium influx is thought to trigger a pathological cascade of events that ultimately results in muscle fiber degeneration. Stretch-activated (or mechanosensitive) ion channels have been proposed to be implicated in the regulation of calcium entry in dystrophic muscle. Although the precise molecular nature of the channels is unclear, emerging evidence suggests that these channels are mediators of altered calcium homeostasis, causing muscle damage and progressive muscle weakness in muscular dystrophy.

INTRODUCTION

Duchenne Muscular Dystrophy (DMD) is a lethal X-linked neuromuscular disease caused by mutations in the dystrophin gene [1, 2]. The estimated incidence of DMD is 1:3500 live male births. The disease is characterized by progressive skeletal muscle degeneration and the subsequent replacement of muscles by fibrosis and adipose cells. This results in progressive skeletal muscle weakness and affected boys usually become wheelchair bound by 10-13 years of age and die from respiratory or cardiac failures in their twenties or thirties [3, 4]. In spite of the extensive research on the molecular and cellular pathogenesis of DMD, there is currently no cure for the disease.

---

[*] Email address: ella.yeung@polyu.edu.hk

Dystrophin is localized at the sarcolemma and forms the dystrophin-associated glycoprotein complex. These associations link the cytoskeletal actin network to the overlying basal lamina. It appears that the dystrophin complex has a dual function. Dystrophin has long been thought to provide a mechanically stabilizing role to reinforce the sarcolemmal membrane; and its absence increases membrane fragility which renders muscle fibers more susceptible to damage during contraction [5, 6]. An alternative view is that DGC acts as a scaffolding protein for ion channel clustering and regulation at the sarcolemma; and its absence leads to abnormal channel function [7-9]. Of particular interest are those channels that allow calcium (Ca^{2+}) entry into muscle fibers.

A consistent feature of the pathology of DMD is that dystrophin-lacking fibers are overloaded with Ca^{2+}. In the *mdx* mouse model of DMD, there is considerable evidence implicating that disturbance of Ca^{2+} homeostasis contribute to the development of the muscle damage and degeneration [10-13]. The route of Ca^{2+} entry leading to increased intracellular calcium $[Ca^{2+}]_i$ in dystrophic muscle is controversial.

Whether it is the transient membrane tears or the altered function of the channels that cause the increase in membrane permeability for Ca^{2+} influx still remain in contention. Our studies thus far point to Ca^{2+} entry via channels rather than membrane tears.

In particular, our findings and others have shown and suggested that the abnormal activity of the Ca^{2+} permeable stretch-activated channels (SACs) contribute to Ca^{2+} influx and involve in the pathogenesis of DMD. It is the intent of this chapter to provide evidence that Ca^{2+} influx through SACs contribute to the elevated $[Ca^{2+}]_i$ that is characteristic of dystrophic skeletal muscle.

INTRACELLULAR CALCIUM IN MUSCULAR DYSTROPHY

There has been some controversy as to whether free resting $[Ca^{2+}]_i$ are higher in muscles lacking dystrophin. Early studies have demonstrated that resting $[Ca^{2+}]_i$ levels are elevated in myotubes and isolated muscle fibers in dystrophic *mdx* mice as well as in cultured myotubes from DMD patients [9, 10, 14, 15]. However several other groups did not confirm this elevation [16-19]. This discrepancy may be due to the methodological differences used to obtain isolated fibers, the fluorescent indicators used to measure $[Ca^{2+}]_i$ and maturation of cultured cells [11]. Of particular significance, dysfunction in Ca^{2+} regulation is directly related to the contractile activity of the muscle. It has been demonstrated on human DMD myotubes that elevated $[Ca^{2+}]_i$ levels is dependent on muscle contractile activity; long term inhibition of contractile activity with tetrodotoxin not only prevents the rise in $[Ca^{2+}]_i$ but returns the resting $[Ca^{2+}]_i$ to normal levels [20]. Similar results have also been observed in cultured myotubes from the *mdx* mice [21]. These findings indicate that muscle contraction is a dominant factor contributing to the disturbance of $[Ca^{2+}]_i$ homeostasis in dystrophin-deficient muscle. Furthermore the abnormal $[Ca^{2+}]_i$ homeostasis observed is a consequence of damage from previous periods of contraction and rather than Ca^{2+} influx during muscle activity.

The simplest explanation for excessive Ca^{2+} influx is greater fragility placed upon the sarcolemma during muscle contraction resulting in membrane tears [6, 22, 23]. The interpretation is based on the observations of elevated levels of serum creatine kinase that leak out of the muscle fibers; or increased uptake of membrane-impermeable dyes such as Evans Blue dye or procion orange found inside the muscle fibers. Alternatively, Ca^{2+} influx can occur through activation of mechanosensitive or stretch-activated channels (SACs) during contraction. We and others have suggested that Ca^{2+} influx through abnormal regulation of SACs as the mechanism for the increased membrane permeability in DMD muscles [13, 24-26]. Consequently, the increase in Ca^{2+} entry through SACs over time results in activation of Ca^{2+}-dependent proteolysis which eventually leads to muscle degeneration.

STRETCH-INDUCED MUSCLE DAMAGE

Muscles are subjected to stress and strain during contractions and there is widespread agreement that muscle damage is more pronounced when the muscles are stretched during contractions (eccentric contractions). For instance, activities such as hiking or downhill running result in greater muscle soreness, stiffness, swelling and damage when compared to isometric or concentric contractions. This muscle damage is characterized by reduced force production, ultrastructural disruption including Z-line streaming and myofibrillar disarray, increased serum creatine kinase activity, inflammatory response and increased proteolytic enzyme activity. In normal individuals these responses are normally short-lived where full recovery is expected within several weeks [27, 28]. Early studies have demonstrated that dystrophic muscles show a greater susceptibility to stretch-induced muscle damage than normal muscles. This has been verified in intact animals and isolated muscle preparations [29-31]. Furthermore, restoration of dystrophin to a portion of *mdx* muscles causes a reduction in the magnitude of the stretch-induced damage [32, 33]. These results suggest that absence of dystrophin renders the sarcolemma more prone to muscle damage.

One obvious feature of muscle damage is the accompanying reduction in tetanic force. The magnitude of damage correlates strongly with the initial muscle length and the amplitude of the stretch [27]. In different muscle preparations and various stretched contraction protocols, force reduction in wild-type muscles is about 10-36 % whereas in *mdx* muscles, force reduction is about to 60-76% [13, 25, 30]. Another characteristic of stretch-induced damage is the increase in the shift in the force-length relationship towards longer muscle length associated with overstretched sarcomeres. We observed a shift to longer lengths in the *mdx* fibers after stretched contractions but this was not different from wild-type fibers [12]. Impaired excitation-contraction coupling is another mechanism that contributes to reduction in force. Several studies have shown that the reduction in Ca^{2+} release in the *mdx* fibers provides an explanation of reduced force. In response to action potential stimulation and voltage clamp pulses, it has been shown that sarcoplasmic reticulum (SR) to release Ca^{2+} is reduced in *mdx* muscle fibers compared with wild-type fibers [34, 35]. One possibility for reduced Ca^{2+} is due to the persisting membrane depolarization observed after stretched contractions. The depolarization could arise from increased permeability to Ca^{2+} and Na^+ either through membrane tears or SACs. SAC blockers prevent the depolarization which strongly suggests that the activation of SACs is responsible for increased membrane permeability [24].

Different biomarkers for quantifying muscle damage have also been employed to assess membrane integrity. Plasma creatine kinase (CK) activity is often used to assess muscle damage following stretch-induced muscle contractions. Elevated CK levels in the serum indicate that intramuscular proteins leak out into the blood; this is interpreted as a loss of membrane integrity. Persistent elevation of CK levels has been well documented in patients with DMD [36]. However CK levels show great variability depending on the rate of release into and clearance of the protein in the circulation. Furthermore there is a delay in the appearance of this protein in blood; the elevated levels may not be evident until 24-72 hours after muscle damage. As such, serum CK levels alone cannot account for the precise temporal relationship between membrane disruption and the eccentric muscle injury event. Also it is noted that CK levels correspond poorly with muscle contractile function [37]. Another method to detect membrane permeability in dystrophic muscle is the use of exogenous membrane-impermeant fluorescent dyes. These large molecular weight proteins do not normally pass through the sarcolemmal membrane. In any conditions under which uptake of these dyes are visually detectable macroscopically [22] or microscopically [38], it indicates that the membrane integrity is compromised. Fluorescent dyes such as Evans blue dye or procion orange have been used in *mdx* mice to assess the susceptibility to stretched contractions [6, 31, 38-40] It has been shown that entry of these dyes in *mdx* muscle is much greater after stretched contractions compared with control. It is important to note that these dye uptake experiments provide evidence of membrane permeability; whether the permeability is caused by mechanical tears or some other mechanism is not certain.

INTACELLULAR CALCIUM CHANGES FOLLOWING STRETCH-INDUCED MUSCLE DAMAGE IN DYSTROPHIC MUSCLE

To establish the causal link between SACs and increase in membrane permeability with concomitant Ca^{2+} influx, an experimental model involving eccentric muscle contraction to induce stress to the sarcolemma is used. We need to show the following: (i) excessive Ca^{2+} influx through SACs is demonstrated in DMD muscle when compared to normal muscle; (ii) aberrant Ca^{2+} entry via SACs occur in dystrophic muscles in different muscle preparations; (iii) the magnitude and the time-course to the rise in $[Ca^{2+}]_i$ resulting from SAC activation or via mechanical tear should be distinctly different; and (iv) the rise in $[Ca^{2+}]_i$ can be attenuated with inhibition of SAC activity/expression which blocks Ca^{2+} influx.

Our approach to study $[Ca^{2+}]_i$ changes during stretched contractions is with isolated single fiber preparations. This approach was first described in 1987 [41]. An advantage of the single fiber preparation is that the length changes imposed on the *mdx* muscle fiber during stretched contractions can be controlled, it also allows simultaneous monitoring of force and $[Ca^{2+}]_i$ fluorescent measurements during muscle activity. In our study [13], the *mdx* muscle fiber was initially subjected to a series of isometric contractions as control, then followed by eccentric contractions at 40% stretch. Compared to wild-type muscle fiber, stretch-induced damage is more pronounced in the *mdx* fiber with a greater reduction in force. Our initial finding was that the resting $[Ca^{2+}]_i$ was higher in *mdx* compared to wild-type fiber. In-situ calibration of fluo-4 fluorescence signals was performed in the *mdx* muscle fiber with the resting $[Ca^{2+}]_i$ estimated to be 261±39 nM (*vs* 109±5.3 nM in wild-type fiber) [42]. It is conceivable that the sustained elevation in resting $[Ca^{2+}]_i$ level exceeding the cell buffering capacity could result in the activation of the Ca^{2+}-sensitive proteolytic cascade.

When the *mdx* muscle fiber was exposed to stretched contractions, the $[Ca^{2+}]_i$ level increased markedly. Of particular importance, there is a slow increase in resting $[Ca^{2+}]_i$ which started after the stretched contractions and reached to a peak of 657 ± 78 nM at 30 min after contractions [42]. This trend was also observed in wild-type fibers but the increase in $[Ca^{2+}]_i$ was significantly larger in *mdx* muscles. The time course and the rate of rise in $[Ca^{2+}]_i$ do not seem to match with mechanical tear(s) hypothesis where one would expect to see an immediate rather than a gradual rise in $[Ca^{2+}]_i$. A previous study that artificially create tears in the membrane with intense laser pulses demonstrated that membrane damage reseals within one minute; and showed that the time to reseal in the *mdx* fibers was no different from that of the wild-type fibers [46]. If the membrane damage is caused by tears, the elevation in $[Ca^{2+}]_i$ should reach its peak almost immediately and when the membrane reseals within a minute, this rise should terminate. Furthermore if such tears were present, one would expect to observe a highly increase Ca^{2+} fluorescence at localized regions just inside the membrane immediately after stretch-induced damage. However, we were not able to identify any 'hot spots' over the course of the experiments [12, 13]. Rather, the distribution of fluorescence was close to uniform. Of course it may be possible that the increase is too small or too transient or too infrequent to be detected by our methods.

For these reasons, we consider that the elevated resting $[Ca^{2+}]_i$ might occur through channels that are permeable to Ca^{2+} ions. One possible contributor to the increased Ca^{2+} entry following stretch-induced muscle damage is the stretch-activated (or mechanosensitive) channels. SACs are Ca^{2+} permeable channels that are activated by membrane stretch [43]. We therefore examined the role of SACs by the blocking of these channels with either streptomycin, Gd^{3+} or the spider venom toxin GsMTx-4 [13]. Our hypothesis was confirmed by the results, showing that SAC blockers completely prevent the rise of $[Ca^{2+}]_i$ after stretched-induced damage and cause a fall in the resting $[Ca^{2+}]_i$ level. In addition, removal of extracellular calcium prevents the rise in resting $[Ca^{2+}]_i$ and ameliorates the decline in force normally produced by stretched contractions. Taken together, these findings suggest that (i) stretch-induced damage is caused by influx of Ca^{2+}; and (ii) both the Ca^{2+} influx and the elevated resting $[Ca^{2+}]_i$ after stretched contractions are mediated by SACs. The elevated $[Ca^{2+}]_i$ level may cause activation of Ca^{2+}-dependent proteases (calpains), phospholipase A_2, and increased reactive oxygen species which cause membrane damage and increased membrane permeability [10, 44].

In a follow up study that investigated the time course of dye uptake in isolated *mdx* muscles, it was demonstrated that procion orange positive fibers progressively increased by 3-4 times from 0 to 60 min after stretched contractions [45]. If mechanical tears had occurred during the stretched contractions, one would predict that the dye rapidly enters the muscle fibers and becomes trapped inside the fibers after membrane resealing within a minute [46]. As such the number of dye-positive fibers would be expected to remain constant over the 60 min period observed. SAC blocker streptomycin reduced the number of procion orange positive fibers significantly at 60 min after stretched contractions, suggesting that inhibiting SAC activity can prevent most of the membrane permeability. These data suggests that dye entry is a secondary consequence of SAC opening resulting in elevation of $[Ca^{2+}]_i$ as observed in our Ca^{2+} imaging study. In other experiments on intact *mdx* mice, two week treatment with streptomycin was shown to improve the reduction of force and minimize muscle damage (as ascertained by reduced EBD uptake) following downhill running [45]. Protection against membrane permeability and stretched-induced muscle damage by SAC blockers has also been documented in several other studies [24, 26, 47].Taken together, the evidence suggests that stretch-induced muscle damage involves rise in $[Ca^{2+}]_i$ level which are attributable to SACs.

STRETCH-ACTIVATED ION CHANNELS IN MUSCULAR DYSTROPHY

Living cells respond and adapt to their physical environments by converting mechanical forces and deformations into biochemical signals. These signals modulate short-term and long-term cellular functions such as migration, growth, differentiation, gene expression, protein synthesis and apoptosis; and have a crucial role in development and homeostasis. In response to mechanical stresses, the activation of cellular signal transduction can occur under physiological and pathological conditions [48, 49]. Stretch-sensitive ion channels change their conformation from closed to open in response to the changes in the physical properties of the lipid bilayer and the cytoskeleton [50]. Traditionally, SACs are characterized by single channel recordings and the simultaneous application of suction to membrane patches [51, 52]. A functional description has been provided for SAC [53, 54]. For a channel to be defined as mechanosensitvie, it has to fulfill the following criteria: the latency of current elicited by the stimulus should be faster than known second-messenger systems (~5 ms); the kinetics of channel activation should depend on the amplitude of the stimulus; there should be a mechanical correlate of channel-gating and a sensory cell or organ that responds in the same range as the channel; and part of the channel and/or associated subunit has to move after a change in mechanical force.

SACs were first observed in chick skeletal muscles [43]. These channels are non-selective cation channels that are permeable to both Na^+ and Ca^{2+} and increase their open probability in response to membrane stretch. Their activities are blocked by SAC inhibitors: streptomycin, Gd^{3+}, the spider venom peptide GsMTx-4. These blockers are thought to inhibit Ca^{2+} entry by blocking the channel pore [55]. GsMTx-4 is the most potent and specific SAC blocker identified to date [56]. It binds to the protein-lipid interface to affect the channel gating [57, 58]. So far the channel proteins are poorly characterized but there is increasing interest in SACs because of their involvement in Duchenne muscular dystrophy. A number of lines of evidence suggest that SACs expression and activity are altered in dystrophic muscle and are responsible for the abnormal influx of Ca^{2+}. SAC channels seem to be expressed more frequently in *mdx* mouse muscle [59] and in human DMD myotubes [60]. Several studies have demonstrated that SACs are more active in *mdx* muscle [52, 61, 62]. These channels have been described as voltage independent, Ca^{2+} permeable and had a conductance of 25 pS. It has also been shown that *mdx* myotubes and muscle fibers exhibit an increase permeability to Ca^{2+} that enter the cell through these channels, and the influx can be blocked by SAC-blocker sensitive pathways [9]. Furthermore, the SAC opening probability in *mdx* fibers is greater than that in the wild-type muscle fibers [61]. Patch-clamp recordings have also confirmed that SAC activity is more active in the human DMD co-cultured myotubes [60]. SACs can become desensitized when exposed to chronic or prolonged high membrane tension. This loss of sensitivity provides mechanoprotection, shielding channels from local stress [63, 64]. This phenomenon was shown in wild-type myotubes where the resting SAC activity decreased over time with repeated suction stimulation, but in contrast, resting SAC activity in *mdx* myotubes was shown to increase over time [65]. Thus, together all these findings suggest that muscles lacking dystrophin have abnormal activity of SACs leading to increase in Ca^{2+} influx.

TRP (TRANSIENT RECEPTOR POTENTIAL) PROTEINS AS STRETCH-ACTIVATED CHANNELS

A very important development in the understanding of the functional significance of SACs has been the identification of genes encoding members of the transient receptor potential (TRP) channel family. Several members of the TRPC, TRPV and TRPM subfamilies of TRP channels are expressed in skeletal muscle as shown by RT-PCR, Western blot and immunohistochemistry [66]. TRPC subfamily, TRPC1, was identified to be mechanosensitive and form the SAC in vertebrate cells [67]. This protein has a molecular weight of 80 kDa and bound antibodies to TRPC1. The investigators expressed human TRPC1 in frog oocytes and CHO-K1 cells and observed a ten-fold increase in SAC expression. Furthermore, TRPC1 is able to translate membrane stretch into cation currents across the plasma membrane and the activity of this channel can be inhibited using TRPC1-specific antisense RNA. Later, two studies with contrasting results were reported which showed that the amplitude of mechanosensitive currents is not significantly altered by both TRPC1 overexpression [68] and ablation [69]. However, it is possible that the expression of tagged-TRP proteins is not localized in the plasma membrane or compensation by other isoforms in TRPC1 KO mice, may affect the proper protein function [70]. TRPC1 is found at the sarcolemma co-localizing with dystrophin [71] and caveolin 3 [72]. It has been recently demonstrated that different levels of TRPC1 are associated to different degrees of the dystrophic phenotype in *mdx* mice [47]. The involvement of TRPC1 in abnormal Ca^{2+} influx in DMD is highlighted in a study which shows increased expression and activity of a channel having biophysical properties identical with SACs in *mdx* muscles [73]. However, this channel was activated by depletion of Ca^{2+} stores with thapsigargin or caffeine. The investigators further showed that repression with antisense oligonucleotides of TRPC1 markedly reduce the activity of the thapsigargin-induced Ca^{2+} channels without affecting the voltage dependent Na^{+} channels. The results from this study suggest that SAC and SOC activity could arise from a channel encoded by TRPC1. A pharmacological study was subsequently performed which shows that SACs and SOCs share similar biophysical and pharmacological properties in adult muscle fibers [52]. Thus, it seems that both SACs and SOCs are closely related and composed of TRPC1.

Apart from the intensely studied TRPC1, other channels of the TRPC family have been implicated in Ca^{2+} entry associated with DMD. For instance, antisense-mediated inhibition directed against TRPC1, 4, and 6 reduced Ca^{2+} leak activity in *mdx* myofibers in culture by 90%, strongly suggesting a role of TRPC channels in Ca^{2+} influx [73]. A recent study demonstrated that overexpression of TRPC3 and the associated increase in TRPC3-dependent Ca^{2+} influx resulted in a phenotype that closely resembles muscular dystrophy. This is evidenced by fibrosis, myofiber degeneration followed by cycles of regeneration and fatty tissue replacement. One important observation in this study is the fact that EBD uptake and serum CK levels were only minimally increased in TRPC3 transgenic mice after exercise despite the observed myofiber degeneration, suggesting that TRPC-dependent Ca^{2+} influx is sufficient to initiate dystrophic phenotype in skeletal muscles without major changes in sarcolemma stability [74]. However, TRPC3 appears to be intracellularly localized.

The possible role of TRPV2 in the pathophysiology of DMD is also demonstrated in several studies [75-77]. TRPV2 is characterized as a Ca^{2+} permeable, stretch-activated and growth factor regulated channel. Its expression is increased in cultured *mdx* myotubes and upon cyclic stretch, TRPV2 translocates to the plasma membrane to become activated and mediates sustained $[Ca^{2+}]_i$ increase. Of significance, removal of extracellular Ca^{2+} or SAC blocker markedly promotes the internalization of TRPV2 in *mdx* myotubes. Furthermore, the elevated $[Ca^{2+}]_i$ levels in muscle fibers could be reduced by dominant-negative inhibition of endogenous TRPV2, suggesting that TRPV2 is a Ca^{2+}-entry route leading to a sustained $[Ca^{2+}]_i$ increase. A recent study extended this observation by showing that *mdx* muscles overexpressing a dominant-negative TRPV2 channel are protected from stretch-induced muscle damage [77]. Results from this study also established that the force deficit and the permeability to procion orange dye were significantly reduced, suggesting that the entry of Ca^{2+} influx following stretched contractions through stretch-sensitive TRPV2 channel, rather than membrane fragility, plays a role in muscle damage in muscular dystrophy.

Although the molecular basis and precise gating mechanism of these TRP channels are largely unknown, the studies mentioned above provide significant insight that dysregulation of mechanosensitive Ca^{2+}-permeable channels plays a key role in the pathogenesis of Duchenne muscular dystrophy.

ROLE OF DYSTROPHIN IN STRETCH-ACTIVATED CHANNEL REGULATION

The absence of dystrophin seems to alter the function of SACs leading to pathological changes in the muscles. Why are stretch-activated channels more active in DMD? One of the possibilities is that dystrophin and its interacting proteins/molecules modulate Ca^{2+} signaling pathways for Ca^{2+} homeostasis. The observations that forced expression of minidystrophin reduces stretch-induced muscle damage [32, 33] and reduces SOC Ca^{2+} influx [78] somewhat support the idea that dystrophin influences Ca^{2+}-permeable channel activity directly or indirectly. The dystrophin-associated glycoprotein (DAG) complex provides a linkage between the actin cytoskeleton and the extracellular matrix. It has been proposed that DAG complex functions as a scaffolding complex for signaling molecules, which regulates activity of channels. Co-immunoprecipitation and co-localization experiments suggest the association of TRPC1 and α-syntrophin to the complex of DAG in cultured myotubes [71, 79]. TRPC1 also co-localizes with the scaffolding protein Homer 1 [80]. Mice lacking Homer 1 exhibit myopathic changes caused by abnormal Ca^{2+} influx from overactive TRPC1 channels. Interestingly, the channel activity can be blocked by the GsMTx4 peptide suggesting it can be carried by a SAC. It was also demonstrated that Homer 1 expression is decreased in dystrophic muscle [80].

We reported the association of TRPC1 with the scaffolding protein caveolin-3, by showing co-localization at the sarcolemma and co-immunoprecipitation of endogenous proteins [72]. We observed increased levels of expression of caveolin-3 together with TRPC1 in *mdx* muscles. Furthermore caveolin-3 co-expression was found to assist trafficking of TRPC1 to the plasma membrane which contributed to Ca^{2+} influx. Taken together, it is possible that TRPC1 channel functions in a costameric macromolecular complex in which dystrophin is linked to the scaffolding proteins (α-syntrophin, Homer 1, and caveolin-3). In the absence of dystrophin, the loss of this molecular assembly leads to aberrant TRPC1 activity and impaired Ca^{2+} homeostasis [81]. Another possibility for the increased sensitivity of the channels in dystrophic muscles may be caused by reactive oxygen species (ROS) damage. We have reported that ROS plays in role in muscle damage which in turn stimulates Ca^{2+} entry through SACs in *mdx* fibers, both at rest and following stretched contractions [72, 82]. These results suggests the missing connection between Ca^{2+} influx and the activation of the longer-term intracellular events causing increased membrane permeability in dystrophic muscles.

CONCLUSION

This chapter presents a mechanism of abnormal Ca^{2+} influx in dystrophin-deficient muscles via SACs causing a rise in $[Ca^{2+}]_i$. Such increase in $[Ca^{2+}]_i$ promotes subsequent proteolysis, causing further increase in membrane permeability and membrane damage.

Although the activation and the regulatory mechanisms of SACs (or TRP) channels are unknown, the fact that muscles lacking dystrophin exhibit abnormal ion channel function suggests their association. Early interventions that can inhibit the increased activity of SACs may be a therapeutic option for Duchenne muscular dystrophy.

ACKNOWLEDGMENTS

The author would like to express her sincere gratitude to Professor David Allen for his guidance and support in the last 13 years. It has been an honor to be his student.

REFERENCES

[1]   E. P. Hoffman, R. H. Brown, Jr. and L. M. Kunkel, *Cell* 51, 919 (1987).
[2]   S. C. Watkins, E. P. Hoffman, H. S. Slayter and L. M. Kunkel, *Nature* 333, 863 (1988).
[3]   H. Moser, *Hum Genet* 66, 17 (1984).
[4]   M. Kohler, C. F. Clarenbach, C. Bahler, T. Brack, E. W. Russi and K. E. Bloch, *J. Neurol. Neurosurg. Psychiatry.* 80, 320 (2009).
[5]   O. F. Hutter, *J. Inherit. Metab. Dis.* 15, 565 (1992).
[6]   B. J. Petrof, J. B. Shrager, H. H. Stedman, A. M. Kelly and H. L. Sweeney, *Proc. Natl. Acad. Sci. USA* 90, 3710 (1993).
[7]   J. M. Gillis, *Acta Physiol. Scand.* 156, 397 (1996).
[8]   C. G. Carlson, *Neurobiol. Dis.* 5, 3 (1998).
[9]   O. Tutdibi, H. Brinkmeier, R. Rudel and K. J. Fohr, *J. Physiol.* 515 (Pt 3), 859 (1999).
[10] P. R. Turner, T. Westwood, C. M. Regen and R. A. Steinhardt, *Nature* 335, 735 (1988).

[11] J. M. Gillis, *J. Muscle Res. Cell Motil.* 20, 605 (1999).

[12] E. W. Yeung, S. I. Head and D. G. Allen, *J. Physiol.* 552, 449 (2003).

[13] E. W. Yeung, N. P. Whitehead, T. M. Suchyna, P. A. Gottlieb, F. Sachs and D. G. Allen, *J. Physiol.* 562, 367 (2005).

[14] P. Y. Fong, P. R. Turner, W. F. Denetclaw and R. A. Steinhardt, *Science* 250, 673 (1990).

[15] T. Mongini, D. Ghigo, C. Doriguzzi, F. Bussolino, G. Pescarmona, B. Pollo, D. Schiffer and A. Bosia, *Neurology* 38, 476 (1988).

[16] P. Gailly, B. Boland, B. Himpens, R. Casteels and J. M. Gillis, *Cell Calcium* 14, 473 (1993).

[17] S. I. Head, *J. Physiol.* 469, 11 (1993).

[18] J. Pressmar, H. Brinkmeier, M. J. Seewald, T. Naumann and R. Rudel, *Pflugers Arch.* 426, 499 (1994).

[19] C. Collet, B. Allard, Y. Tourneur and V. Jacquemond, *J. Physiol.* 520 Pt 2, 417 (1999).

[20] N. Imbert, C. Cognard, G. Duport, C. Guillou and G. Raymond, *Cell Calcium* 18, 177 (1995).

[21] F. W. Hopf, P. R. Turner, W. F. Denetclaw, Jr., P. Reddy and R. A. Steinhardt, *Am. J. Physiol.* 271, C1325 (1996).

[22] V. Straub, J. A. Rafael, J. S. Chamberlain and K. P. Campbell, *J. Cell Biol.* 139, 375 (1997).

[23] B. J. Petrof, *Mol. Cell Biochem.* 179, 111 (1998).

[24] T. A. McBride, B. W. Stockert, F. A. Gorin and R. C. Carlsen, *J. Appl. Physiol.* 88, 91 (2000).

[25] N. P. Whitehead, E. W. Yeung and D. G. Allen, *Clin. Exp. Pharmacol. Physiol.* 33, 657 (2006).

[26] T. Sonobe, T. Inagaki, D. C. Poole and Y. Kano, *Am. J. Physiol. Regul. Integr. Comp. Physiol.* 294, R1329 (2008).

[27] D. G. Allen, *Acta Physiol. Scand.* 171, 311 (2001).

[28] U. Proske and D. L. Morgan, *J. Physiol.* 537, 333 (2001).

[29] S. I. Head, D. A. Williams and D. G. Stephenson, *Proc. Biol. Sci.* 248, 163 (1992).

[30] P. Moens, P. H. Baatsen and G. Marechal, *J. Muscle Res. Cell Motil.* 14, 446 (1993).

[31] V. Brussee, F. Tardif and J. P. Tremblay, *Neuromuscul. Disord.* 7, 487 (1997).

[32] N. Deconinck, T. Ragot, G. Marechal, M. Perricaudet and J. M. Gillis, *Proc. Natl. Acad. Sci. U S A* 93, 3570 (1996).

[33] V. Brussee, F. Merly, F. Tardif and J. P. Tremblay, *Biochem. Biophys. Res. Commun.* 250, 321 (1998).

[34] C. E. Woods, D. Novo, M. DiFranco and J. L. Vergara, *J. Physiol.* 557, 59 (2004).

[35] C. E. Woods, D. Novo, M. DiFranco, J. Capote and J. L. Vergara, *J. Physiol.* 568, 867 (2005).

[36] E. Ozawa, Y. Hagiwara and M. Yoshida, *Mol. Cell Biochem.* 190, 143 (1999).

[37] J. Fridén and R. L. Lieber, *Scand. J. Med. Sci. Sports* 11, 126 (2001).

[38] P. W. Hamer, J. M. McGeachie, M. J. Davies and M. D. Grounds, *J. Anat.* 200, 69 (2002).

[39] R. Matsuda, A. Nishikawa and H. Tanaka, *J. Biochem.* 118, 959 (1995).

[40] C. N. Pagel and T. A. Partridge, *J. Neurol. Sci.* 164, 103 (1999).

[41] J. Lannergren and H. Westerblad, *J. Physiol.* 390, 285 (1987).

[42]   D. G. Allen, O. L. Gervasio, E. W. Yeung and N. P. Whitehead, *Can. J. Physiol. Pharmacol.* 88, 83 (2010).

[43]   F. Guharay and F. Sachs, *J. Physiol.* 352, 685 (1984).

[44]   P. Gailly, F. De Backer, M. Van Schoor and J. M. Gillis, *J. Physiol.* 582, 1261 (2007).

[45]   N. P. Whitehead, M. Streamer, L. I. Lusambili, F. Sachs and D. G. Allen, *Neuromuscul. Disord.* 16, 845 (2006).

[46]   D. Bansal, K. Miyake, S. S. Vogel, S. Groh, C. C. Chen, R. Williamson, P. L. McNeil and K. P. Campbell, *Nature* 423, 168 (2003).

[47]   C. Y. Matsumura, A. P. Taniguti, A. Pertille, H. Santo Neto and M. J. Marques, *Am. J. Physiol. Cell Physiol.* 301, C1344 (2011).

[48]   O.P. Hamill and B. Martinac, *Physiol. Rev.* 81, 685 (2001).

[49]   D. E. Jaalouk and J. Lammerding, *Nat. Rev. Mol. Cell Biol.* 10, 63 (2009).

[50]   J. B. Lansman and A. Franco-Obregon, *Mechanosensitivity in Cells and Tissues. Moscow: Academia,* 390 (2005).

[51]   Franco-Obregon and J. B. Lansman, *J. Physiol.* 539, 391 (2002).

[52]   T. Ducret, C. Vandebrouck, M. L. Cao, J. Lebacq and P. Gailly, *J. Physiol.* 575, 913 (2006).

[53]   A.P. Christensen and D. P. Corey, *Nat. Rev. Neurosci.* 8, 510 (2007).

[54]   B. Nilius, *EMBO Rep.* 11, 902 (2010).

[55]   O.P. Hamill and D. W. McBride Jr., *Pharmacol. Rev.* 48, 231 (1996).

[56]   T. M. Suchyna, J. H. Johnson, K. Hamer, J. F. Leykam, D. A. Gage, H. F. Clemo, C. M. Baumgarten and F. Sachs, *J. Gen. Physiol.* 115, 583 (2000).

[57]   T. M. Suchyna, S. E. Tape, R. E. Koeppe 2nd, O. S. Andersen, F. Sachs and P. A. Gottlieb, *Nature* 430, 235 (2004).

[58]   K. Kamaraju, P. A. Gottlieb, F. Sachs and S. Sukharev, *Biophys. J.* 99, 2870 (2010).

[59]   C. M. Haws and J. B. Lansman, *Proc. Biol. Sci.* 245, 173 (1991).

[60]   C. Vandebrouck, G. Duport, C. Cognard and G. Raymond, *Neuromuscul. Disord.* 11, 72 (2001).

[61]   Franco-Obregon Jr. and J. B. Lansman, *J. Physiol.* 481 (Pt 2), 299 (1994).

[62]   P. Gailly, *Curr. Opin. Pharmacol.* (in press).

[63]   C. E. Morris, *Cell. Mol. Biol. Lett.* 6, 703 (2001).

[64]   E. Honore, A. J. Patel, J. Chemin, T. Suchyna and F. Sachs, *Proc. Natl. Acad. Sci. USA* 103, 6859 (2006).

[65]   T. M. Suchyna and F. Sachs, *J. Physiol.* 581, 369 (2007).

[66]   H. Brinkmeier, *Adv. Exp. Med. Biol.* 704, 749 (2011).

[67]   R. Maroto, A. Raso, T. G. Wood, A Kurosky, B. Martinac and O. P. Hamill, *Nat. Cell Biol.* 7, 179 (2005).

[68]   P. Gottlieb, J. Folgering, R. Maroto, A. Raso, T. G. Wood, A. Kurosky, C. Bowman, D. Bichet, A. Patel, F. Sach, B. Martinac, O. P. Hamill and E. Honoré, *Pflugers Arch.* 455, 1097 (2008).

[69]   Dietrich, H. Kalwa, U. Storch, M. Mederos y Schnitzler, B. Salanova, O. Pinkenburg, G. Dubrovska, K. Essin, M. Gollasch, L. Birnbaumer and T. Gudermann, *Pflugers Arch.* 455, 465 (2007).

[70]   S. Alfonso, O. Benito, S. Alicia, Z. Angélica, G. Patricia, K. Diana and L. Vaca. *Cell Calcium* 43, 375 (2008).

[71] Vandebrouck, J. Sabourin, J. Rivet, H. Balghi, S. Sebille, A. Kitzis, G. Raymond, C. Cognard, N. Bourmeyster and B. Constantin, *FASEB J.* 21, 608 (2007).

[72] O.L. Gervasio, N. P. Whitehead, E. W. Yeung, W. D. Phillips and D. G. Allen, *J. Cell Sci.* 121, 2246 (2008).

[73] C. Vandebrouck, D. Martin, M. Colson-Van Schoor, H. Debaix and P. Gailly, *J. Cell Biol.* 158, 1089 (2002).

[74] D. P. Millay, S. A. Goonasekera, M. A. Sargent, M. Maillet, B. J. Aronow and J. D. Molkentin, *Proc. Natl. Acad. Sci. U S A* 106, 19023 (2009).

[75] Y. Iwata, Y. Katanosaka, Y. Arai, K. Komamura, K. Miyatake and M. Shigekawa, *J. Cell Biol.* 161, 957 (2003).

[76] Y. Iwata, Y. Katanosaka, Y. Arai, M. Shigekawa and S. Wakabayashi, *Hum. Mol. Genet.* 18, 824 (2009).

[77] N. Zanou, Y. Iwata, O. Schakman, J. Lebacq, S. Wakabayashi and P Gailly, *FEBS Lett.* 583, 3600 (2009).

[78] Vandebrouck, T. Ducret, O. Basset, S. Sebille, G. Raymond, U. Ruegg, P. Gailly, C. Cognard and B. Constantin, *FASEB J.* 20, 136 (2006).

[79] J. Sabourin, C. Lamiche, A. Vandebrouck, C. Magaud, J. Rivet, C. Cognard, N. Bourmeyster and B. Constantion, *J. Biol. Chem.* 284, 36248 (2009).

[80] J. A. Stiber, Z. S. Zhang, J. Burch, J. P. Eu, S. Zhang, G. A. Truskey, M. Seth, N. Yamaguchi, G. Meissner, R. Shah and P. F. Worley, *Mol. Cell Biol.* 28, 2637 (2008).

[81] J. Sabourin, C. Cognard and B. Constantin, *J. Muscle Res. Cell Motil.* 30, 289 (2009).

[82] N. P. Whitehead, E. W. Yeung, S. C. Froehner and D. G. Allen, *PLoS One* 5, e15354 (2010).

In: Skeletal Muscle
Editor: Mark Willems

ISBN: 978-1-62417-271-7
© 2013 Nova Science Publishers, Inc.

Chapter 9

CELLULAR AND MOLECULAR MECHANISMS REGULATING THE HYPERTROPHY AND ATROPHY OF SKELETAL MUSCLE

***Kunihiro Sakuma**[*1] **and Akihiko Yamaguchi**[2]*
[1]Research Center for Physical Fitness, Sports and Health,
Toyohashi University of Technology, Toyohashi, Japan
[2]School of Dentistry, Health Sciences University of Hokkaido, Kanazawa,
Ishikari-Tobetsu, Hokkaido, Japan

ABSTRACT

Beyond skeletal muscle's primary function as a force generator for locomotion, there is a growing recognition of the important role skeletal muscle plays in overall health through its impact on whole-body metabolism as well as directly influencing quality of life issues with chronic disease and aging. Gain or loss of skeletal muscle mass occurs in situations of altered use such as strength training, aging, denervation, or immobilization. Skeletal muscle is a highly adaptable tissue with well-known sensitivities to environmental cues such as growth factors, cytokines, nutrients, and mechanical loading. Over the last decade, extensive progress has been made with regard to our understanding of the molecules that regulate skeletal muscle mass. Not surprisingly, many of these molecules are intimately involved in the regulation of protein synthesis and protein degradation. This chapter examines our current understanding of the cellular and molecular events involved in the control of muscle mass under conditions of muscle use and disuse. DNA content, a critical determinant of protein synthesis by providing the amount of DNA necessary to sustain gene transcription, can be either increased (activation of satellite cells) or decreased (apoptosis) depending on neuromuscular activity. In addition, several transcription factors are sensitive to functional demand and may control muscle-specific protein expression to promote or repress myofiber enlargement. The control of skeletal muscle mass is also markedly mediated by the regulation of transduction pathways that promote the synthesis and/or degradation of proteins. Insulin-like growth factor-I (IGF-I) plays a key role in this balance by activating the Akt/tuberous sclerosis complex

---

[*] E-mail address: ksakuma@las.tut.ac.jp

2/mammalian target of rapamycin pathway. Stimulation of this pathway leads to the concomitant activation of initiation and elongation factors resulting in an increase in protein translation and decrease in the expression of ubiquitin proteasome components through Forkhead box O (FOXO). A serum response factor (SRF)-dependent pathway transduces the mechanical signaling to the myonucleus to activate the transcription of muscle-specific molecules. In contrast, various negative regulators of muscle mass exist such as ubiquitin proteasome system (UPS), autophagy, and caspase-dependent signaling. In this chapter we will attempt to summarize the current understanding of how these molecules regulate skeletal muscle mass.

LIST OF ABBREVIATIONS

ActRIIB = activin type IIB receptor
AIF = apoptosis-inducing factor
ALK = activin receptor-like kinase
ALS = amyotrophic lateral sclerosis
AR = androgen receptor
Caspases = cysteine-aspartic proteases
COPD = chronic obstructive pulmonary disease
CsA = cyclosporine A
DM = Myotonic dystrophy
eIF = elongation initiation factor
eIF3-f = eIF 3 subunit 5
ER = endoplasmic reticulum
ERK = extracellular signal-regulated kinase
4E-BP = eukaryotic initiation factor 4E binding protein
FAK = focal adhesion kinase
FOXO = forkhead box O
GASP-1 = growth and differentiation factor-associated serum protein-1
GH = growth hormone
GSK-3β = glycogen synthase kinase-β
HGF = hepatocyte growth factor
HSP = heat shock protein
IFN = interferon
IGF-I = insulin-like growth factor-I
IκB = inhibitor of NFκB
IκBαSR = super-repressor mutant of NF-κB inhibitory protein IκBα
IKK = IκB kinases
IL = interleukin
JAK = Janus kinase
LC3 = microtubule-associated protein light chain 3
LIF = leukemia inhibitory factor
MAPK = mitogen activated protein kinase
MHC = myosin heavy chain
MRTF-A = myocardin-related transcription factor-A
mTOR = mammalian target of rapamycin

MtDNA = mitochondrial DNA
MuRF = muscle ring finger
NF-κB = nuclear factor-kappaB
NO = nitric oxide
NOS = nitric oxide synthase
PGC1α = peroxisome proliferator-activated receptor γ coactivator 1α
PI3-K = phosphatidylinositol-3-kinase
PLD = phospholipase D
PPAR = peroxisome proliferator-activated receptor
RM = repetition maximum
ROS = reactive oxygen species
SRF = serum response factor
STARS = striated muscle activators of Rho signalng
STAT = signal transducers and activators of transcription
TGF-β = transforming growth factor-β
TNF-α = tumor necrosis factor-α
mTORC1 = mTOR signaling complex 1
TSC = tuberous sclerosis complex
TWEAK = TNF-like weak promoter of apoptosis
TRAF = TNF receptor associated factor
UPS = ubiquitin proteasome system

1. INTRODUCTION

In humans, skeletal muscle is the most abundant tissue in the body comprising 40-50% of body mass and playing vital roles in locomotion, heat production during periods of cold stress, and overall metabolism. Skeletal muscle is composed of bundles of muscle fibers called fascicles. The cell membrane surrounding the muscle cell is the sarcolemma, and beneath the sarcolemma lies the sarcoplasm, which contains the cellular proteins, organelles, and myofibrils: the thin actin filament and the thicker myosin filament. The arrangement of these protein filaments gives skeletal muscle its striated appearance.

Skeletal muscle is capable of remarkable adaptations in response to altered activity. These adjustments to mechanical and metabolic demands elicit marked modifications of gene expression that could lead to gain (hypertrophy) or loss (atrophy) of muscle mass. Whereas endurance training leads to minor changes in skeletal muscle mass, strength training induces marked hypertrophy of exercising muscles. Histochemical analyses clearly show a 10 to 30% increase in muscle fiber cross-sectional area after 10-12 weeks of resistance training in sedentary subjects [1].

Satellite cells are resident myogenic stem cells found in postnatal skeletal muscle, accounting for 3-9% of the sublaminal nuclei associated with adult normal muscle fiber [2], with the variation widely depending on animal species, age, muscle fiber type, and longitudinal location of the cell along the fiber [3, 4]. Satellite cells, existing between the basal lamina and the sarcolemma of the fiber [5], are normally found in a mitotically and metabolically quiescent or dormant state most of the time in adult muscles. When muscle is injured or mechanically stretched, satellite cells activate to enter the cell cycle from a protracted G1 state (often referred to as G0). Activated satellite cells have been shown to migrate to the damaged site where they replicate DNA, divide, differentiate, and fuse with the adjacent muscle fiber or form new fibers [4, 6].

It has been reported that satellite cells are activated in compensatory hypertrophy [6, 7], and the addition of a new nuclei to the growing fiber seems to be required for extreme hypertrophy. Since the myonuclear domain is constant in hypertrophied muscle after mechanical overloading [8, 9], many satellite cells must be incorporated adjacent to muscle fibers. In fact, irradiation of satellite cells followed by a loading stimulus results in an attenuated increase in skeletal muscle mass and protein content [10]. Therefore, it is necessary for consecutive processes (the activation, proliferation, and differentiation of satellite cells) to elicit muscle hypertrophy in the case of mechanical overloading as well as normal growth. However, several researchers recently suggested satellite cell-independent muscle hypertrophy during mechanical overloading. In addition, some have debated whether the contribution of satellite cells to fiber hypertrophy in adult muscle is minor [11, 12]. In hypertrophied muscle, increasing protein synthesis and decreasing protein degradation are also important events. Phosphatidylinositol-3-kinase (PI3-K)/Akt/ mammalian target of rapamycin (mTOR) signaling has been shown to be crucial to protein synthesis [13, 14]. Mechanical stretching *in vivo* and *in vitro* activates serum response factor (SRF)-dependent signaling in skeletal muscle similar to smooth and cardiac muscles [15, 16]. In contrast, several possible mediators for muscle atrophy have been described. Many negative regulators are proposed to induce muscle atrophy by inhibiting protein synthesis and enhancing protein degradation in skeletal muscle. For example, the ubiquitin proteasome system (UPS) is thought to be a major contributor to many structural proteins [17]. The autophagy-lysosome system has been largely ignored despite the evidence that lysosomal degradation contributes to protein breakdown in atrophying muscles [18, 19]. Recent studies demonstrated that autophagy is an important pathway for appropriate protein degradation in several neuro-muscular disorders [20]. The group of Sandri et al. [21, 22] has shown that the autophagy-lysosome and UPS are coordinately regulated during muscle wasting. Furthermore, specific expression of mutant SOD1^{G93A} in skeletal muscle caused muscle atrophy and weakness mainly via autophagy activation [23]. On the other hand, myostatin is a potent inhibitor for muscle growth, and therefore, a therapeutic target for muscle wasting including cachexia and sarcopenia, muscular dystrophy, or amyotrophic lateral sclerosis (ALS) [24, 25]. In this chapter, we summarize possible candidates for proteins that regulate muscle hypertrophy and atrophy. In addition, we describe the possible modulators for switching, proliferation, and differentiation of satellite cells, a possible contributor to muscle hypertrophy.

2. MUSCLE HYPERTROPHY

2.1. Satellite Cell

Muscle fibers are multinucleated cells, and during postnatal muscle growth the number of nuclei present in muscle fibers increase because of the proliferation of satellite cells and their incorporation into the fibers [26]. An increase in myonuclear number during overload hypertrophy resulting from the proliferation and subsequent incorporation into muscle fibers of satellite cells has been documented by electron microscope autoradiography following ^{3}H-thymidine labeling [7]. However, it is still debatable whether the activation and fusion of satellite cells is necessary to adapt to changes in muscle fiber size, when muscle hypertrophy is induced in adult skeletal muscles, such that the ratio of cytoplasm to nuclei, the myonuclear domain, remains constant [9, 11, 27]. In contrast, another study has shown that the myonuclear domain is not constant from species to species and between different fiber types [28]. In addition, the finding that nuclear domains may not be tightly regulated as previously thought [29], and the lack of elimination of myonuclei during atrophy observed by *in vivo* imaging [30] also questions the idea that myonuclei are added to muscle fibers during hypertrophy. More recently, it has been shown that in transgenic myostatin knockout mice, satellite cells do not contribute to the significant hypertrophy seen in these muscles [31]. Although ablation of satellite cells by γ-irradiation of muscles blunts the hypertrophic response due to overloading or insulin-like growth factor-I (IGF-I) overexpression [32], γ-irradiation may affect muscle fibers themselves, and not only satellite cells. Thus, γ-irradiation for skeletal muscle would damage intramuscular capillary, endothelial, immune, and fibroblast cells, which have an important role in the hypertrophic process. Intriguingly, irradiation of the anterior latissimus dorsi of quail did not block stretch-induced hypertrophy [33]. It was argued that molecular evidence suggests that there is an early increase in protein synthesis (per nucleus), and that addition of nuclei follows later if at all. When mouse extensor digitorum longus muscle was overloaded by synergistic ablation, the addition of new nuclei occurred mainly after 6-8 days, while the increase in size occurred after 8-12 days [34]. This sequence of events strongly suggests that the increased number of nuclei is a major cause of hypertrophy, but it does not preclude that at least some hypertrophy could occur in the absence of new nuclei under experimental conditions where satellite cells are eliminated. In hypertrophic muscles *in vivo*, several growth factors [hepatocyte growth factor (HGF), IGF-I, etc], cytokines [leukemia inhibitory factor (LIF)], and transcription factors (MyoD) modulate the activity of satellite cells. Next, we introduce important modulators of the switching, proliferation, and differentiation of satellite cells, a possible contributor to muscle hypertrophy.

2.2. Modulators of Satellite Cell Activity

2.2.1. Hepatocyte Growth Factor (HGF) and Neuronal Nitric Oxide Synthase

Two factors have been demonstrated to activate quiescent satellite cells. The first is HGF. Early experiments using single muscle fibers with associated quiescent satellite cells have shown that growth factors, such as IGF-I and fibroblast growth factor (FGFs), do not activate satellite cells in fibers [35, 36]. Although IGF-I and FGFs are reported to activate satellite cells, the studies involved typically used cultures of muscle cells that were not quiescent; IGF-I and FGFs increase the proliferative activity of satellite cells once they are activated, even when that activation results during the cell isolation process, i.e. prior to the plating of cells or fibers for culture. Moreover, platelet-derived growth factor BB, transforming growth factor-β (TGF-β), and epidermal growth factor do not stimulate quiescent cells to enter the cell cycle *in vitro* [37, 38]. Therefore, HGF is the only growth factor that has been established to have the ability to stimulate quiescent satellite cells to enter the cell cycle early in a culture assay and *in vivo* [39, 40]. HGF is localized to the extracellular domain of un-injured skeletal muscle fibers through a possible association with glycosaminoglycan chains of proteoglycans that are essential components of the extracellular matrix, and following injury, quickly associates with satellite cells [41].

The second component shown to be involved in satellite cell activation is nitric oxide (NO), a short-lived free radical that is well known as a freely diffusible and ubiquitous molecule produced by nitric oxide synthase (NOSs) from the L-arginine of substrates. In skeletal muscle, neuronal NOS (nNOS, also called NOS-1) is localized to the sarcolemma of muscle fibers by association at its amino terminus with alpha1-syntrophin linked to the dystrophin cytoskeleton [42]. The NO radical is normally produced in very low level pulses by muscles under conditions where satellite cells are quiescent [43], and the expression and activity of constitutive NOS (nNOS and eNOS) are up-regulated by exercise, loading injury, shear force, and mechanical stretch. Therefore, the NO radical is an exciting subject in the 'mechanobiology' of skeletal muscles that can respond to mechanical stimuli and initiate the molecular programs that ensure tissue growth, adaptation, and regeneration.

Studies *in vitro* and *in vivo* using rodent muscle have shown HGF and NO to regulate the activity of many satellite cells [40, 41, 44, 45]. Although several studies have demonstrated an important role for HGF in satellite cells during muscle hypertrophy *in vivo*, only a few [46, 47] have found HGF mRNA expression in the plantaris muscle to be up-regulated after functional overload. Just one study has indicated a role for HGF in human hypertrophied muscle. Shelmadine et al. [48] found an increase in the amount of phosphorylated c-Met, the receptor for HGF, after 28 days of resistance training (a program consisting of nine exercises such as bench press, shoulder shrugs, chest flies, biceps curls, i.e., 3x10RM, 4 times/week). Strangely, whether resistance training modulates the expression pattern of HGF in human muscle and/or satellite cells has not been investigated. In contrast, the amount of nNOS mRNA and protein has been demonstrated to be increased in rodent muscles after swim training (60 min twice/day for 3-4 weeks) [49] and acute eccentric exercise (downhill running) [50] and in human muscle after short-term (10 days) endurance training [51]. Similarly, it is unknown whether resistance training increases nNOS expression in human skeletal muscle. Therefore, the functional role of HGF and nNOS during muscle hypertrophy in humans needs to be clarified by further studies. It may be that the extent of mechanical stimulation by resistance training affects the expression of HGF and nNOS in human muscle.

2.2.2. Leukemia Inhibitory Factor (LIF)

LIF is a newly discovered myokine [52], originally identified by its ability to induce the terminal differentiation of myeloid leukemic cells. Today, LIF is known to have a wide array of functions, including acting as a stimulus for platelet formation, the proliferation of hematopoietic cells, bone formation, neural survival and formation, muscle satellite cell proliferation and acute phase production by hepatocytes [53]. LIF is a long chain four α-helix bundle cytokine, which is highly glycosylated and may be present with a weight of 38-67 kDa, which can be deglycosylated to ~20 kDa [54, 55]. Several tissues, including skeletal muscle, express LIF. LIF is constitutively expressed at a low level in type I muscle fibers [56, 57] and is implicated in conditions affecting skeletal muscle growth and regeneration [56-58]. Production of the LIF protein is augmented in mechanically overloaded rat plantaris muscle and in denervated rat muscles [57], thus endogenous LIF production is modulated by factors influencing muscle activity. Furthermore, LIF restored the hypertrophic response to increased loading in LIF (-/-) mice, and has been denoted as an important factor in skeletal muscle hypertrophy [59].

In 1991, Austin and co-workers demonstrated that LIF stimulated myoblast proliferation in culture [60], thereby showing that LIF functions as a mitogenic growth factor when added to muscle precursor cells *in vitro*. To date, different groups have confirmed this finding and shown that LIF induces satellite cell and myoblast proliferation, while preventing premature differentiation, by activating a signaling cascade involving Janus kinase (JAK) 1, signal transducers and activators of transcription (STAT) 1 and STAT3 [61, 62]. In line with this, the specific LIF receptor is primarily expressed by satellite cells and not by mature muscle fibers [56]. Thus, it seems that LIF has the potential to affect satellite cells rather than mature muscle fibers.

Broholm and Pedersen [63] proposed the intriguing hypothesis that the primary function of LIF, as a contraction-induced myokine, is that of a mitogenic growth factor affecting nearby satellite cells in a paracrine fashion. Indeed, in primary human skeletal myocytes, Broholm et al. [52] observed a marked increase in both LIF mRNA and protein in response to a Ca^{2+} ionophore, ionomycin. Although Broholm et al. [52] also demonstrated that neuromuscular activity caused by endurance exercise (3 hours on a cycle ergometer at ~ 60% of VO$_2$max) elicited LIF mRNA expression in vastus lateralis muscle 0-3 hours postexercise, they did not observe a similar significant increase at the protein level. In contrast, our previous study [57] found that chronic mechanical overloading of rat plantaris muscle induced by ablation of synergists was followed by an increase in LIF protein from 2 to 14 days of overload. Broholm et al. [52] proposed that repetitive bouts of exercise are necessary to induce accumulation of the LIF protein in skeletal muscle. Our finding may support this that no LIF protein was detected in hypertrophied muscle on the first day [57]. It is unknown whether increased neuromuscular activity after resistance training up-regulates LIF mRNA and protein expression in human skeletal muscle. The functional role of LIF during muscle hypertrophy needs to be clarified further.

2.2.3. IGF-I

The anabolic effects of IGF-I have been demonstrated in both muscle cell lines and *in vitro* animal models [64-67]. For example, the addition of IGF-I to cultured myotubes results in an enlargement of myotube diameters and a higher protein content [64, 65], while the delivery of IGF-I through either osmotic pumps or genetic overexpression results in increased muscular mass in rodents [64, 68]. Mechanical loading also results in skeletal muscle synthesis of IGF-I [69, 70] *in vivo*, which has led investigators to conclude that IGF-I is a critical factor involved in skeletal muscle hypertrophy. Resistance training also results in increased IGF-I mRNA expression in human skeletal muscle [71, 72].

IGF-I is thought to induce muscle growth through the increased proliferation of satellite cells and the enhancement of protein translation resulting in an increase in the rate of protein synthesis [73]. In addition to stimulating the proliferation of myoblasts, IGF-I stimulates their differentiation [74]. IGF-I also alters the transcription and translation of muscle factors that regulate myocyte growth or differentiation [75]. For example, IGF-I inhibits production of myogenin, a protein that stimulates muscle cell differentiation, thus allowing increased myoblast proliferation. However, its effects are biphasic because prolonged exposure to IGF-I results in increased myogenin expression, thus facilitating differentiation [76]. IGF-I also modulates the expression of MyoD, myocyte enhancer factor 2 (MEF2) and p21 to control differentiation.

Many investigators hypothesize that IGF-I expression induced by mechanical overloading is crucial to the hypertrophic process [77-79]. However, several studies indicated earlier activation of mTORC1 than enhancement of PI3-K/Akt [80] or Akt/mTOR activation not through IGF-I receptor-dependent signaling [81]. Although the possibility of IGF-I-dependent muscle hypertrophy after mechanical overloading can not be excluded, further study is needed to elucidate the true role of IGF-I in muscle hypertrophy *in vivo*. Figure 1 provides an overview of several modulators of satellite cells in hypertrophied muscle.

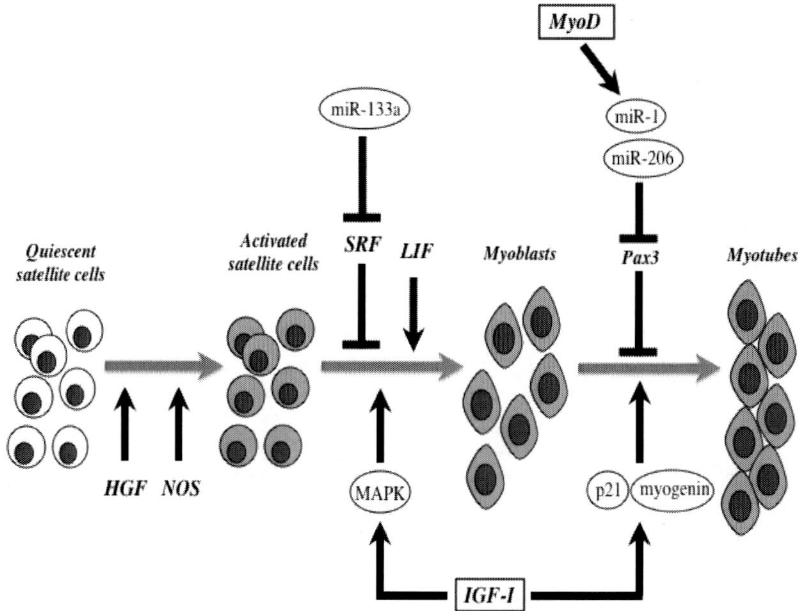

Figure 1. HGF and nNOS coordinately regulate switching of satellite cells from quiescence to activation. IGF-I and LIF enhance the proliferation of satellite cells via MAPK- and JAK1-STAT3-dependent pathway, respectively. During the proliferation phase, miR-133a inhibits the functional role of SRF. In myogenesis, Pax3 is involved in maintaining proliferation, preventing precocious differentiation, and protecting against apoptosis [82]. Therefore, down-regulation of Pax3 is essential for the terminal differentiation. In the differentiation phase, both miR-1 and miR-206, which are induced by MyoD, block the multiple roles of Pax3 [83]. IGF-I also promotes myogenic differentiation via p21 and myogenin. In hypertrophic muscle after mechanical overloading, the differentiating myotubes seem to be incorporated into the existing muscle fibers ultimately. HGF; hepatocyte growth factor, IGF-I; insulin-like growth factor-I, LIF; leukemia inhibitory factor, MAPK; mitogen-activated protein kinase, NOS; nitric oxide synthase.

2.3. PI3-K/Akt/mTOR Pathway

A central pathway involved in hypertrophy is regulated at the translational level by the serine/threonine kinase Akt. In muscle, Akt is activated by the upstream PI3-K, induced either by receptor binding or by integrin-mediated activation of focal adhesion kinase (FAK), such as in cardiac myocytes [84, 85]. PI3-K activates Akt, which then has the ability to phosphorylate and change the activity of many signaling molecules. Possible downstream regulators of Akt, mTOR and glycogen synthase 3-β (GSK-3β) play a crucial role in the regulation of translation [86]. Akt activates mTOR via phosphorylation and inactivation of tuberous sclerosis complex (TSC)-2 [87]. Subsequently, mTOR phosphorylates and activates the 70kDa ribosomal protein S6 kinase (p70S6K), which results in increased translation either directly or indirectly by activating initiation and elongations, elongation initiation factor (eIF)-2, eIF-4E [through eukaryotic initiation factor 4E binding protein (4E-BP)] and eEF-2 [14]. In addition, Akt also phosphorylates and inactivates GSK-3β, thereby activating translation via the initiation factor eIF-2B [88]. Other functions of Akt include the negative regulation of protein degradation by inhibiting forkhead box O (FOXO)-mediated proteasome activity [89].

2.3.1. Akt

Disruption of the Akt1 gene causes growth retardation and apoptosis [90, 91], whereas deletion of Akt2 causes defects in glucose metabolism but not altered growth [92]. When both the Akt1 and Akt2 genes were deleted, skeletal muscle atrophy at embryonic day 18.5 was observed, together with dwarfism, impaired skin and bone development, and reduced adipogenesis [93]. The striking effect of Akt1 on muscle size was demonstrated by the transient transfection of a constitutively active inducible Akt1 transgene in skeletal muscle *in vivo* [21, 94-96]. Muscle hypertrophy was rapidly achieved in all cases when Akt1 expression was induced in the adult animal for a period ranging from 1 to 3 weeks. Downstream mediators of protein synthesis (p70S6K, S6) were activated, but no incorporation of satellite cells was observed [96]. Akt1 hypertrophic muscles showed increased strength, demonstrating that a functional hypertrophy was induced [95, 96]. Moreover, muscle mass was completely preserved in denervated transgenic Akt mice [97]. Akt activation causes multiple changes in muscle gene expression, documented by microarray analyses [95, 96, 98]. Several different downstream effectors, including GSK-3β, mTOR and FOXO, can mediate the effects of Akt on muscle mass regulation.

2.3.2. mTOR and mTORC (mTOR Signaling Complex) 1

More recently, it has been shown that mTOR exists in two functionally distinct multi-protein signaling complexes, mTORC1 and mTORC2 [99]. In general, only signaling by mTORC1 is inhibited by rapamycin, and thus the growth regulatory effects of rapamycin are believed to be primarily exerted through the mTORC1 complex [100]. It is now widely accepted that signaling by mTORC1 is involved in the regulation of several anabolic processes including protein synthesis, ribosome biogenesis, and mitochondrial biogenesis, as well as catabolic processes such as autophagy [100, 101]. Two of the most studied mTORC1 targets are the 4E-BP1 and p70S6K, which both play important roles in the initiation of mRNA translation.

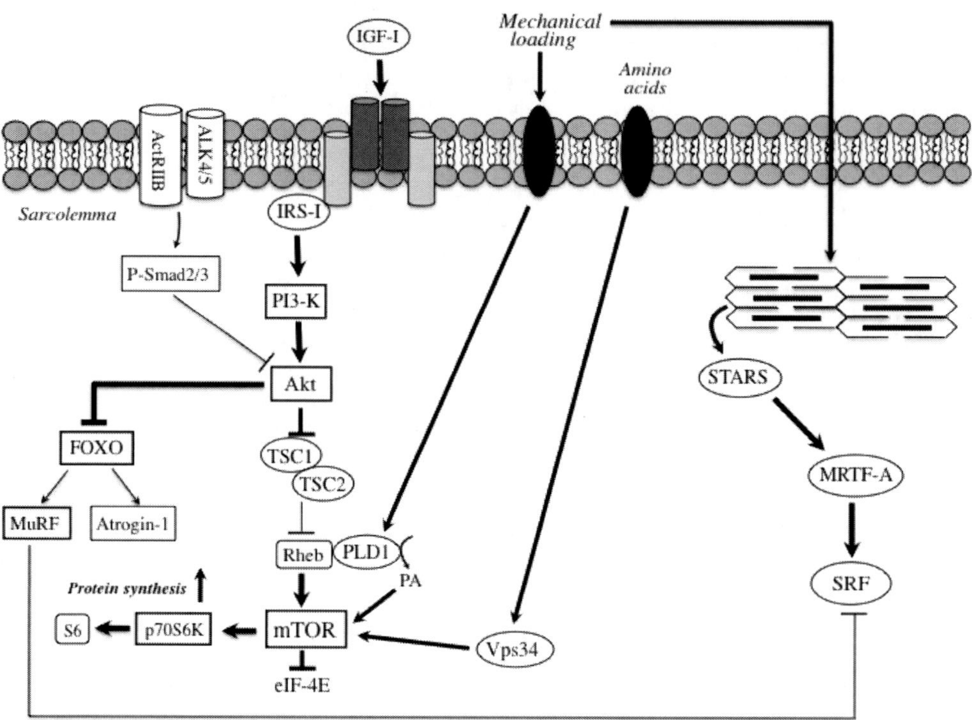

Figure 2. Anabolic pathway regulating skeletal muscle mass. The major anabolic pathway regulating protein synthesis in skeletal muscle is mTOR/TORC1 signaling. Upstream triggers (IGF-I, mechanical loading, amino acids) activate mTOR signaling through a number of different intermediary proteins such as Rheb, PLD1 and its metabolite PA, and Vps34 [121]. Although myostatin signals through the ActRIIB-ALK4/5 heterodimer activate Smad2/3, reduced stimulation of myostatin in the presence of IGF-I and mechanical loading cannot block the functional role of Akt. Myosin-actin interaction by mechanical loading activates STARS /MRTF-A/SRF signaling [16]. The accumulation of MuRF in muscle tissue during inactivity (hindlimb suspension, immobilization, etc) is known to inhibit SRF-dependent transcription of muscle-specific genes [122]. However, the functional role of SRF is not abrogated under such conditions, which lower MuRF expression because of marked inhibition of FOXO by abundant Akt. ActRIIB; activin receptor IIB, ALK4/5; activin-like kinase 4/5, eIF; eukaryotic initiation factor, FOXO; Forkhead box O, IGF-I; insulin-like growth factor-I, IRS-1; insulin receptor substrate-1, MRTF-A; mycardin-related transcription factor-A, mTOR; mammalian target of rapamycin, MuRF; muscle ring-finger protein, PA; phosphatidic acid, PI3-K; phosphatidylinositol 3-kinase, p70S6K; 70 kDa ribosomal protein S6 kinase, Rheb; Ras homolog enriched in brain, SRF; serum response factor, STARS; striated muscle activators of Rho signalng, TORC1; a component of TOR signaling complex 1, TSC; tuberous sclerosis complex.

In skeletal muscle, signaling by mTORC1 has been shown to be regulated by a variety of different stimuli that control skeletal muscle mass. For example, signaling by mTORC1 is activated in response to hypertrophic stimuli such as increased mechanical loading, feeding and growth factors [66, 102, 103]. In fact, hypertrophy induced by mechanical loading, IGF-I and clenbuterol has been shown to be significantly, if not completely, blocked by rapamycin [66, 104]. In addition, overexpression of constitutively active Akt activates mTORC1 signaling and induces hypertrophy through a rapamycin-sensitive mechanism [105]. These findings support the hypothesis that signaling through mTORC1 is sufficient to induce hypertrophy, however, the hypertrophic stimuli employed in these studies also induce signaling through PI3-K and Akt. Since signaling through PI3-K/Akt can regulate mTOR-independent growth regulatory molecules such as GSK-3β, tuberin (TSC-2) and the FOXO transcription factors [106-108], it was not clear if signaling by mTORC1 was sufficient, or simply permissive, for the induction of hypertrophy.

To address this issue, overexpression of Rheb (Ras homolog enriched in brain) was recently used as a means to induce a PI3-K/Akt-independent activation of mTORC1 [109]. Marked increases in protein synthesis and hypertrophy have been recognized in several muscles of Rheb abundant mice [109, 110]. Hornberger et al. [111] found that stretch-induced activation of mTOR signaling was not abolished in the skeletal muscle of Akt1-/- mice, showing that Akt is not necessary for mechanically induced signaling through mTOR. Furthermore, Akt-independent stimulation of mTOR may be positively or negatively regulated by phosphorylation of TSC-2. For instance, TSC-2 is inhibited by focal adhesion kinase (FAK) in 293T cells [112] suggesting that up-regulation of FAK expression with increased loading could stimulate protein synthesis via TSC-2 inhibition. All these regulatory influences may explain the rise in the level of phosphorylated p70S6K [113]. These results suggested that the activation of mTORC1 is indeed sufficient to induce hypertrophy, at least in part by increasing protein synthesis.

Although one of the most well characterized upstream triggers of mTOR signaling in skeletal muscle is IGF-I, mechanical loading has been shown to activate mTOR by an IGF-I independent pathway involving phospholipase D (PLD) via its metabolite phosphatidic acid [114]. More recently, Hornberger et al. [115] extended these initial findings by showing mTOR activation following eccentric contractions via PLD synthesis but not PI3-K-Akt activity. It has been shown that PLD1, but not PLD2, was a downstream effector of Rheb's activation of mTOR [116]. In contrast, Vps34 is a class III PI3-K previously shown to mediate amino acid activation of p70S6K by mTOR [117]. In skeletal muscle, MacKenzie et al. [118] reported that high-resistance contractions increased Vps34 activity possibly in response to increased intramuscular leucine levels. In addition to Vps34, two groups reported the exciting discovery that the Rag family of GTPases was necessary and sufficient for amino acid activation of the mTOR pathway [119, 120].

Therefore, mTOR is currently thought to be the major hub for the integration of an array of upstream signaling pathways which, when activated, ultimately result in increased translational efficiency [14, 121]. Figure 2 summarizes the anabolic pathway (PI3-K/Akt/mTOR and SRF-dependent) regulating skeletal muscle mass.

2.4. Serum Response Factor (SRF)

SRF is an ubiquitously expressed member of the MADS (MCM1, Agamous, Deficiens, SRF) box transcription factor family, sharing a highly conserved DNA-binding/dimerization domain, which binds the core sequence of SRF/CArG boxes [CC (A/T)6 GG] as homodimers [123]. Functional CArG boxes have been found in the cis-regulatory regions of various muscle-specific genes, such as the skeletal α-actin [124], muscle creatine kinase, dystrophin, tropomyosin, and MLC (myosin light chain) 1/3 [125] genes. SRF-dependent signaling plays a major role in a variety of physiological processes, including cell growth, migration, and cytoskeletal organization [126]. Recent results obtained with specific SRF knock-out models by the Cre-LoxP system, emphasize a crucial role for SRF in postnatal skeletal muscle growth and regeneration by modulating interleukin (IL)-4 and IGF-I mRNA expression [127]. SRF also enhances the hypertrophic process in muscle fibers after mechanical overloading [16, 128] as well as muscle differentiation and MyoD gene expression *in vitro* [129]. For example, we demonstrated frequent co-localization between SRF and MyoD or myogenin in myoblast-like cells in hypertrophied muscle of rodent. In contrast, a marked reduction in levels of SRF was recognized in the muscle of unweighted hindlimbs in mice [15] and in merosin-deficient dystrophic mice [130]. In addition, using RT-PCR, crude and fractionated homogenates, and immunofluorescence, we found a blunted expression of SRF mRNA and protein in the quadriceps and triceps brachii in aged mice [131]. More recently, a decrease of SRF expression achieved by a transgenic approach using the Cre-LoxP system was found to accelerate the atrophic process in muscle fibers with age [132]. These lines of evidence clearly show that mechanical loading for skeletal muscle is a determinant of SRF expression.

2.4.1. SRF-Dependent Signaling

It is proposed that the transcriptional activity of SRF is regulated by muscle ring finger (MuRF)-2 [122] and striated muscle activators of Rho signaling (STARS) [133, 134]. At the M-band, the mechanically modulated kinase domain of titin interacts with a complex of the protein products of the atrogenes NBR1, p62/SQSTM-1 and MuRFs [122, 135]. This complex dissociates under mechanical arrest, and MuRF-1 and MuRF-2 translocate to the cytoplasm and the nucleus [122, 136]. One of the probable nuclear targets of MuRFs is SRF [122], suggesting that the MuRF-induced nuclear export and transcriptional repression of SRF may contribute to amplifying the transcriptional atrophy program [137]. Thus, it is possible that the synergistic transactivation of SRF and SRF-linked molecules is abrogated by MuRF-2 *in vivo*. On the other hand, SRF activity is exquisitely sensitive to the state of actin polymerization. G-actin monomers inhibit SRF activity, whereas polymerization of actin occurs in response to serum stimulation and RhoA signaling. In this pathway, signal inputs lower the ratio of globular actin to fibrillar actin thereby liberating the binding of myocardin-related transcription factor-A (MRTF-A) to globular actin resulting in the nuclear accumulation of MRTF-A and subsequent SRF-dependent gene expression [138]. It has been well established that overexpression of STARS contributes to the nuclear translocation of MRTF-A and MRTF-B [134, 139], and these factors activate SRF transcription.

2.4.2. The Functional Role of SRF during Muscle Hypertrophy

SRF also enhances the hypertrophic process in muscle fibers after mechanical overloading [128, 140, 141]. For example, Flück et al. [141] utilized a stretch-induced hypertrophic model, in which a weight equal to 10% of body weight was attached to the left wing of a rooster to induce enlargement of the anterior latissimus dorsi muscle. Gordon et al. [15] also indicated a significant increase in SRF protein in the soleus and plantaris muscles after 8 days of functional overload caused by surgical ablation of the gastrocnemius muscle in rats. In humans, Lamon et al. [142] demonstrated that 8 weeks of resistance training (leg presses, squats, and leg extensions) induced increases in SRF mRNA (3-fold) and nuclear protein (1.25-fold) in the vastus lateralis muscle. In the same training period, they also observed a similar increase in the mRNA levels of several SRF-targeted molecules (α-actin [120], myosin heavy chain (MHC) IIa [143], and IGF-I [127]). They proposed the induction of these molecules by SRF in human hypertrophied muscle, though they did not provide any direct evidence such as transcriptional activation by increased binding of SRF to the promoter region of α-actin, MHC IIa, and IGF-I. Although SRF would regulate proliferation and differentiation using different pathways, it would mainly activate the differentiation of satellite cells during muscle hypertrophy. Indeed, we [128] showed that, in mechanically overloaded muscles of rats, the SRF protein co-localized with MyoD and myogenin in myoblast-like cells during the active differentiation phase. In this study, we did not detect abundant SRF protein at 2 days after mechanical overloading, when many proliferating satellite cells and/or myoblasts are expected to exist. In addition, the location of the SRF protein did not correspond with that of BrdU-positive satellite cells or ED1-positive macrophages in the hypertrophied plantaris muscle.

As indicated above, by reducing the cytoplasmic concentration of monomeric G-actin, STARS promotes the nuclear translocation of SRF transcriptional co-activator-A and -B (MRTF-A and MRTF-B), resulting in an increase in SRF-mediated gene transcription [139]. A real-time PCR analysis conducted by Lamon et al. [142] demonstrated that increased mechanical loading from resistance training in humans caused significant increases in the upstream modulators of SRF (STARS mRNA; 3.4-fold, MRTF-A mRNA; 2.5-fold, MRTF-B mRNA; 3.6-fold, and RhoA protein; 2-fold).

SRF seems to regulate the transcriptional facilitation of the α-actin promoter by the androgen receptor (AR) during muscle hypertrophy. Using male adult Sprague-Dawley rats, Lee et al. [144] showed an increase in AR protein of 106 and 279% after 7 and 21 days in mechanically overloaded plantaris muscles by surgical ablation of two synergistic muscles. Co-overexpression of either SRF or active RhoA with AR indicated a synergistic 36- and 28-fold increase in the skeletal α-actin promoter activity. In contrast, co-transfection of AR, SRF, and active RhoA induced a 180-fold increase in skeletal α-actin promoter activity. Therefore, it is possible that intimate linkages among these three modulators induce α-actin expression in hypertrophied muscle *in vivo*. Intriguingly, experiments using C2C12 cells indicated that this AR coactivation for the α-actin promoter requires a co-expressed full-length serum response factor and SRF-binding site but not AR's direct binding to GRE sites [145].

2.5. Calcineurin

During hypertrophy, the binding of IGF-I to its receptor initiates a cascade of signaling events which result in muscle growth. The major pathway believed to be involved in the onset of hypertrophy is PI3-K/Akt/mTOR/p70S6K, as eloquently reviewed by Glass [13, 14]. Recently however, a Ca^{2+}-dependent pathway and its downstream mediators have also been implicated in the adaptive responses seen during hypertrophy. It has been demonstrated that inhibition of calcineurin by cyclosporine A (CsA) results in inhibition of the rapid growth of all fiber types in the overloaded muscle of the mouse [146] through the downstream targets myogenin, utrophin A, the Id family, and myostatin [147-151].

The potent effect of calcineurin upon hypertrophy has further been demonstrated in cell culture. In cell lines which received CsA, satellite cell differentiation was inhibited. Conversely, in cells administered a constitutively active calcineurin, the differentiation process was enhanced [152]. These results would suggest that calcineurin-dependent signaling is capable of directly targeting satellite cells. However, while satellite cell activity is undoubtedly of importance in overload-related muscle adaptations, it does not appear to be dependent solely upon calcineurin activity [153].

Calcineurin-dependent signaling has been shown to a major modulator for the hypertrophy in slow-type soleus muscle [154-156]. During muscular hypertrophy *in vivo*, potential downstream targets for the effects of calcineurin are myostatin and MEF2C. Studies in mice have found decreased myostatin mRNA levels in animals with compromised calcineurin signaling compared to wild-type controls [157]. The hypertrophy-defective soleus muscle treated with CsA possessed a lower level of the MEF2C but not MEF2D protein in the subsarcolemmal region in a group of myotubes and/or myofibers during an active-differentiation period (4 days postsurgery) [158]. Two recent findings [159, 160] clearly showed that MEF2C is required for thick filaments to form in nascent muscle fibers and for the integrity of the sarcomere and M-line during postnatal muscle growth, by directly regulating several muscle structural genes such as the genes for myomesin, MHC and MLC.

Endurance training has been shown to increase the expression of calcineurin and calcineurin-downstream candidates [161-163], whereas it has not been studied until recent years whether resistance training modulates calcineurin expression levels in mammalian skeletal muscles. Using twenty-nine physically active male subjects, Lamas et al. [164] tested whether the hypertrophy of skeletal muscle caused by strength and power training included an increase in calcineurin expression. In their study, the strength group trained with an intensity of between 4 and 10 RM (repetition maximum), while the power training group trained with an intensity of between 30% and 60% of 1RM. Eight-week progressive training regimens elicited a significant increase in strength and muscle fiber hypertrophy, but there was no change in calcineurin mRNA levels in the vastus lateralis muscle after the training. These data may support our hypothesis of a limited role for calcineurin in the hypertrophy of slow-type soleus muscle.

2.6. PGC1α

Peroxisome proliferator-activated receptor (PPAR)-γ coactivator-1α (PGC1α) was originally identified as a transcriptional co-activator of PPAR-γ induced by cold exposure in brown adipose tissue [165]. It has since been shown to co-activate several nuclear hormone receptors and transcription factors (e.g. estrogen-related receptorα, PPAR-α, nuclear receptor factor-1, and MEF2) which leads to an upregulation of mitochondrial gene expression and an increase in mitochondrial DNA in tissues such as skeletal muscle [166]. The expression of PGC1α in skeletal muscle is greatly influenced by levels of physical activity with endurance exercise increasing PGC1α expression [167] and physical inactivity leading to decreased expression [168].

PGC1α has been implicated in the regulation of skeletal muscle mass, particularly under conditions of muscle atrophy [169]. For example, PGC1α expression decreases in a multitude of different muscle atrophic models such as denervation, fasting, and cancer cachexia [169-173], although some models have shown increases in PGC1α expression [23, 174]. Recently, it was proposed that a decrease in PGC1α mRNA, prior to the induction of denervation-induced muscle atrophy, may contribute to ensuing muscle atrophy [169]. To date, however, no studies have demonstrated that a decrease in muscle-specific knockout of PGC1α does not exacerbate denervation-induced atrophy, nor increase the expression of Atrogin-1 (Atrophy gene-1) and MuRF-1 [175]. Thus, notwithstanding potential compensatory adaptations in these transgenic models, to date, there is little support for the hypothesis that the loss of PGC1α expression *per se* is sufficient to induce muscle atrophy.

Significant evidence exists that increased expression of PGC1α is sufficient to inhibit increases in protein degradation and protect skeletal muscle mass from various atrophic stimuli. For example, transient overexpression of PGC1α *in vivo* blocks c.a.-FOXO3-induced muscle fiber atrophy, and transgenic muscle-specific PGC1α overexpression protects against denervation- and fasting-induced atrophy, an effect associated with a blunting of the expression of Atrogin-1, MuRF-1 and cathepsin L [169]. Moreover, PGC1α overexpression also protects aging muscle from sarcopenia and is associated with reduced apoptotic markers and suppression of autophagy and ubiquitin-proteasome system genes [176]. The anti-atrophic effects of PGC1α may be due, at least in part, to the suppression of FOXO3's binding to, and activation of, genes such as the Atrogin-1 gene [169, 170, 177]. NF-κB, a transcription factor also implicated in skeletal muscle atrophy [178], has recently been shown to be profoundly inhibited by PGC1α overexpression *in vivo* [176], which could explain some of PGC1α's effect.

In contrast to the findings of Sandri et al., [169] Miura et al. [179] found that overexpression of PGC1α increased the oxidative capacity of skeletal muscle by increasing mitochondrial biogenesis in 13-to-15-wk-old mice. Although mitochondria from PGC1α-overexpressing mice exhibited a two- to three-fold increase in oxygen consumption, respiration was uncoupled for ATP synthesis and ATP content was markedly decreased in PGC1α transgenic mice [180]. When PGC1α-overexpressing mice were followed for longer periods of time (25 wk), the mice developed selective atrophy of the quadriceps and gastrocnemius (fast glycolytic type IIB fibers) but not in the soleus, and these muscles exhibited increased infiltration of adipocytes [180]. The dramatically different conclusion that can be reached by both studies may be due to differences in the promoters used to drive PGC1α expression (muscle creatine kinase vs the human α-skeletal actin), and differences in the functional role (pure inducers of muscle hypertrophy such as IGF-I and growth hormone (GH) vs atrophy-inhibitors by the formation of neuromuscular junction and mitochondrial induction).

2.7. Testosterone

Androgens play crucial physiological roles in establishing and maintaining the male phenotype. Their actions are essential for the differentiation and growth of the male reproductive organs, initiation and regulation of spermatogenesis, and control of male sexual behavior. In addition, androgens also have anabolic actions on several extragenital structures including muscle and bone [181]. Indeed, testosterone, the main androgen in skeletal muscle [182], increases muscle size and strength in both young [183] and older men [184]. The testosterone-induced increase in muscle mass is partly due to muscle fiber hypertrophy, reflected by an increase in myonuclear number and cross-sectional area of both type I and type II muscle fibers [185]. The responsiveness of skeletal muscle to androgens could potentially be exploited clinically in the treatment of muscle wasting [e.g. cancer cachexia, chronic obstructive pulmonary disease (COPD)] [186]. Another important health issue associated with testosterone deficiency is the age-related increase in sarcopenia and frailty in elderly men [187]. Indeed, testosterone administration to frail elderly men may increase muscle strength [188].

Androgens exert their effects largely by binding to the nuclear AR (androgen receptor). The AR is a ligand-inducible transcription factor that binds to specific DNA sequences called androgen response elements and recruits coactivators, which help affect the transcription of target genes [189]. Binding sites for the MEF2 family are also abundant in these areas [190], indicating that at least for part of the target genes, AR could be recruited indirectly by tethering via MEF2 factors. Similar AR tethering has been described via other transcription factors such as SRF [191] and T cell factor [192]. AR-binding regions were found near genes encoding androgen-regulated microRNAs, as well as, e.g., the MEF2C gene, which controls muscle differentiation by regulating the expression of other muscle-specific genes [190]. AR binding was also observed near genes encoding factors involved in sarcomere integrity and muscle contraction, like myomesin and myozenin [190].

2.7.1. Testosterone-Dependent Signaling

Several clinical studies have demonstrated that testosterone therapy augments GH secretion [193, 194], which in turn correlates with an increase in serum IGF-I [195]. It seems to be mediated centrally, since mice selectively lacking AR in the nervous system show a twofold reduction in serum IGF-I levels [196]. However, testosterone increases total body weight and levator ani muscle mass even in hypophysectomized rats that are deficient in GH and low in IGF-I serum levels [197]. In addition, administration of high doses of dihydrotestosterone to orchidectomized rats did not change serum IGF-I concentrations although levator ani weight was restored to levels in sham-operated rats [198]. Collectively, these results suggest that circulating GH and IGF-I play only a minor role in mediating the anabolic effects of androgens.

The muscle hypertrophy observed in myostatin knockout mice is more pronounced in males than females [199] and, conversely, muscle-specific myostatin overexpression lowers muscle mass more in male than female mice [200]. This gender-based difference suggests crosstalk between androgens and myostatin. Moreover, androgens regulate myostatin expression at the gene level. Indeed, several studies support the hypothesis that androgens enhance β-catenin signaling, thereby increasing the expression of target genes including follistatin, resulting in myostatin inhibition [201]. In addition, co-immunoprecipitation assays showed direct interaction between AR and β-catenin, which might stabilize β-catenin and prevent its degradation [201, 202]. In addition, the activation of adenosine monophosphate-activated kinase by androgens might further contribute to the stabilization of β-catenin via phosphorylation at Ser552 [202]. The myostatin-mediated effects are antagonized by follistatin [203-205], a protein whose expression is regulated through β-catenin signaling [201]. Follistatin antagonizes myostatin via direct interaction, which prevents myostatin from binding to its receptor [203].

Notch's regulation by androgens was proposed to be involved in the protective effect of androgens on age-associated muscle degradation [205, 206]. Testosterone-induced muscle hypertrophy in mice is accompanied by an upregulation of the Notch ligand Delta1 and an activation of Notch signaling, as evidenced by the increase in activated forms of Notch1 and Notch2 [207]. Moreover, testosterone treatment inhibited c-Jun NH2-terminal kinase and activated p38 mitogen-activated protein kinase (MAPK), two factors that are critical for the activation of Notch signaling [208]. Enhancement of Notch activation could also play a role in the effects on androgens on satellite cell proliferation and myogenic progression [209, 210]. In addition, a study exploring androgen effects on aged muscle revealed that testosterone treatment can restore Notch signaling in old mice and reverse the age-associated increase in p21, a downstream member of the Notch cascade, which is known to interfere with satellite cell regenerative capacity [208].

2.8. Growth Hormone (GH)

GH is a single-chain peptide of 191 amino acids produced and secreted mainly by the somatotrophs of the anterior pituitary gland. GH secretion occurs in a pulsatile manner with a major surge at the onset of slow-wave sleep and less conspicuous secretory episodes a few hours after meals [211], and is controlled by the action of two hypothalamic factors, GH releasing hormone, which stimulates GH secretion, and somatostatin, which inhibits GH secretion [212]. Different factors, including gender, age, adiposity, sleep, diet, and exercise, affect the frequency and magnitude of the GH pulses. The secretion of GH is maximal at puberty accompanied by very high circulating IGF-I levels [213], with a gradual decline during adulthood. Indeed, in aged men, daily GH secretion is 5- to 20-fold lower than that in young adults [214].

GH coordinates the postnatal growth of multiple target tissues, including skeletal muscle [215]. With respect to the somatomedin hypothesis, the growth-promoting actions of GH are mediated by circulating or locally produced IGF-I [216]. GH-induced muscle growth may be mediated in an endocrine manner by circulating IGF-I derived from liver [213] and/or in an autocrine/paracrine manner by direct expression of IGF-I from target muscle via GH receptors on muscle membranes [212]. Several studies have shown that GH treatment increases IGF-I mRNA levels in skeletal muscle tissues as well as in the myoblast cell line C2C12 [215, 217]. The critical step in initiating GH signaling is the activation of receptor-associated JAK2, which induces cross-phosphorylation of tyrosine residues in the kinase domain of JAK2 and GH receptor. Phosphorylated residues in JAK2 and GH receptor form docking sites for the membranes of the STAT family of transcription factor [218]. Phosphorylation of the STATs by JAK2 results in their dissociation from the receptor and translocation to the nucleus, with subsequent binding to DNA and regulation of gene expression. Among the targets, GH regulates the expression of suppressor of cytokine signaling, a family of negative regulators that terminate the GH signaling cascade [219]. In contrast, Sotiropoulos et al. [220] clearly showed a direct GH-mediated effect on the growth of muscle cells in primary cultures by facilitating the addition of myoblasts with nascent myotubes via NFATc2 independent of IGF-I. They demonstrated that GH does not regulate IGF-I expression in myotubes. In addition, they suggesed that the hypertrophic effects of GH and IGF-I are additive and rely on different signaling pathways (GH via NFATc2 vs IGF-I via Akt-mTOR). Taken together, GH may enhance the fusion of muscle cells not via IGF-I in target muscle.

3. NEGATIVE REGULATORS OF SKELETAL MUSCLE MASS

At least four major proteolytic pathways (lysosomal, Ca^{2+}-dependent, caspase-dependent, and ubiquitin-proteasome-dependent) operate in skeletal muscle and may be altered during aging, thus contributing to sarcopenia. The lysosomal and proteasomal systems lead to an exhaustive degradation of cell proteins into amino acids or small peptides, whereas the Ca^{2+}-dependent and caspase systems can perform only limited proteolysis, owing to their restricted specificity.

The endosome-lysosome system is relatively nonselective and mostly involved in the degradation of long-lived proteins [221]. Lysosomal autophagic degradation has been reported to be induced in skeletal muscle by a 6-h nutrient starvation [222], whereas autophagins, a class of cysteine proteases putatively involved in the formation of autophagosomes, are particulary abundant in the skeletal muscle [223]. Recently, autophagy-related genes have been shown to be hyperexpressed during muscle atrophy induced by denervation or fasting [21, 224, 225].

The UPS, initially described as relevant to the catabolism of regulatory or damaged proteins, is also involved in bulk protein degradation, at least in skeletal muscle. In particular, the identification of muscle-specific components of E3 ubiquitin-ligases has improved knowledge of the regulation of the ubiquitin-proteasome system in skeletal muscle [226].

The Ca^{2+}-dependent system comprises several cystein proteases called calpains and a physiological inhibitor named calpastatin. Calpains affect only a limited proteolysis on their substrates (i.e. protein kinase C, calcineurin, and titin), resulting in irreversible modifications that lead to changes in activity or to degradation via other proteolytic pathways [227, 228].

Finally, caspases, a family of cysteine protease, are mostly known for their role in the execution of apoptosis. Whether caspases are relevant to the regulation of skeletal muscle mass still remains to be elucidated [229, 230].

3.1. Ubiquitin-Proteasome System (UPS)

The ATP-dependent UPS is essential for regulating protein degradation [231]. In eukaryotic cells, most proteins in the cytosol and nucleus are degraded via the UPS. The degradation of a protein via the UPS involves two steps: (1) tagging of the substrate by covalent attachment of multiple ubiquitin molecules and (2) degradation of the tagged protein by the 26S proteasome complex with the release of a free and reusable ubiquitin [232]. In a variety of conditions such as cancer, diabetes, denervation, uremia, sepsis, disuse, and fasting, skeletal muscles undergo atrophy through degradation of myofibrillar proteins via the UPS [233]. For instance, studies with animal models of cancer cachexia, as well as in cancer patients, suggest that the UPS plays a prominent role in the degradation of myofibrillar proteins, particulary in patients with a weight loss of >10% [234]. The expression of mRNA for both α- and β-proteasome subunits was increased in gastrocnemius muscle of weight-losing mice bearing the MAC16 adenocarcinoma [235]. The fact that ubiquitin and proteasomes have been implicated in many disease states suggests that these inhibitors have therapeutic potential.

3.1.1. Ubiquitin and Proteasomes

Ubiquitin, composed of 76 amino acids, is an 8.45-kDa protein that is highly conserved in nearly all eukaryotes. The ubiquitination of proteins is regulated by at least three enzymes: ubiquitin-activating enzyme (E1); ubiquitin-conjugating enzyme (E2); and ubiquitin ligase (E3) [236]. Kwak et al. [237] suggested that the 14-kDa ubiquitin-conjugating enzyme $E2_{14K}$ and the ubiuitin ligase E3 are particulary important for the degradation of muscle proteins. Indeed, recent findings assert that muscle atrophy in these conditions shares a common mechanism in the induction of the muscle-specific E3 ubiquitin ligase Atrogin-1 and MuRF-1 [226, 238, 239]. The labeled proteins are then fed into the cells' "waste disposers", the so-called proteasomes, where they are chopped into small pieces and destroyed [240].

The proteasome is a highly conserved intracellular multicatalytic proteinase complex [241], which can catalyze the cleavage of peptide bonds on the carboxyl side of basic, acidic, and hydrophobic amino acid residues in both natural peptides and synthetic substrates [242]. The 26S proteasome is composed of a 19S capped-shape regulatory particle and a 20S catalytic core particle. X-ray crystallography of the 26S proteasome revealed the following: (1) the 19S proteasome is made up of 18 subunits, which control the recognition, deubiquitinylation and unfolding of protein substrates prior to their translocation into the 20S proteasome [243] and (2) the cylindrically shaped 20S proteasome is composed of two copies of 14 different subunits, arranged as $(\alpha1\text{-}\alpha7, \beta1\text{-}\beta7)_2$ with four stacked rings.

3.1.2. Characteristics of Atrogin-1 and MuRF-1

Atrogin-1 is a member of the Skp1, Cullin 1 and F-box-containing proteins (SCF) complex, which bind together to establish their E3 Ub-protein ligase activity, and features an approximately 40-amino acid motif known as an F-box through which Atrogin-1 mediates its interaction with Sp1 [244]. It also contains a leucine-charged residue-rich domain that binds several proteins as well as a predicted PDZ domain, cytochrome c binding site and two nuclear localization signals [245]. MuRF-1 contains a canonical N-terminal RING domain characteristic of RING-containing E3 ligases [246] followed by a MuRF family conserved region, zinc-finger domain (B-box) and leucine-rich coiled-coil domains, which allows it to form homo- and heterodimers with other MuRF proteins [247], and an acidic C' terminal tail [248].

Atrogin-1 and MuRF-1 were first identified following transcript profiling in fasting and immobilization models of rodent muscle atrophy [226, 244]. Consistent increases in their gene expression have been observed in a wide range of *in vivo* models of skeletal muscle atrophy including diabetes, cancer, renal failure, denervation, unweighting and glucocorticoid or cytokine treatment [226, 244, 249], and their expression is primarily confined to striated muscle [244, 250]. The importance of these atrophy-regulated genes or "atrogenes" in muscle wasting was confirmed through knockout studies in mice where an absence of Atrogin-1 or MuRF-1 reduced denervation-induced muscle atrophy [226]. Yeast two-hybrid analysis had identified eIF 3 subunit 5 (eIF3-f) and MyoD as interactors of Atrogin-1, and all were found subsequently to be ubiquitylated and degraded in skeletal muscle [251-253]. eIF3-f appears to be a key effector of Atrogin-1 as targeted increases and decreases in eIF3-f levels cause skeletal muscle hypertrophy and atrophy, respectively [251]. The Atrogin-1-MyoD interaction occurs in the nucleus. Overexpression of Atrogin-1 results in the polyubiquitination of MyoD and an inhibition of MyoD-induced myotube differentiation and formation [254]. Conversely, the knockdown of Atrogin-1 reversed endogenous MyoD proteolysis and the overexpression of a mutant MyoD, unable to be ubiquinated, prevented muscle atrophy *in vivo* [254]. These results confirmed MyoD as a substrate of Atrogin-1 resulting in its polyubiquitination and subsequent degradation during dexamethasone-induced myotube atrophy [255]. In contrast to Atrogin-1, it appears that MuRF-1 mainly interacts with structural proteins. MuRF-1 binds to titin and potentially affects titin signaling [248]. It also binds to and degrades MHC proteins following the treatment of skeletal muscle with dexamethasone. Additionally, MuRF-1 degrades myosin-binding protein C and MLC 1 and 2 during denervation and fasting conditions [256]. These last two studies identified these proteins *in vivo* using RING-deleted MuRF-1 mutant transgenic mice while other *in vitro* ubiquitination studies have confirmed the targeted depletion of MHC isoforms [257] and an oxidized form of creatine kinase [258]. These studies suggest that while numerous stimuli can activate both Atrogin-1 and MuRF-1, the downstream pathways affected may be separate for each protein. It is possible that Atrogin-1 targets molecules affecting protein synthesis and muscle growth whereas MuRF-1 may act to control the level of protein degradation while perhaps also contributing to skeletal muscle metabolism. The identification of substrate targets of Atrogin-1 and MuRF-1 will greatly aid in delineating the roles they may play in skeletal muscle remodeling and function.

3.1.3. Adapatative Changes in Atrogin-1 and MuRF-1 in Muscle Atrophy Models

Atrogin-1 and MuRF-1 mRNA levels rapidly increase in numerous models of muscle atrophy, suggesting that they may contribute to the initiation of an atrophy program. For example, several experimental models of cancer cachexia (AH-130, C26, Lewis Lung calcinoma etc) showed increased UPS activity, as well as upregulation of both Atrogin-1 and MuRF-1 expression [259, 260]. The expression of MuRF-1 in muscle after cancer cachexia has been recently shown to be regulated by tumor necrosis factor (TNF) receptor-associated factor (TRAF) 6-dependent signaling [261]. It has been widely accepted during the last decade that these two atrogenes mainly regulate protein degradation through the UPS. However, few studies have investigated the actual functional role of Atrogin-1 and MuRF-1, making it unclear if these genes are critically required for muscle wasting. While considerable research has used animal models of human muscle wasting to investigate the role and regulation of Atrogin-1 and MuRF-1, few investigations have been performed in humans with muscle atrophy [262]. In addition, the manner of adaptation in these atrogenes in human studies did not correspond to those of animal atrophy models [262, Table 1]. This incomplete comparison between what is known about the role and regulation of Atrogin-1 and MuRF-1 in animal versus human models of skeletal muscle atrophy confounds their validity as potential therapeutic targets. Indeed, to our knowledge, there is no patent or original paper showing the effectiveness of pharmaceutics and/or supplements targeting these atrogenes in spite of an abundance of patents and original papers on blockers [MYO-029, PF-354, etc] of myostatin, another mediator of muscle atrophy [24].

With recent studies revealing new substrates, and hence, specific new roles for Atrogin-1 and MuRF-1, it appears that there is more to their previously known functions. It is possible that these ubiquitin E3 ligases play additional roles in skeletal muscle, potentially contributing to muscle remodeling activity and metabolism. Inhibitor studies have also highlighted that the attenuation of muscle atrophy under catabolic conditions may be independent of reduced Atrogin-1 and MuRF-1 mRNA. These observations suggest that the activities of Atrogin-1 and MuRF-1 are altered without expression levels changing or that the primary role of

Atrogin-1 and MuRF-1 is not to enhance muscle atrophy. Therefore, caution is needed when using the expression levels of these genes as indicators of accelerated muscle protein breakdown. How to assess changes in their activity levels and understanding their contribution to muscle atrophy *in vivo* is a challenging prospect. Overexpression, knockout, knockdown and inhibitor studies in animals and *in vitro* can assist in determining the specific contribution that Atrogin-1 and MuRF-1 make to each atrophy process. With this in mind, and due to considerable species differences, there is a need for more *in vitro* studies to be performed in human primary muscle cells. Furthermore, the identification of further transcriptional modifications of Atrogin-1 and MuRF-1, in addition to understanding their interplay with deubiquitinating enzymes in the ubiquitination of potential target substrates [274], will aid in the understanding of their regulation and function. Finally, the identification of novel substrates for these atrogenes presents multiple challenges in terms of the turnover of the substrates by E3 ligase-substrate interactions.

Table 1. The adaptine changes in ubiquitin ligase (Atrogin-1, MuRFI, etc) and FOXO in human atrophy model

Author (year)	References	Model	analysis	Outcomes
Jones et al. 2004 [263]	FASEB J. 18, p 1025-1027	Immobilization (14 days)	Real-time PCR	Atrogin-1 mRNA ↑ MuRF-1 and E3 ligase mRNA ⇔
Whitman et al. 2005 [264]	Pflügers Arch. 450, p 437-446	Sarcopenia	Real-time PCR	Atrogin-1 and MuRF-1 mRNA ⇔
Ogawa et al. 2006 [265]	Muscle Nerve 34, p 463-469	Bed rest (20 days)	Real-time PCR	Atrogin-1 and cbl-b mRNA ↑ MuRF-1 mRNA ⇔
Leger et al. 2006 [266]	J. Physiol. 576, p 923-933	detraining (8 weeks)	Real-time PCR, Western blot	Atrogin-1, MuRF1, and UBB mRNA ↓ FOXO1 protein ⇔
Leger et al. 2006 [267]	FASEB J. 20, p 583-585	ALS	Real-time PCR, Western blot	Atrogin-1 mRNA and protein ↑ FOXO1 and FOXO3A mRNA ⇔ FOXO1 and FOXO3A protein ⇔
Urso et al. 2007 [268]	J. Physiol. 579, p 877-892	Spinal cord injury (2 and 5 days)	Real-time PCR, Western blot	Atrogin-1, MuRF-1, UBE3C, and PSMD11 mRNA ↑ PSMD11 protein ↑ Atrogin-1, MuRF-1, and UBE3C protein ⇔
Salanova et al. 2008 [269]	J. Anat. 212, p 306-318	bed rest (60 days)	Western blot, immunofluorescence	Soleus ••• MuRF-1 protein ↑ Vastus lateralis ••• MuRF1 protein ⇔
Leger et al. 2008 [270]	Rejuvenat. Res. 11, p 163B-175B	Sarcopenia	Real-time PCR, Western blot (fractionated homogenate)	Atrogin-1, MuRF1 mRNA ⇔ FOXO1 and FOXO3A protein (Nuclear fraction) ↓
Leger et al. 2009 [271]	Muscle Nerve 40, p 69-78	Chronic spinal cord-injured	Real-time PCR, Western blot	Atrogin-1 and MuRF1 mRNA ↓ FOXO1, FOXO3A, and Atrogin-1 protein ↓
Sakuma et al. 2009 [272]	Acta Physiol. 197, p 151-159	Lower limb unloading (20 days)	RT-PCR, Western blot (total and fractionated homogenates,)	Atrogin-1 and FOXO1 mRNA ⇔ FOXO1 and FOXO3A protein ⇔
Gustafsson et al. 2010 [273]	J. Appl. Physiol. 109, p 721-727	Unilateral lower limb suspension (3 days)	Real-time PCR, Western blot	Soleus ••• Atrogin-1 and MuRF-1 mRNA ⇔ Vastus lateralis ••• Atrogin-1 and MuRF-1 mRNA ↑ FOXO1A and FOXO3A protein ⇔

3.2. Autophagy-Dependent Signaling

3.2.1. Autophagy

Autophagy (derived from Greek and meaning "to eat oneself ") occurs in all eukaryotic cells and is evolutionarily conserved from yeast to humans [275]. Autophagy is a ubiquitous catabolic process that involves the bulk degradation of cytoplasmic components through a lysosomal pathway [276-279]. This process is characterized by the engulfment of part of the cytoplasm inside double-membrane vesicles called autophagosomes. Autophagosomes subsequently fuse with lysosomes to form autophagolysosomes in which the cytoplasmic cargo is degraded and the degradation products are recycled for the synthesis of new molecules [280]. Turnover of most long-lived proteins, macromolecules, biological membranes, and whole organelles, including mitochondria, ribosomes, the endoplasmic reticulum and peroxisomes, is mediated by autophagy [281].

Three major mechanisms of autophagy have been described: (1) microautophagy, in which only a small portion of the cytoplasm, proteins and maybe glycogen in mammals are destroyed [282]; (2) chaperone-mediated autophagy, in which soluble proteins with a particular pentapeptide motif are recognized and transported across the lysosomal membrane for degradation; and (3) macroautophagy, in which a portion of the cytoplasm, including subcellular organelles, is sequestered within a double membrane-bound vacuole that ultimately fuses with a lysosome [283].

3.2.2. Autophagy-Linked Molecule

In the autophagy system, small ubiquitin-like molecules [microtubule-associated protein light chain 3 (LC3), GABARAP, GATE16, and Atg12] are transferred from the conjugation system to membranes for their growth and commitment to become a double-membrane vesicle (autophagosome) that engulfs portions of the cytoplasm [20, 284].

This reaction requires the recruitment and assembly of different components of the autophagy machinery on phospholipids, but only the ubiquitin-like components LC3, GABARAP, and GATE16 are covalently bound to phosphatidylethanolamine [285, 286]. This covalent bond occurs on both the outer and inner membranes of the autophagosome. Sequestered organelles and proteins are then docked to the lysosome for their degradation.

The fusion of the outer membrane of the autophagosome with the lysosomal membrane also determines the degradation of the inner membrane and of the proteins that are associated with it. Because of the transient nature of autophagosomes, the life span of LC3 and its homologs is rather short. p62/SQSTM1 interacts with the autophagosome membane LC3 protein, which facilitates delivery of its polyubiquitinated cargo for lysosomal degradation [287, 288].

Interestingly, that the UPS and the lysosomal-autophagy system in skeletal muscle are interconnected was suggested by Mammucari et al. [21], and Zhao et al. [22]. Both studies identified FOXO3 as a regulator of the lysosomal and proteasomal pathways in muscle wasting (Figure 3). FOXO3 is a transcriptional regulator of the ubiquitin ligases MuRF-1 and Atrogin-1. It has now been linked to the expression of autophagy-related genes in skeletal muscle *in vivo* and C2C12 myotubes [22]. More recently, Masiero et al. [289] found an intriguing characteristic using muscle-specific autophagy knockout mice. The atrophy, weakness and mitochondrial abnormalities in these mice are also features of sarcopenia. Autophagy-dependent protein degradation seems to be also modulated by TRAF6 and PGC1α. Paul et al. [261] demonstrated that depletion of TRAF6 prevented tumor-induced muscle loss and the upregulation of LC3B and Beclin-1 mRNA expression. In addition, Wenz et al. [176] recognized no significant age-related increase in the ratio of LC3-II to LC3-I in MCK-PGC1α mice. Therefore, PGC1α would attenuate the autophagic process probably through increased anti-oxidant defense and mitochondrial biogenesis, and then preserved muscle integrity (ex. protection from age-associated fibrosis).

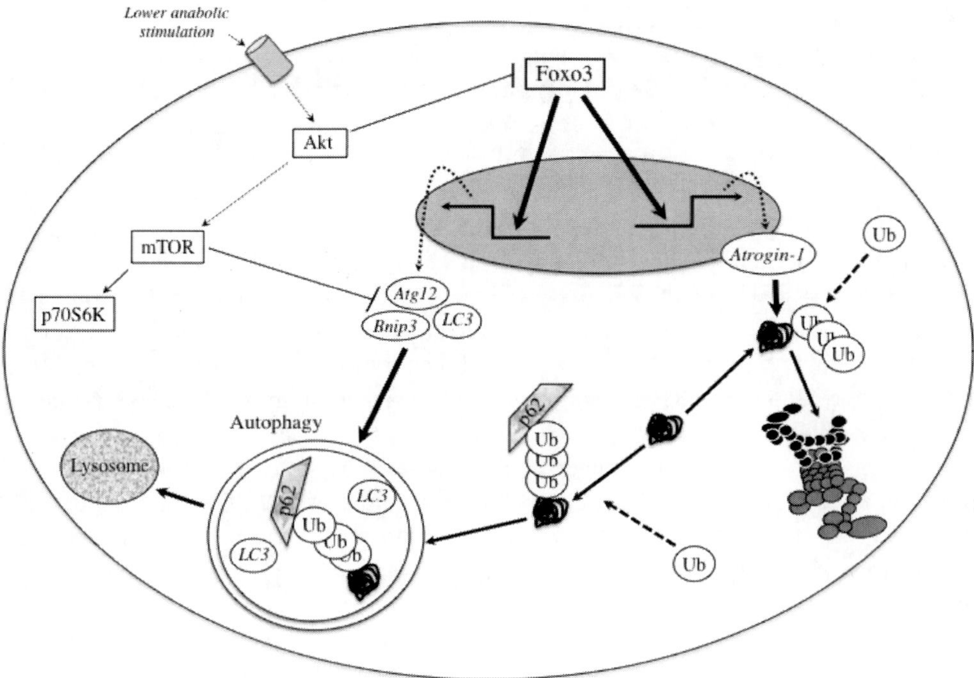

Figure 3. Contribution of the proteolytic pathways to muscle atrophy during catabolic conditions. In catabolic conditions such as denervation, cancer, and fasting, an atrophy program is induced to degrade muscle proteins and organelles. Proteins can have a double fate, being recognized and removed by the proteasome or docked to the autophagosome. In the latter case the chains of polyubiquitins are interacting with the p62. These proteins have also a domain for the interaction with LC3 therefore bringing the ubiquitinated proteins to the growing autophagosome. Less anabolic stimulation (IGF-I, mechanical loading, amino acids etc) reduces the amount of activated Akt, not promoting protein synthesis by activating the mTOR/p70S6K pathway. Lower Akt activity also does not block the nuclear translocation of FOXO3 to enhance the expression of autophagy-related genes (Bnip3, LC3, Atg12) and Atrogin-1 and the consequent protein degradation. FOXO; forkhead box O, IGF-I; insulin-like growth factor-I, LC3; microtubule-associated protein light chain 3, mTOR; mammalian target of rapamycin, p70S6K; 70 kDa ribosomal protein S6 kinase, Ub; ubiquitin.

3.2.2. Adaptation of Autophagy-Linked Signaling during Muscle Atrophy

Several recent studies have dealt with the changes of autophagy with age in mammalian skeletal muscle [176, 290-292]. Compared to those in young male Fischer 344 rats, amounts of Beclin-1 were significantly increased in the plantaris muscles of senescent rats [292]. In contrast, aging did not influence the amounts of Atg7 and Atg9 proteins in rat plantaris muscle [292]. Indeed, Western blot analysis by Wohlgemuth et al. [292] clearly showed a marked increase in the amount of LC3 in muscle during aging. However, they could not demonstrate an aging-related induction of the ratio of LC3-II to LC3-I, a better biochemical marker to assess ongoing autophagy. In contrast, Wenz et al. [173] recognized a significant increase in the ratio of LC3-II to LC3-I during aging (3 vs 22 months) in the biceps femoris muscle of wild-type mice. None of the studies determining the transcript level of autophagy-linked molecules found a significant increase with age [290-292]. Therefore, not all contributors to autophagy signaling seem to change similarly at both the mRNA and protein levels in senescent skeletal muscle.

Many recent studies have dealt with the changes of autophagy during cachexia in mammalian skeletal muscle. Using biopsy samples, Plant et al. [293] demonstrated that there was no difference in the levels of Beclin-1 and LC3 transcripts in the quadriceps muscle of patients with COPD compared with control individuals. In contrast, endotoxin-induced cachexia seems to include the upregulation of autophagic signaling. A significant increase in the Atg7 and Atg12 proteins was observed in the soleus and white vastus lateralis muscles as early as 8 hours after the injection of Lipopolysaccharide (1 mg/Kg). The amount of Beclin-1 and the ratio of LC3-II to LC3-I also increased in the soleus and white vastus lateralis muscles, respectively. In addition, intraperitoneal administration of dexamethasone (600 µg/Kg) for 5 days significantly increased both the ratio of LC3-II/LC3-I and Atrogin-1 mRNA but not MuRF-1 mRNA along with decreased total protein concentrations in rat soleus muscle [294].

Autophagy-linked molecules appear to be modulated by several neuromuscular diseases. Noglaska et al. [295] suggested the prominent accumulation of p62/SQSTM1 in muscle fibers of patients with sporadic myositis. Myotonic dystrophy (DM) type-1 is characterized by muscle distress with myotonia, progressive muscle weakness and wasting as well as a high neonatal mortality, mental retardation and respiratory failure [296, 297]. Immunofluorescence clearly shows that muscle cells from DM1 patients possess abundant LC3-positive vesicles *in vitro* differentiation phase [298]. Limb girdle muscular dystrophy type 2B and Miyoshi myopathy are both caused by mutations of the gene encoding dysferlin [299], a type II transmembrane protein expressed primarily in muscle sarcolemma. Although wild-type dysferlin in the endoplasmic reticulum (ER) is degraded primarily by the UPS, mutant dysferlin, which spontaneously aggregates in the ER, is degraded primarily by the autophagy/lysosomal system [300]. In cells expressing mutant dysferlin, autophagy inhibition due to Atg5 depletion or a mutation in ER-stress-induced eIF2a-mediated autophagy increases aggregation [301]. Therefore, the enhancement of autophagy signaling acts positively to improve the muscular disorder in this case.

3.3. Myostatin

Growth and differentiation factor 8, otherwise known as myostatin, was first discovered during screening for novel members of the TGF-β superfamily, and shown to be a potent negative regulator of muscle growth [302]. Indeed, mutations in myostatin can lead to massive hypertrophy and/or hyperplasia in developing animals, as evidenced by knockout experiments in mice. Like other TGF-β family members, myostatin is synthesized as a precursor protein that is cleaved by furin proteases to generate the active C-terminal dimer. When produced in Chinese hamster ovary cells, the C-terminal dimer remains bound to the N-terminal propeptide, which remains in a latent, inactive state [303, 304]. Most, if not all, of the myostatin protein that circulates in the blood also appears to exist in an inactive complex with a variety of proteins, including the propeptide [305]. The latent form of myostatin seems to be activated *in vitro* by dissociation of the complex with either acid or heat treatment [304,

306] or by proteolytic cleavage of the propeptide with members of the bone morphogenic protein-1/tolled family of metalloproteases [304].

Myostatin binds to and signals through a combination of ActRIIA/B receptors on the cell membrane, but has higher affinity for activin type IIB receptor (ActRIIB). On binding to ActRIIB, myostatin forms a complex with a second surface type I receptor, either activin receptor-like kinase (ALK) 4 or ALK5 to stimulate the phosphorylation of receptor Smad and the Smad2/3 transcription factors in the cytoplasm. This leads to the assembly of Smad2/3 with Smad4 to form a heterodimer that is able to translocate to the nucleus and activate the transcription of target genes [306, 307]. Myostatin circulates in the blood in a latent complex with non-covalently bound propeptide at the N-terminus [303, 304], crucial for the correct folding of the protein [302]. The propeptide was shown to bind the active myostatin *in vivo* and *in vitro*, and its overexpression in mice results in an increase of muscle mass [303, 308].

Specific inhibitors can prevent binding of myostatin to ActRIIB in serum. One of them is follistatin, an extracellular cysteine-rich glycoprotein with a structure not similar to members of the TGF-β family. Follistatin binds myostatin and inhibits its activity by preventing its binding to the receptor [303, 305]. Mice with a homozygous mutation in the Follistatin gene showed a decrease in muscle mass suggesting an important role for follistatin in the regulation of myogenesis [309]. Follistatin-like 3 protein has more than 30% homology with follistatin and was also shown to bind circulating myostatin [310].

Another inhibitor of myostatin is growth and differentiation factor-associated serum protein-1 (GASP-1). Unlike members of the follistatin family, GASP-1 does not have affinity for activin [311]. GASP-1 differs in structure from previously described inhibitors. Interestingly, GASP-1 can also bind to the myostatin propeptide and possibly regulate the activation of myostatin through proteolytic cleavage [311]. Moreover, myostatin can be inactivated upon covalently binding to latent TGF-β binding protein 3 which is used for storing myostatin in the extracellular matrix [312]. Taken together, myostatin is regulated by at least four different inhibitors via binding of the active or latent form.

The role of myostatin in the regulation of adipose tissue is not well understood. Myostatin and its receptor were shown to be expressed in adipose tissue at low levels [313]. Several studies showed a negative effect of myostatin on preadipocyte differentiation and proliferation [314]. Conversely, promotion of differentiation leading to increased adipogenesis was observed in pluripotent mesenchymal cells [315]. Moreover, Myostatin null mice had a decreased amount of adipose tissue [313, 316]. It is still unclear whether the effect of myostatin on adipose tissue is the direct result of regulation or an indirect consequence of skeletal muscle growth. The decreased amount of adipose tissue in Myostatin null mice could be an indirect drawback of significant increases in muscle mass. Such an effect can be mediated through a leptin-adipose-specific hormone regulating food intake and energy homeostasis, whose level was shown to be lower in myostatin null mice [316].

3.3.1. Adaptive Changes in Myostatin during Muscle Wasting

Myostatin levels increase with muscle atrophy due to unloading in mice and humans [53, 272, 317], and with severe muscle wasting in patients with cancer cachexia, chronic heart failure, COPD, AIDS, and diabetes [24]. For example, our study [272], using healthy university students, showed significantly increased levels of myostatin mRNA in muscle after the unilateral suspension of a hindlimb for two weeks. Although many researchers consider myostatin levels to increase with age, studies using sarcopenic muscles have yielded conflicting results [270, 318, 319]. For example, Léger et al. [270] found that myostatin mRNA and protein levels were significantly elevated by 2 and 1.4 fold in young (20 ± 0.2 years) compared to older males (70 ± 0.3 years). Haddad et al. [319] showed a lower expression of myostatin mRNA in aged (30-month-old) than young (6-month-old) rats. Intriguingly, Carlson et al. [318] showed enhanced levels of Smad3 but not myostatin in sarcopenic muscles of mice. Therefore, it seems that myostatin- dependent signaling is activated in sarcopenic mammalian muscles, although several researchers have found various age-dependent results regarding myostatin levels.

The administration of myostatin to adult mice induces a profound muscle loss analogous to that seen in human cachexia syndromes [320]. *In vitro* studies showed that myostatin-induced cachexia was caused by the activation of the UPS through FOXO1-dependent signaling [321]. Myostatin induction was also recognized in an animal model of cardiac cachexia (rats with volume overload induced by aortacaval shunt) [322]. In support of these findings, Heineke et al. [323] demonstrated that cardiomyocyte-specific overexpression of myostatin significantly reduced both cardiac and skeletal muscle mass. Furthermore, several recent reports suggested that up-regulation of myostatin expression was involved in the development of muscle atrophy associated with COPD [324, 325]. For example, Hayot et al. [326] found increased skeletal muscle myostatin expression in hypoxemic COPD patients (PaO_2 = 64 ± 2 mm Hg) compared with healthy controls. Therefore, abundant myostatin protein further worsens the muscle loss in various cases of cachexia.

3.3.2. Myostatin-Depedent Signaling

Studies indicate that myostatin inhibits cell cycle progression and reduces levels of myogenic regulatory factors (MRFs), thereby controlling myoblastic proliferation and differentiation during developmental myogenesis [327-329]. One of the known downstream targets of Smad signaling is MyoD. Interestingly, myostatin downregulates MyoD expression in an NFκB-independent way [330, 331]. Myostatin also inhibits Pax3 expression, which is possibly an upstream target of MyoD [331, 332].

TGF-β family members were shown to activate MAPKs, particularly p38 and extracellular signal-regulated kinase (ERK) 1/2 [333]. p38 MAPK is responsible for the cell response to stress factors and was shown to be activated by myostatin via a TGF-β activated kinase 1/MAPK kinase cascade. The role of ERK1/2 in the regulation of muscle mass is controversial. ERK1/2 has been demonstrated to activate satellite cell proliferation [334] and induce protein synthesis [335]. On the other hand, the administration of myostatin significantly activated ERK1 in C2C12 cells and muscle of mice [336]. It seems likely that myostatin mediates its signaling at least partially through ERK1/2, which is activated by MEK1.

Akt plays a significant role in different metabolic processes in the cell, particularly in the hypertrophic response to insulin and IGF [105]. Akt is the crossing point between the IGF-I/myostatin pathways. The results of several studies suggest that the genetic loss of myostatin leads to an increase in Akt activity in skeletal muscle *in vivo* and *in vitro* [337]. In contrast, a decreased level of phosphoryated Akt is associated with the incubation of myotubes with myostatin [328]. Akt regulates the activity of FOXO1 and FOXO3 by phosphorylating them and thereby retaining them in the cytoplasm [331]. In myotube cultures, the level of FOXO1 in the nucleus was shown to increase after treatment with myostatin [239, 331]. Recently, it was found that FOXO1 and Smad synergistically increase the expression of myostatin mRNA and its promoter activity in C2C12 myotubes but via different pathways [338]. Taken together, myostatin-mediated signaling activates FOXO, and this leads to the expression of ubiquitin ligases, MuRF-1 and Atrogin-1, which participate in protein degradation. mTOR is a kinase downstream of Akt, which phosphorylates 4E-BP1 and p70S6K thereby initiating protein synthesis [14]. There are two mTOR complexes called TORC1 and TORC2, containing the proteins RAPTOR and RICTOR, respectively [339]. Blockade of the Akt/mTOR/p70S6K pathway using siRNA to RAPTOR facilitates myostatin's inhibition of muscle differentiation because of an increase in Smad2 phosphorylation [340].

3.3.3. Use of Myostatin to Treat Muscle Wasting

Taking note of the functional role of myostatin in muscle mass, many researchers have examined the possibility of inhibiting myostatin to treat muscle disorders. The use of neutralizing antibodies to myostatin improved muscle disorders in rodent models of Duchenne muscular dystrophy (*mdx*), limb girdle muscular dystrophy 2F (Sgcg-/-), and amyotrophic lateral sclerosis (SOD1^{G93A} transgenic mouse) [341-343]. Indeed, myostatin inhibition using MYO-029 was tested in a phase I/II trial in 116 adults with muscular dystrophy such as Becker muscular dystrophy and limb-girdle muscular dystrophy [344]. Long-term administration of MYO-029 at 1-30 mg/Kg showed good safety and tolerability except for cutaneous hypersensitivity in spite of no improvement in muscle function.

Although the functional role of myostatin in sarcopenia is unclear, many researchers have investigated the effect of inhibiting myostatin to counteract sarcopenia using animals [345, 346]. Lebrasseur et al. [346] reported several positive effects of 4 weeks of treatment with PF-354 (24 mg/Kg), a drug for inhibiting myostatin, in aged mice. They found that PF-354-treated mice exhibited significantly greater muscle mass (by 12%) probably due to decreased levels of phosphorylated Smad3 and MuRF-1 in muscle. More recently, Murphy et al. [347] showed, with once-weekly injections, that a lower dose of PF-354 (10 mg/Kg) significantly increased the fiber cross-sectional area (by 12%) and *in situ* muscle force (by 35%) of aged mice (21-month-old). The attenuation of myostatin-dependent signaling markedly improves the muscle wasting of cachexia. Zhou et al. [348] indicated the therapeutic potential of blocking the myostatin receptor, ActRIIB, on cancer cachexia. They utilized various models of cancer cachexia, including colon 26 tumor-bearing mice and nude mice bearing human TOV-21G ovarian carcinoma xenografts. Blockade of the ActRIIB pathway using a decoy receptor (sActRIIB) completely reversed the pathology and muscle atrophy caused by cancer cachexia. Treatment with sActRIIB abolished the activation of the UPS probably due to decreased phospho-Smad2 and increased inactivated (phospho)-FOXO3 and phospho-Akt proteins in cachexic muscles.

3.4. TNF-α and NF-κB (Nuclear Factor-KappaB) Signaling

3.4.1. The Structure of the NF-κB Signaling Pathway

NF-κB refers to structurally related Rel family eukaryotic transcription factors which regulate a variety of cellular responses [349]. The NF-κB family constitutes five members which can be further divided in two groups. One group includes RelA (p65), RelB, and c-Rel, which are synthesized as mature proteins and characterized by the presence of an N-terminal

Rel homology domain essential for dimerization and DNA-binding and a C terminal transcriptional activation domain. The second group consists of the NF-κB1 (p50) and NF-κB2 (p52) proteins, which are synthesized as the larger precursors p105 and p100, respectively, containing an N-terminal ankyrin repeat domain. Proteolytic processing of p105 and p100 at the C terminus gives rise to p50 and p52, respectively. Both p50 and p52 contain the N-terminal Rel homology domain but lack the transcriptional activation domain at the C terminus [349].

Different members of the NF-κB family dimerize to facilitate the binding of NF-κB to DNA. Among them, p50 and p65 are the most prototypical heterodimers, present in almost all cell types and responsible for the increased expression of a number of pro-inflammatory and cell survival genes. However, homodimers or heterodimers of p50 and p52, which lack transcriptional activation domains, can still bind to NF-κB consensus sites in DNA and act as transcriptional repressors by blocking the consensus sites [350]. Prior to activation, most NF-κB dimers are retained in the cytoplasm by binding to specific inhibitors - the inhibitors of NF-κB (IκBs). The interaction with IκBs masks the nuclear localization sequence in the NF-κB complex, thus preventing nuclear translocation and maintaining NF-κB in an inactive state in the cytoplasmic compartment [350].

3.4.2. Mechanisms of NF-κB's Action in Skeletal Muscle

Systemic inflammation is the primary cause of skeletal muscle wasting and fatigue that is characteristic of normal aging, AIDS, chronic heart failure, COPD, and cancer cachexia. Pro-inflammatory cytokines such as TNF-α, TNF-like weak promoter of apoptosis (TWEAK), IL-Iβ, interferon-γ, and IL-6 are some of the most important inducers of muscle wasting in chronic disease states [351]. Elevated serum levels of several of these pro-inflammatory cytokines have been consistently observed in humans and animal models of chronic disorders involving muscle wasting [351, 352]. Increased levels of TNF-α or IL-1β have been shown to inhibit the expression and activity of hormones such as GH and IGF-I, the two most important contributors to postnatal growth [353]. In addition, the expression of several chemokines, cell adhesion molecules, and extracellular proteases can augment the infiltration of macrophages, lymphocytes and neutrophils in skeletal muscle tissue. Interestingly, NF-κB regulates the expression of a number of inflammatory molecules [354]. In addition, several NF-κB-regulated molecules, especially pro-inflammatory cytokines (e.g. TNF-α), are also potent activators of NF-κB.

A wealth of research has clearly shown the importance of NF-κB signaling in regulating skeletal muscle mass during periods of disease and inactivity [355]. While the necessity of NF-κB signaling during muscle atrophy is unequivocal, current research has focused on defining the contribution of pathway components to muscle atrophy. Van Gammeren and coworkers extended earlier work in the mouse by showing that both IκB kinases (IKK), IKKα and IKKβ, are necessary and sufficient for muscle atrophy [238, 356]. To arrive at this conclusion, these researchers overexpressed in the rat soleus muscle dominant negative or constituitively active forms of each IKK and found, respectively, that disuse-induced muscle atrophy was reduced by 50% (individually) and 70% (in combination) or a ~ 40% reduction in fiber cross-sectional area under weight-bearing conditions. An interesting question for future research will be whether or not the ability to blunt muscle atrophy with the dominant negative forms of IKK is observed under muscle wasting conditions associated with cachexia.

Some candidates such as heat shock protein (HSP) and reactive oxygen species (ROS) seem to interact with NF-κB-dependent signaling. In a series of studies, the Judge laboratory has found that HSP27 and HSP70 can prevent immobilization-elicited muscle atrophy by modulating the signaling activities of NF-κB and FOXO pathways [357-359]. A 5-fold overexpression of HSP70 in the rat soleus muscle completely prevented the loss in fiber area in response to seven days of immobilization and attenuated the increase in Atrogin-1 and MuRF-1 mRNA expression by ~ 70%. Further examination revealed that HSP70 inhibited the transcriptional activities of both FOXO3 and NF-κB during immobilization [357]. In a similar fashion, the overexpression of HSP27 was able to ameliorate the loss in fiber area with immobilization and inhibit NF-κB. Collectively, these studies indicate that HSP27 and HSP70 can significantly modify the activity of catabolic pathways involved in muscle atrophy and highlight the need to identify the direct transcriptional targets of NF-κB during muscle atrophy [360].

3.4.3. The NF-κB Adaptation in Muscle Atrophy

The majority of muscle atrophy in the general population results from disuse. Hunter et al. [361] demonstrated that hindlimb unloading leads to a drastic increase in the NF-κB DNA binding and reporter genes activities in rat soleus muscle. Compared to weight-bearing rats, there was a significant increase in the nuclear levels of NF-κB/IκB family proteins in unloaded soleus muscle [361]. Analysis of the NF-κB/DNA complex showed that it mainly contained p50, c-Rel, and Bcl-3, but not other members of the NF-κB/IκB family [361]. The role of p50, c-Rel, and Bcl-3 proteins in unloading-induced muscle atrophy was later evaluated using knockout mice [362, 363]. Genetic ablation of either NF-κB1 or Bcl-3 in mice prevented the unloading-induced NF-κB reporter gene activity in the soleus muscle [11]. The same group of investigators also reported that the c-Rel subunit was not involved in unloading-induced muscle atrophy because of no difference in the NF-κB transcriptional activity and fiber area in c-Rel knockout mice.

The role of NF-κB in denervation-induced skeletal muscle atrophy was elucidated by two independent studies using transgenic approaches [238, 364]. Mourkioti et al. [364] demonstrated that skeletal-muscle-restricted NF-κB inhibition in mice through targeted deletion of the IKK2 subunit of the IKK complex causes an increase in total cross-sectional area and twitch and tetanic forces in soleus muscle. Cai et al. [238] studied the role of NF-κB in denervation-induced skeletal muscle atrophy through transgenic overexpression of super-repressor mutant of NF-κB inhibitory protein IκBα (IκBαSR) in the skeletal muscle of mice. There was a marked decrease in NF-κB activity in the denervated muscle of IκBαSR mice (1.5 fold vs 9 fold in wild-type mice). IκBαSR mice showed a blunted atrophy (muscle mass and fiber size) after denervation, suggesting that NF-κB signaling is involved in the denervation-induced muscle atrophy.

There is growing evidence that higher levels of inflammatory markers [TNF-α, IL-1β, etc] are associated with physical decline in older persons [365]. An age-related disruption in the intracellular redox balance appears to be a primary causal factor of a chronic state of low-grade inflammation. More recently, Chung et al. [366] hypothesized that abundant NF-κB protein induced age-related increases in IL-6 and TNF-α. Moreover, ROS also appear to function as second messengers for TNF-α in skeletal muscle, activating NF-κB either directly or indirectly [360]. Indeed, marked production of ROS has been documented in muscle of the elderly [367, 368]. However, it is not clear whether NF-κB signaling is enhanced with age.

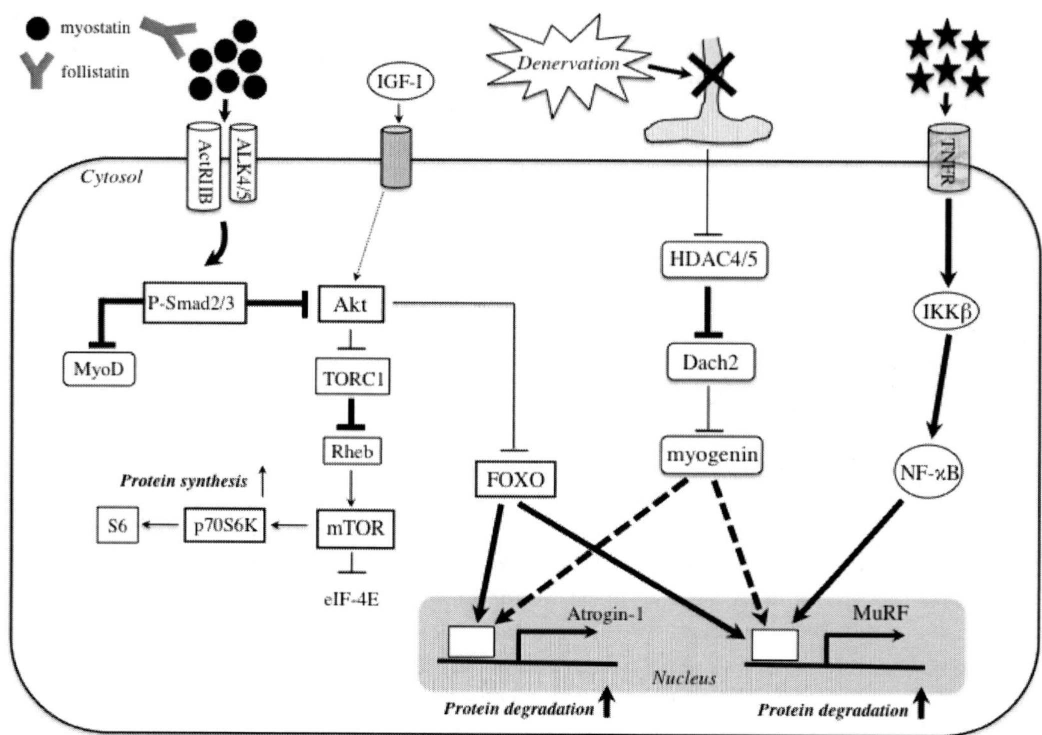

Figure 4. Catabolic pathway regulating protein degradation in skeletal muscle. Muscle wasting upregulates myostatin expression. Myostatin signals through the ActRIIB-ALK4/5 heterodimer activate Smad2/3 while blocking MyoD transactivation via an autoregulatory feedback loop. In addition, Smad3 sequesters MyoD in the cytoplasm to prevent it from entering the nucleus and activating the stem cell population. Recent findings [97, 340] suggest that the myostatin-Smad pathway inhibits protein synthesis probably by blocking the functional role of Akt. The reduction in mechanical loading decreases the amount of follistatin, a myostatin blocker, and of IGF-I. Less IGF-I stimulation reduces the amount of activated Akt, not promoting protein synthesis by activating the mTOR/p70S6K pathway. Lower Akt activity also does not block the nuclear translocation of FOXO to enhance the expression of Atrogin-1 and MuRF and the consequent protein degradation. Neuromuscular inactivity (denervation) results in the upregulation of HDAC4 and HDAC5, which represses Dach2, a negative regulator of myogenin, upregulating myogenin expression [370]. Myogenin activates protein degradation by inducing the expression of MuRF-1 and Atrogin-1. The blood and muscle levels of TNF-α were increased in muscle wasting, and the elevation of IKKβ/NF-κB induced MuRF-1 expression. ActRIIB; activin receptor IIB, ALK4/5; activin-like kinase 4/5, eIF; eukaryotic initiation factor, FOXO; forkhead box O, HDAC; histone deacetylase, IGF-I; insulin-like growth factor-I, IKKβ; inhibitor of κB kinase β, mTOR; mammalian target of rapamycin, MuRF; muscle ring-finger protein, NF-κB; nuclear factor of kappa B, p70S6K; 70 kDa ribosomal protein S6 kinase, Rheb; Ras homolog enriched in brain, TORC1; a component of TOR signaling complex 1, TNF-α; tumor nectosis factor-α, TNFR; TNF receptor.

Despite some evidence supporting enhanced NF-κB signaling in type I fibers of aged skeletal muscle, direct evidence for increased activation and DNA binding of NF-κB is lacking [352, 369]. For example, Phillips and Leeuwenburgh [352] found that neither p65 protein expression nor the binding activity of NF-κB was significantly altered in the vastus lateralis muscles of 26-month-old rats despite the marked upregulation of TNF-α expression in both blood and muscle. In aging mammals, up-regulated TNF-α expression in serum and muscle seems to enhance apoptosis in mitochondria resulting in a loss of muscle fibers [330, 352] but not NF-κB signaling. Considering these findings, NF-κB dependent-signaling is not always similarly upregulated in various musclular atrophy models. Figure 4 provides an over view of the catabolic pathway regulating protein degradation in skeletal muscle.

3.5. Apoptosis Signaling

Apoptosis is an evolutionary conserved process of programmed cell death, which is performed via a systematic set of morphological and biochemical events, resulting in cellular self-destruction without inflammation or damage to the surrounding tissue. Execution of apoptosis in skeletal muscle displays unique features due to the multinucleated nature of myofibers. Therefore, apoptosis may result in the elimination of individual myonuclei (myonuclear apoptosis) and a portion of sarcoplasm, without demise of the entire fiber [371]. Importantly, skeletal muscle, similar to the heart, contains two bioenergetically and structurally [372] distinct mitochondrial subpopulations: subsarcolemmal mitochondria and intermyofibrillar mitochondria arranged in parallel rows between the myofibrils. These two subpopulations differ in their susceptibility to apoptotic stimuli [373] and may therefore be differentially involved in the pathogenesis of sarcopenia and other muscle atrophying conditions.

Cysteine-aspartic proteases (caspases) are the executioner enzymes that carry out the dismantling of the cell and are normally present as inactive zymogens (procaspases) [374]. Upon apoptotic stimuli, initiator caspases (i.e., caspase-8) are activated, subsequently leading to the activation of effector caspases (i.e., caspase-3) that perform the actual cellular degradation [375]. Effector caspases can be activated through extrinsic and intrinsic pathways [376]. The extrinsic apoptotic signaling is initiated by the activation of death receptors on cell membranes, such as the Fas receptor and TNF receptor [376]. Activation of these receptors by death-stimulating ligands recruits adaptor proteins such as Fas-associated-death domain that engage procaspase-8. Intrinsic pathways of caspase activation include several internal cellular stimuli mediated by the endoplasmic reticulum or the mitochondria [374]. When intracellular calcium homeostasis is disrupted, calpain-mediated activation of caspase-12 can occur, leading to caspase-3's activation [377].

Mitochondria are essential for proper cellular functioning and viability, being the main sites for energy production and playing an essential role in the maintenance of redox homeostasis [378]. MtDNA (mitochondrial DNA) is especially prone to oxidative damage [379] due to its proximity to the electron transport chain, the lack of protective histones, and a less efficient repair system compared to nuclear DNA [380]. The mitochondria play an important part in apoptosis by activating caspases [381]. Mitochondrial dysfunction can lead to mitochondrial cytochrome c release [381]. In the cytoplasm, cytochrome c, Apaf-1, caspase-9, and dATP form an apoptosome, which can activate caspase-3, a key cell death protease.

3.5.1. The Role of Mitochondria-Mediated Apoptosis in Disuse Muscle Atrophy

Several mechanisms are thought to be involved in the execution of mitochondria-driven apoptosis in skeletal muscle [374]. However, it appears that altered mitochondrial function precedes and is required for the initiation of apoptosis in disuse muscle atrophy [381].

Enhanced mitochondrial apoptotic signaling has been observed in experimental models of muscle denervation and in humans affected by neuromuscular disorders. Elevated caspase-9 expression was detected in patients with muscle atrophy due to peripheral neuropathy [382]. In the same disorder, Tews et al. [383] also identified increased expression of Smac/DIABLO, which was exclusively observed in atrophied fibers. In addition, changes in the expression of pro-apoptotic Bax and anti-apoptotic Bcl-2 were detected in skeletal muscles of patients with spinal muscular atrophy [384], polyneuropathy and ALS [385]. Denervation resulted in the up-regulation of Bax expression by 115% and a decrease in Bcl-2 by 89%, producing a 16-fold increase in the Bax-to-Bcl-2 ratio. Similarly, Siu and Always [386] found an increased Bax-to-Bcl-2 ratio in gastrocnemius muscle 2 weeks after tibial nerve transection. This coincided with enhanced apoptotic DNA fragmentation and activation of mitochondria-dependent activation apoptosis, as evidenced by increased cytosolic levels of cytochrome c, Smac/DIABLO and apoptosis-inducing factor (AIF).

Several reports indicate that apoptosis is enhanced in aged skeletal muscle, likely contributing to the development of sarcopenia [387]. In the white gastrocnemius and soleus muscles of Fischer 344 rats, it was demonstrated that aging significantly increased DNA fragmentation, and the amounts of cleaved caspase-3 and proapoptotic Bax, and decreased antiapoptotic Bcl-2 protein content [388]. MtDNA is especially prone to oxidative damage [383], particularly in aged mammalian muscles [389, 390]. Aside from mitochondria-mediated apoptosis, other pathways of myonuclear apoptosis may be involved at an advanced age. In particular, the death receptor-mediated pathway, triggered by TNF-α, appears to be upregulated in aged rodent muscles [352, 387]. Recent findings suggest that apoptotic signaling is required for and precedes protein degradation during muscle atrophy [391]. Although a clear mechanistic link between myonuclear apoptosis and sarcopenia is yet to be established, data from animal models suggest this link exists [387, 392]. Moreover, other rodent studies indicate an association between myocyte apoptosis and declines in muscle mass and strength [393]. Despite that most studies with animal models appear to support a key role for apoptosis in age-related muscle atrophy, evidence in humans is still scarce. A recent report indicated increases in the expression of the mitochondrial caspase-independent mediator AIF in semitendinousus muscle of middle-aged men relative to younger controls [394]. This finding is consistent with previously described data from animal experiments and supports the involvement of mitochondria-driven apoptosis in the development of sarcopenia.

The involvement of myonuclear apoptosis in disuse muscle atrophy has recently been challenged by Gundersen's group [395, 396], who reported no apoptotic loss of myonuclei in murine skeletal muscles atrophied for up to 28 days. Myonuclei were transfected with green fluorescent protein containing a nuclear localization tag and tracked using *in vivo* time-lapse microscopy. Unexpectedly, no myonuclear loss was observed in either slow- or fast-twitch muscles, despite a significant reduction in muscle cross-sectional area. Furthermore, terminal deoxynucleotidyl transferase biotin-dUTP nick-end labeling (TUNEL) analysis of approximately 27,000 myonuclei revealed only an exiguous number to be apoptotic. In contrast, a large number of studies have suggested that myonuclei undergo apoptosis during atrophy [384, 397-403]. However, as pointed out by Gundersen and Bruusgaard [396], several of the papers indicating myonuclei apoptosis are based only on finding molecular markers of apoptosis in muscle homogenates [384, 399, 401, 402], and the majority of the literature is based on light-microscopy studies [397, 398, 400, 403]. Nuclear labeling combined with the labeling of laminin and dystrophin is generally utilized to distinguish between myonuclei,

nuclei belonging to satellite cells and nuclei of stromal cells in light microscopy. Using such techniques, Gundersen and Bruusgaard [396] observed that denervation and nerve impulse block both led to high levels of apoptotic nuclei in muscle tissue. These were, however, unlikely to be myonuclei, since virtually none of them had their mass centre inside the dystrophin ring of any muscle fibers. In addition, basal lamina staining suggested that apoptosis occurred both in stromal cells and in satellite cells probably due to an increased turnover rather than net loss of nuclei. It was shown that after denervation mitotic activity increased to a similar extent as the rate of apoptosis [404]. Considering these findings, a more descriptive analysis using several direct and accurate techniques would be needed to clarify whether myonuclear apoptosis contributes to muscle atrophy under conditions of denervation, hindlimb suspension, and sarcopenia.

3.6. Glucocorticoids

The catabolic effects of glucocorticoids have been well known for many years. Either as drugs used to treat medical conditions or as endocrine hormones released in response to stress, glucocorticoids appear to cause skeletal muscle atrophy. Many pathological conditions characterized by muscle atrophy (sepsis, cachexia, starvation, etc) are associated with increases in circulating glucocorticoids levels [405], suggesting that these hormones could trigger the muscle atrophy observed in these situations.

In skeletal muscle, glucocorticoids decrease the rate of protein synthesis and increase the rate of protein breakdown [406] contributing to atrophy. The inhibitory effect on protein synthesis results from different mechanisms. First, glucocorticoids inhibit the transport of amino acids into the muscle [407], which could limit the protein synthesis. Second, glucocorticoids inhibit the stimulatory action of insulin, IGF-I, and amino acids (in particular leucine), on the phosphorylation of eIF-4E-BP1 and the ribosomal protein S6 kinase 1 (S6K1), two factors that play a key role in the protein synthesis machinery by controlling the initiation step of mRNA translation [408, 409]. Finally, there is also evidence that glucocorticoids cause muscle atrophy by inhibiting myogenesis through the downregulation of myogenin, a transcription factor mandatory for the differentiation of satellite cells into muscle fibers [410].

The muscle cell catabolism caused by glucocorticoids is thought to be mediated by FOXO, GSK-3β, and p300-C/EBPβ. For example, overexpression of a dominant negative form of FOXO3a prevents muscle cell atrophy together with Atrogin-1 induction caused by glucocorticoids *in vitro* [239]. Recent *in vitro* data suggest that glucocorticoid-induced muscle proteolysis is at least in part regulated by p300-histone acetyl transferase activity. Indeed, p300 protein levels and activity are increased, in a time- and dose-dependent manner, in dexamethasone-treated myotubes [411]. Intriguingly, Myostatin knockout mice did not develop a reduction of muscle mass or fiber cross-sectional area after receiving glucocorticoids, different to wild-type mice [412]. This observation indicates that myostatin is mandatory for the atrophic effects of glucocorticoids on muscle.

3.7. microRNAs

The human genome contains thousands of non-coding RNAs, the best-studied class of which are microRNAs (miRNAs) [413], which regulate gene expression at the transcriptional and post-transcriptional levels. miRNAs suppress gene expression through their complementarity to the sequence of one or more RNAs, usually at a site in the 3' untranslated region. The formation of a miRNA-target complex results either in inhibition of protein translation or in degradation of the mRNA transcript through a process similar to RNA interference [414]. There is no doubt that the formation, maintenance, and physiological and pathophysiological responses of skeletal muscles, with all their complex regulatory circuits, are subject to regulation by non-coding RNAs. In fact, the increase of complexity provided by the extent of genomic non-coding sequences provides a satisfying explanation for the intricate layers of regulation found in skeletal muscle cells.

Many miRNAs are expressed in skeletal and cardiac muscle. Some of them are found specifically, or at least are highly concentrated, in skeletal and/or cardiac muscle, suggesting specific roles in myogenesis [415]. The expression of the muscle-specific miRNAs miR-1, miR-133, miR-206 and miR-208 seems to be under the control of a core muscle transcriptional network, which involves the pleiotropic SRF, MyoD, and the bHLH transcription factor Twist in cooperation with MEF2 [416-418]. Chromatin immunoprecipitation followed by microarray analysis indicated that MyoD and myogenin bind sequences upstream of miR-1 and miR-133 [417]. miR-133a increases myoblast proliferation, via its repression of SRF [419], while miR-1 stimulates myoblast differentiation via its inhibition of histone deacetylase 4 (HDAC4) [419]. In addition, MyoD has been demonstrated to utilize miRNAs, including miR-1 and miR-206, to suppress downstream gene expression [419, 420]. More recently, Hirai et al. [83] have demonstrated that miR-1 and miR-206 bind to two miR-1/miR-206-binding sequences within the Pax3-3'UTR and suppress Pax3 expression. Since Pax3 expression increases cell survival and suppresses myogenic differentiation in myoblasts, down-regulation of Pax3 has been shown to elicit proper myogenic differentiation along with an increase in apoptosis [83]. An analogus role was described for the regulation of Pax7, which is repressed by miR-1 and miR-206 [419, 421]. In contrast, miR-221 and miR-222 are downregulated during the transition from proliferation to differentiation [422]. Decreases in these miRNAs are associated with increased expression of the cell cycle inhibitor p27. Overexpression of miR-221 and miR-222 in differentiating myotubes delays cell cycle withdrawal and differentiation, a response associated with a reduction in sarcomeric protein [422].

3.7.1. The Functional Role of MicroRNA during Muscle Hypertrophy and Atrophy

Mechanical overloading for plantaris muscle has been shown to induce marked reductions in miR-1 and miR-133a [423]. The decrease in these mature miRNAs may contribute to muscle hypertrophy. miR-1 and miR-133a may play a role in muscle hypertrophy by the removal of their transcriptional inhibition of growth factor gene targets, including HGF, LIF, IGF-I and SRF [423]. Increased levels of endogenous miR-1 in differentiating C2C12 myotubes resulted in a reduction in the activation of an IGF-I luciferase reporter construct and a decrease in IGF-I protein levels. Conversely, IGF-I treatment reduced miR-1 levels in C2C12 myotubes. As previously indicated, IGF-I stimulates C2C12 myotube hypertrophy via the activation of Akt signaling and the inhibition of FOXO [66]. miR-1 levels are reduced by active Akt and increased by active FOXO3a [424] demonstrating a regulatory loop whereby IGF-I regulates miR-1 via an Akt/FOXO3a pathway.

Following 11 days of spaceflight, miR-206 was significantly decreased while miR-1 and miR-133a showed a trend towards a reduction in mouse gastrocnemius muscle [425]. miR-206 is upregulated via MyoD, a protein which can be degraded by Atrogin-1 [426]. TWEAK down-regulated several grown-related myomiRs, including miR-1, miR-23, miR-133a, miR-133b, and miR-206 in C2C12 myotubes. However, it only reduced

**Table 2. The adaptive changes in miR-1, miR-133 and miR-266
in various atrophy model**

References	Model	Outcomes
Jeng et al. 2009 [428]	Denervation (mouse)	miR-1 ↓
Rau et al. 2010 [429]	Entrapment neuropathy (rat)	miR-1 ↓
McCarthy et al. 2009 [430]	Hindlimb suspension (mouse)	miR-206 ⇔
Allen et al. 2009 [425]	Space flight (mouse)	miR-1 ⇔ , miR-133 ⇔ , miR-206 ↓
Lewis et al. 2012 [431]	COPD (human)	miR-1 ↓ , miR-133 ⇔ , miR-206 ⇔
Panguluri et al. 2010 [427]	TWEAK treatment (*in vitro*) TWEAK treatment (*in vivo*, mouse)	miR-1 ↓ , miR-133 (a & b) ↓ , miR-206 ↓ miR-1 ↓ , miR-133 (a & b) ↓
McCarthy et al. 2007 [423]	*mdx* mouse	miR-133a soleus ↓ / plantaris ⇔ / diaphragm ⇔ ; miR-206 soleus ⇔ / plantaris ↓ / diaphragm ↑
Yuasa et al. 2008 [432]	*mdx* mouse	miR-1 ⇔ miR-206 ↑
Greco et al. 2002 [433]	DMD (human), mdx mouse	miR-1 ↓ miR-206 ↑
Gambardella et al. 2010 [434]	Myotonic dystrophy type 1 (human)	miR-1 ⇔ , miR-133 (a & b) ⇔ , miR-206 ↑
Perbellini et al. 2011 [435]	Myotonic dystrophy type 1 (human)	miR-1 ↑ , miR-133 (a & b) ⇔ , miR-206 ⇔
Williams et al. 2009 [436]	ALS mouse model	miR-1 ⇔ , miR-133a ⇔ / miR-133b ↑ miR-206 ↑
Drummond et al. 2008 [437]	Age-related muscle wasting (human)	miR-1 ⇔ , miR-133a ⇔

miR-1, miR-133a and miR-133b in mouse skeletal muscle *in vivo* [427]. While treatment with TWEAK regulates several muscle miRNAs involved in muscle growth, it is not known whether their regulation is a cause of muscle wasting or a response to prevent further muscle wasting. Table 2 provides an overview of the adaptive changes in miR-1, miR-133, and miR-266 in the skeletal muscle under various atrophic conditions.

CONCLUSION AND PERSPECTIVES

Advances in our understanding of muscle biology have led to new approaches to muscle wasting. For example, resistance training combined with amino acid-containing supplements [438, 439] would be the best way to prevent age-related muscle wasting and weakness including sarcopenia, in which older people suffer various functional defects including decreased muscle strength, decreased mobility and function, and a greater risk of falls. A combination of exercise and nutrition (e.g. amino acids) is widely accepted to be a major promoter for muscle mass in muscular wasting probably due to upregulation of the Akt/mTOR/p70S6K pathway. Recent progress has significantly expanded our understanding of the molecular mechanisms that regulate skeletal muscle protein synthesis and degradation. Despite this, considerably more research is required to fully elucidate the many different mechanisms that potentially regulate these two processes. Successful identification of common regulatory molecules/pathways will greatly aid our understanding of how different types of stimuli promote changes in skeletal muscle mass. The Akt/mTOR/p70S6K pathway and SRF-dependent signaling have been considered to be major contributors to protein synthesis and muscle-specific transcription, respectively [14, 16]. Over the past decade, studies using rodent muscles have indicated that Atrogin-1 and MuRF contribute to the protein degradation in muscular wasting [226]. More recent studies using human muscle do not necessarily support such a role for these atrogenes [262]. In contrast, myostatin functions to negatively regulate the hypertrophy of muscles, but a role in the induction of muscle loss was observed in muscle wasting diseases and cachexia associated with severe illness. The pharmacological inhibition by myostatin seems to occur mostly in animal models of muscle wasting (muscular dystrophy, cachexia, sarcopenia etc) [24, 25], but lacked efficacy in phase I clinical trials [344]. The mechanism of myostatin signaling is complex and involves the activation of several downstream pathways. In addition, more recent studies indicated crosstalk between a myostatin dependent-pathway and autophagy [440] or TNF-α/NF-κB signaling [441]. Future findings may lead to the development of novel therapeutics for use in a range of clinical conditions associated with muscle atrophy/wasting, metabolic disease and reduced mobility, and thus help to improve quality of life.

ACKNOWLEDGMENTS

This work was supported by a research Grant-in-Aid for Scientific Research C (No. 23500778) from the Ministry of Education, Culture, Sports, Science and Technology of Japan.

REFERENCES

[1]   R. S. Staron, M. J. Leonardi, D. L. Karapondo, E. S. Malicky, J. E. Falkel, F. C. Hagerman and R. S. Hikida, *J. Appl. Physiol.* 70, 631 (1991).

[2]   F. P. Moss and C. P. Leblond, *Anat. Rec.* 170, 421 (1971).

[3]   R. E. Allen and D. E. Goll, In: C. G. Scanes (ed.), *Biology of Growth of Domestic Animals*, pp. 148, Iowa State Press, Iowa (2003).

[4]   R. Bischoff, In: A. G. Engel and C. Franzini-Armstrong (ed.), *Myology, 2nd edn*, pp. 97, McGraw-Hill, New York, (1994).

[5]   A. Mauro, *J. Biophys. Biochem.* 9, 493 (1961).

[6]   K. M. McCormick and E. Schultz, *Dev. Biol.* 150, 319 (1992).

[7]   S. Schiaffino, S. P. Bormioli and M. Aloisi, *Virchows Arch. B Cell Pathol.* 21, 113 (1973).

[8]   D. L. Allen, R. R. Roy and V. R. Edgerton, *Muscle Nerve* 22, 1350 (1999).

[9]   R. R. Roy, S. R. Monke, D. L. Allen and V. R. Edgerton, *J. Appl. Physiol.* 87, 634 (1999).

[10]   J. D. Rosenblatt and D. J. Parry, *J. Appl. Physiol.* 73, 2538 (1992).

[11]   J. J. McCarthy and K. A. Esser, *J. Appl. Physiol.* 103, 1100 (2007).

[12]   R. S. Hikida, *J. Appl. Physiol.* 103, 1104 (2007).

[13]   D. J. Glass, *Int. J. Biochem. Cell Biol.* 37, 1974 (2005).

[14]   D. J. Glass, *Curr. Top. Microbiol. Immunol.* 346, 267 (2010).

[15]   S. E. Gordon, M. Flück and F. W. Booth, *J. Appl. Physiol.* 90, 1174 (2001).

[16]   K. Sakuma and A. Yamaguchi, In: S. G. Pandalai (ed.), *Recent. Res. Devel. Life Sci. 5th edn*, pp. 13, Research Signpost, Kerala, India (2012).

[17]   U. Cheema, R. Brown, V. Mudera, S. Y. Yang, G. McGrouther and G. Goldspink, *J. Cell. Physiol.* 202, 67 (2005).

[18]   K. Furuno, M. N. Goodman and A. L. Goldberg, *J. Biol. Chem.* 265, 8550 (1990).

[19]   S. Schiaffino and V. Hanzlikova, *J. Ultrastruct. Res.* 39, 1 (1972).

[20]   B. Levine and G. Kroemer, *Cell* 132, 27 (2008).

[21]   C. Mammucari, G. Milan, V. Romanello, E. Masiero, R. Rudolf, P. Del Piccolo, S. J. Burden, R. Di Lisi, C. Sandri, J. Zhao, A. L. Goldberg, S. Schiaffino and M. Sandri, *Cell Metab.* 6, 458 (2007).

[22]   J. Zhao, J. J. Brault, A. Schild, P. Cao, M. Sandri, S. Schiaffino, S. H. Lecker and A. L. Goldberg, *Cell Metab.* 6, 472 (2007).

[23]   G. Dobrowolny, M. Aucello, E. Rizzuto, S. Beccafico, C. Mammucari, S. Boncompagni, S. Belia, F. Wannenes, C. Nicoletti, Z. Del Prete, N. Rosenthal, M. Molinaro, F. Protasi, G. Fanò, M. Sandri and A. Musarò, *Cell Metab.* 8, 425 (2008).

[24]   K. Sakuma and A. Yamaguchi, *Recent Pat. Regen. Med.* 1, 284 (2011).

[25]   K. Tsuchida, *Curr. Opin. Drug Discov. Devel.* 11, 487 (2008).

[26]   C. B. Mantilla, R. V. Sill, B. Aravamudan, W. Z. Zhan and G. C. Sieck, *J. Appl. Physiol.* 104, 787 (2008).

[27]   R. S. O'Connor, G. K. Pavlath, J. J. McCarthy and K. A. Esser, *J. Appl. Physiol.* 103, 1107 (2007).

[28]   J. X. Liu, A. S. Hoglund, P. Karlsson, J. Lindblad, R. Qaisar, S. Aare, E. Bengtsson and L. Larsson, *Exp. Physiol.* 94, 117 (2009).

[29]   K. Gundersen and J. C. Bruusgaard, *J. Physiol.* 586, 2675 (2008).

[30]   J. C. Bruusgaard and K. Gundersen, *J. Clin. Invest.* 118, 1450 (2008).

[31]   H. Amthor, A. Otto, A. Vulin, A. Rochat, J. Dumonceaux, L. Garcia, E. Mouisel, C. Hourde, R. Macharia, M. Friedrichs, F. Relaix, P. S. Zammit, A. Matsakas, K. Patel and T. Partridge, *Proc. Natl. Acad. Sci. U. S. A.* 106, 7479 (2009).

[32] E. R. Barton-Davis, D. I. Shoturma and H. L. Sweeney, *Acta Physiol. Scand.* 167, 301 (1999).

[33] D. A. Lowe and S. E. Alway, *Cell Tissue Res.* 296, 531 (1999).

[34] J. C. Bruusgaard, I. B. Johansen, I. M. Egner, Z. A. Rana and K. Gundersen, *Proc. Natl. Acad. Sci. U. S. A.* 107, 15111 (2010).

[35] R. Bischoff, *Dev. Biol.* 115, 140 (1986).

[36] R. Bischoff, *J. Cell Biol.* 111, 201 (1990).

[37] R. Bischoff, *Dev. Biol.* 115, 129 (1986).

[38] S. E. Johnson and R. E. Allen, *Exp. Cell Res.* 219, 449 (1995).

[39] R. E. Allen, C. J. Temm-Grove, S. M. Sheehan and G. M. Rice, *Methods Cell Biol.* 52, 155 (1997).

[40] R. Tatsumi, J. E. Anderson, C. J. Nevoret, O. Halevy and R. E. Allen, *Dev. Biol.* 195, 114 (1998).

[41] J. E. Anderson, *Mol. Biol. Cell* 11, 1859 (2000).

[42] J. E. Brenman, D. S. Chao, H. Xia, K. Aldape and D. S. Bredt, *Cell* 82, 743 (1995).

[43] J. G. Tidball, E. Lavergne, K. S. Lau, M. J. Spencer, J. T. Stull and M. Wehling, *Am. J. Physiol. Cell Physiol.* 275, C260 (1998).

[44] K. J. Miller, D. Thaloor, S. Matteson and G. K. Pavlath, *Am. J. Physiol. Cell Physiol.* 278, C174 (2000).

[45] M. Yamada, R. Tatsumi, K. Yamanouchi, T. Hosoyama, S. Shiratsuchi, A. Sato, W. Mizunoya, Y. Ikeuchi, M. Furuse and R. E. Allen, *Am. J. Physiol. Cell Physiol.* 298, C465 (2010).

[46] A. Yamaguchi, H. Ishii, I. Morita, I. Oota and H. Takeda, *Pflügers Arch.* 448, 539 (2004).

[47] Y. Tanaka, A. Yamaguchi, T. Fujikawa, K. Sakuma, I. Morita and K. Ishii, *Acta Physiol. (Oxf.)* 194, 149 (2008).

[48] B. Shelmadine, M. Cooke, T. Buford, G. Hudson, L. Redd, B. Leutholtz and D. S. Willoughby, *J. Int. Soc. Sports Nutr.* 6, 16 (2009).

[49] R. Tatchum-Talom, R. Schulz, J. R. McNeill and F. H. Khadour, *Am. J. Physiol. Heart Circ. Physiol.* 279, H1757 (2000).

[50] E. Lima-Cabello, M. J. Cuevas, N. Garatachea, M. Baldini, M. Almar and J. González-Gallego, *J. Appl. Physiol.* 108, 575 (2010).

[51] G. K. McConell, S. J. Bradley, T. J. Stephens, B. J. Canny, B. A. Kingwell and R. S. Lee-Young, *Am. J. Physiol. Regul. Integr. Comp. Physiol.* 293, R821 (2007).

[52] C. Broholm, O. H. Mortensen, S. Nielsen, T. Akerstrom, A. Zankari, B. Dahl and B. K. Pedersen, *J. Physiol.* 586, 2195 (2008).

[53] D. Metcalf, *Stem Cells* 21, 5 (2003).

[54] M. G. Hinds, T. Mauer, J. G. Zhang, N. A. Nicola and R. S. Norton, *J. Biomed. NMR* 9, 113 (1997).

[55] C. H. Schmelzer, L. E. Burton and C. M. Tamony, *Protein Expr. Purif.* 1, 54 (1990).

[56] K. Kami and E. Semba, *Muscle Nerve* 21, 819 (1998).

[57] K. Sakuma, K. Watanabe, M. Sano, I. Uramoto and T. Totsuka, *Biochm. Biophys. Acta Mol. Cell Res.* 1497, 77 (2000).

[58] P. Gregorevic, D. A. Williams and G. S. Lynch, *Muscle Nerve* 25, 194 (2002).

[59] E. E. Spangenberg, and F. W. Booth, *Cytokine* 34, 125 (2006).

[60] L. Austin and A. W. Burgss, *J. Neurol. Sci.* 101, 193 (1991).

[61] Y. Diao, X. Wang and Z. Wu, *Mol. Cell. Biol.* 29, 5084 (2009).

[62] L. Sun, K. Ma, H. Wang, F. Xiao, Y. Gao, W. Zhang, K. Wang, X. Gao, N. Ip and Z. Wu, *J. Cell Biol.* 179, 129 (2007).

[63] C. Broholm and B. K. Pedersen, *Exerc. Immunol. Rev.* 16, 77 (2010).

[64] G. R. Adams and S. A. McCue, *J. Appl. Physiol.* 84, 1716 (1998).

[65] M. V. Chakravarthy, B. S. Davis and F. W. Booth, *J. Appl. Physiol.* 89, 1365 (2000).

[66] C. Rommel, S. C. Bodine, B. A. Clarke, R. Rossman, L. Nunez, T. N. Stitt, G. D. Yancopoulos and D. J. Glass, *Nat. Cell Biol.* 3, 1009 (2001).

[67] D. R. Vyas, E. E. Spangenburg, T. W. Abraha, T. E. Childs and F. W. Booth, *Am. J. Physiol. Cell Physiol.* 283, C545 (2002).

[68] A. Musaró, K. McCullagh, A. Paul, L. Houghton, G. Dobrowolny, M. Molinaro, E. R. Barton, H. L. Sweeney and N. Rosenthal, *Nat. Genet.* 27, 195 (2001).

[69] D. L. Devol, P. Rotwein, J. L. Sadow, J. Novakofski and P. J. Bechtel, *Am. J. Physiol.* 259, E89 (1990).

[70] K. Sakuma, K. Watanabe, T. Totsuka, I. Uramoto, M. Sano and K. Sakamoto, *Acta Neuropathol. (Berl.)* 95, 123 (1998).

[71] M. M. Bamman, J. K. Petrella, J. S. Kim, D. L. Mayhew and J. M. Cross, *J. Appl. Physiol.* 102, 2232 (2007).

[72] C. D. Willborn, L. W. Taylor, M. Greenwood, R. B. Kreider and D. S. Willoughby, *J. Strength Cond. Res.* 23, 2179 (2009).

[73] G. R. Adams, *J. Appl. Physiol.* 93, 1159 (2002).

[74] S. Adi, B. Bin-Abbas, N. Y. Wu and S. M. Rosenthal, *Endocrinology* 143, 511 (2002).

[75] V. Jacquemin, G. S. Butler-Browne, D. Furling and V. Mouly, *J. Cell Sci.* 120, 670 (2007).

[76] J. Tureckova, E. M. Wilson, J. L. Cappalonga and P. Rotwein, *J. Biol. Chem.* 276, 39264 (2001).

[77] D. R. Clemmons, *Trends Endocrinol. Metabol.* 20, 349 (2009).

[78] K. Gundersen, *Biol. Rev. Camb. Philos. Soc.* 86, 564 (2011).

[79] A. Philippou, M. Maridaki, A. Halapas and M. Koutsilieris, *In Vivo* 21, 45 (2007).

[80] M. Miyazaki, J. J. McCarthy, M. J. Fedele and K. A. Esser, *J. Physiol.* 589, 183 (2011).

[81] E. E. Spangenburg, D. L. Roith, C. W. Ward and S. C. Bodine, *J. Physiol.* 586, 283 (2008).

[82] F. Relaix, D. Montarras, S. Zaffran, B. Gayraud-Morel, D. Rocancourt, S. Tajbakhsh, A. Mansouri, A. Cumano and M. Buckingham, *J. Cell Biol.* 172, 91 (2006).

[83] H. Hirai, M. Verma, S. Watanabe, C. Tastad, Y. Asakura and A. Asakura, *J. Cell Biol.* 191, 347 (2010).

[84] K. G. Franchini, A. S. Torsoni, P. H. A. Soares and M. J. A. Saad, *Circ. Res.* 87, 558 (2000).

[85] K. Sakamoto, M. F. Hirshman, W. G. Aschenbach and L. J. Goodyear, *J. Biol. Chem.* 277, 11910 (2002).

[86] D. A. E. Cross, D. R. Alessi, P. Cohen, M. Andjelkovich and B. A. Hemmings, *Nature* 378, 785 (1995).

[87] B. D. Manning, A. R. Tee, M. N. Logsdon, J. Blenis and L. C. Cantley, *Mol. Cell* 10, 151 (2002).

[88] L. S. Jefferson, J. R. Fabian and S. R. Kimball, *Int. J. Biochem. Cell Biol.* 31, 191 (1999).

[89] T. N. Stitt, D. Drujan, B. A. Clarke, F. Panaro, Y. Timofeyva, W. O. Kline, M. Gonzalez, G. D. Yancopoulos and D. J. Glass, *Mol. Cell* 14, 395 (2004).

[90] H. Cho, J. L. Thorvaldsen, Q. Chu, F. Feng and M. J. Birnbaum, *J. Biol. Chem.* 276, 38349 (2001).

[91] W. S. Chen, P. Z. Xu, K. Gottlob, M. L. Chen, K. Sokol, T. Shiyanova, I. Roninson, W. Weng, R. Suzuki, K. Tobe, T. Kadowaki and N. Hay, *Genes Dev.* 15, 2203 (2001).

[92] H. Cho, J. Mu, J. K. Kim, J. L. Thorvaldsen, Q. Chu, E. B. Crenshaw, K. H. Kaestner, M. S. Bartolomei, G. I. Shulman and M. J. Birnbaum, *Science* 292, 1728 (2001).

[93] X. D. Peng, P. Z. Xu, M. L. Chen, A. Hahn-Windgassen, J. Skeen, J. Jacobs, D. Sundararajan, W. S. Chen, S. E. Crawford, K. G. Coleman and N. Hay, *Genes Dev.* 17, 1352 (2003).

[94] K. M. Lai, M. Gonzalez, W. T. Poueymirou, W. O. Kline, E. Na, E. Zlotchenko, T. N. Stitt, A. N. Economides, G. D. Yancopoulos and D. J. Glass, *Mol. Cell. Biol.* 24, 9295 (2004).

[95] Y. Izumiya, T. Hopkins, C. Morris, K. Sato, L. Zeng, J. Viereck, J. A. Hamilton, N. Ouchi, N. K. Lebrasseur and K. Walsh, *Cell Metab.* 7, 159 (2008).

[96] B. Blaauw, M. Canato, L. Agatea, L. Toniolo, C. Mammucari, E. Masiero, R. Abraham, M. Sandri, S. Schiaffino and C. Reggiani, *FASEB J.* 23, 3896 (2009).

[97] R. Sartori, G. Milan, M. Patron, C. Mammucari, B. Blaauw, R. Abraham and M. Sandri, *Am. J. Physiol. Cell Physiol.* 296, C1248 (2009).

[98] B. Blaauw, C. Mammucari, L. Toniolo, L. Agatea, R. Abraham, M. Sandri, C. Reggiani and S. Schiaffino, *Hum. Mol. Genet.* 17, 3686 (2008).

[99] E. Jacinto, R. Loewith, A. Schmidt, S. Lin, M. A. Ruegg, A. Hall and M. N. Hall, *Nat. Cell Biol.* 6, 1122 (2004).

[100] R. Zoncu, A. Efeyan and D. M. Sabatini, *Nat. Rev. Mol. Cell Biol.* 12, 21 (2011).

[101] E. A. Dunlop and A. R. Tee, *Cell. Signal.* 21, 827 (2009).

[102] M. J. Drummond, C. S. Fry, E. L. Glynn, H. C. Dreyer, S. Dhanani, K. L. Timmerman, E. Volpi and B. B. Rasmussen, *J. Physiol.* 587, 1535 (2009).

[103] E. I. Glover, B. R. Oates, J. E. Tang, D. R. Moore, M. A. Tarnopolsky and S. M. Phillips, *Am. J. Physiol. Regul. Integr. Comp. Physiol.* 295, R604 (2008).

[104] T. A. Hornberger, T. J. McLoughlin, J. K. Leszczynski, D. D. Armstrong, R. R. Jameson, P. E. Bowen, E. S. Hwang, H. Hou, M. E. Moustafa, B. A. Carlson, D. L. Hatfield, A. M. Diamond and K. A. Esser, *J. Nutr.* 133, 3091 (2003).

[105] S. C. Bodine, T. N. Stitt, M. Gonzalez, W. O. Kline, G. L. Stover, R. Bauerlein, E. Zlotchenko, A. Scrimgeour, J. C. Lawrence, D. J. Glass and G. D. Yancopoulos, *Nat. Cell Biol.* 3, 1014 (2001).

[106] M. Sandri, *Physiology* 23, 160 (2008).

[107] R. A. Frost and C. H. Rang, *J. Appl. Physiol.* 103, 378 (2007).

[108] D. L. Hamilton, M. G. MacKenzie and K. R. Baar, In: J. Magalhaes and A. Ascensao (ed.), *Muscle plasticity – Advances in Biochemical and Physiological Research*, pp. 45, Research Signpost, Kerala, India (2009).

[109] C. A. Goodman, M. H. Miu, J. W. Frey, D. M. Mabrey, H. C. Lincoln, Y. Ge, J. Chen and T. A. Hornberger, *Mol. Biol. Cell* 21, 3258 (2010).

[110] C. A. Goodman, D. M. Mabrey, J. W. Frey, M. H. Miu, E. K. Schmidt, P. Pierre and T. A. Hornberger, *FASEB J.* 25, 1028 (2011).

[111] T. A. Hornberger, R. Stuppard, K. E. Conley, M. J. Fedele, M. L. Fiorotto, E. R. Chin and K. A. Esser, *Biochem. J.* 380, 795 (2004).

[112] B. Can, Y. Yoo and J. L. Guan, *J. Biol. Chem.* 281, 37321 (2006).

[113] V. G. Coffey, Z. Zhong, A. Shield, B. J. Canny, A. V. Chibalin, J. R. Zierath and J. A. Hawley, *FASEB J.* 20, 190 (2006).

[114] T. A. Hornberger, W. K. Chu, Y. W. Mak, J. W. Hsiung, S. A. Huang and S. Chien, *Proc. Natl. Acad. Sci. U. S. A.* 103, 4741 (2006).

[115] T. K. O'Neil, L. R. Duffy, J. W. Frey and T. A. Hornberger, *J. Physiol.* 587, 3691 (2009).

[116] Y. Sun, Y. Fang, M. S. Yoon, C. Zhang, M. Roccio, F. J. Zwartkruis, M. Armstrong, H. A. Brown and Wittinghofer, *Proc. Natl. Acad. Sci. U. S. A.* 105, 8286 (2008).

[117] M. P. Byfield, J. T. Murray and J. M. Backer, *J. Biol. Chem.* 280, 33076 (2005).

[118] M. G. MacKenzie, D. L. Hamilton, J. T. Murray, P. M. Taylor and K. Baar, *J. Physiol.* 587, 253 (2009).

[119] E. Kim, P. Goraksha-Hicks, L. Li, T. P. Neufeld and K. L. Guan, *Nat. Cell Biol.* 10, 935 (2008).

[120] Y. Sankak, T. R. Peterson, Y. D. Shaul, R. A. Rindquist, C. C. Thoreen, L. Bar-Peled and D. M. Sabatini, *Science* 320, 1496 (2008).

[121] M. Miyazaki and K. A. Esser, *J. Appl. Physiol.* 106, 1367 (2009).

[122] S. Lange, F. Xiang, A. Yakovenko, A. Vihola, P. Hackman, E. Rostkova, J. Kristensen, B. Brandmeier, G. Franzen, B. Hedberg, L. G. Gunnarsson, S. M. Hughes, S. Marchand, T. Sejersen, I. Richard, L. Edström, E. Ehler, B. Udd and M. Gautel, *Science* 308, 1599 (2005).

[123] R. Treisman, *EMBO J.* 6, 2711 (1987).

[124] G. E. Muscat, T. A. Gustafson and L. Kedes, *Mol. Cell. Biol.* 8, 4120 (1988).

[125] H. Ernst, K. Walsh, C. A. Harrison and N. Rosenthal, *Mol. Cell. Biol.* 11, 3735 (1991).

[126] G. C. Pipes, E. E. Creemers, and E. N. Olson, *Genes Dev.* 20, 1545 (2006).

[127] C. Charvet, C. Houbron, A. Parlakian, J. Giordani, C. Lahoute, A. Bertrand, A. Sotiropoulos, L. Renou, A. Schmitt, J. Melki, Z. Li, D. Daegelen and D. Tuil. *Mol. Cell. Biol.* 26, 6664 (2006).

[128] K. Sakuma, J. Nishikawa, R. Nakao, H. Nakano, M. Sano, K. Watanabe and T. Totsuka, *Histochem. Cell Biol.* 119, 149 (2003).

[129] C. Gauthier-Rouviére, M. Vandromme, D. Tuil, N. Lautredou, M. Morris, M. Soulez, A. Kahn, A. Fernandez and N. Lamb, *Mol. Biol. Cell* 7, 719 (1996).

[130] K. Sakuma, R. Nakao, S. Inashima, M. Hirata, T. Kubo and M. Yasuhara, *Acta Neuropathol. (Berl.)* 108, 241 (2004).

[131] K. Sakuma, M. Akiho, H. Nakashima, H. Akima and M. Yasuhara, *Biochim. Biophys. Acta Mol. Basis Dis.* 1782, 453 (2008).

[132] C. Lahoute, A. Sotiropoulos, M. Favier, I. Guillet-Deniau, C. Charvet, A. Ferry, G. Butler-Browne, D. Metzger, D. Tuil and D. Daegelen, *PLoS One* 3, e3910 (2008).

[133] A. Arai, J. A. Spencer and E. N. Olson, *J. Biol. Chem.* 277, 24453 (2002).

[134] K. Kuwahara, T. Barrientos, G. C. Pipes, S. Li and E. N. Olson, *Mol. Cell. Biol.* 25, 3173 (2005).

[135] E. M. Puchner, A. Alexandrovich, A. L. Kho, U. Hensen, L. V. Schäfer, B. Brandmeier, F. Gräter, H. Grubmüller, H. E. Gaub and M. Gautel, *Proc. Natl. Acad. Sci. U. S. A.* 105, 13385 (2008).

[136] J. Ochala, A. M. Gustafson, M. L. Diez, G. Renaud, M. Li, S. Aare, R. Qaisar, V. C. Banduseela, Y. Hedström, X. Tang, B. Dworkin, G. C. Ford, K. S. Nair, S. Perera, M. Gautel and L. Larsson, *J. Physiol.* 589, 2007 (2011).

[137] J. A. Spencer, S. Eliazer, R. L. Ilaria Jr, J. A. Richardson and E. N. Olson, *J. Cell Biol.* 150, 771 (2000)

[138] F. Miralles, G. Posern, A. I. Zaromytidou and R. Treisman, *Cell* 113, 329 (2003).

[139] K. Kuwahara, G. C. Teg Pipes, J. McAnally, J. A. Richardson, J. A. Hill, R. Bassel-Duby and E. N. Olson, *J. Clin. Invest.* 117, 1324 (2007).

[140] J. A. Carson and L. Wei, *J. Appl. Physiol.* 88, 337 (2000).

[141] M. Flück, J. A. Carson, R. J. Schwartz and F. W. Booth, *J. Appl. Physiol.* 86, 1793 (1999).

[142] S. Lamon, M. A. Wallace, B. Léger and A. P. Russell, *J. Physiol.* 587, 1795 (2009).

[143] D. L. Allen, C. A. Sartorius, L. K. Sycuro and L. A. Leinwand, *J. Biol. Chem.* 276, 43254 (2001).

[144] W. J. Lee, R. W. Thompson, J. M. McClung and J. A. Carson, *Am. J. Physiol. Regul. Integr. Comp. Physiol.* 285, R1076 (2003).

[145] S. Vlahopoulos, W. E. Zimmer, G. Jenster, N. S. Belaguli, S. P. Bals, A. O. Brinkmann, R. B. Lanz, V. C. Zoumpourlis and R. J. Schwartz, *J. Biol. Chem.* 260, 7786 (2005).

[146] S. E. Dunn, J. L. Burns and R. N. Michel, *J. Biol. Chem.* 274, 21908 (1999).

[147] R. N. Michel, E. R. Chin, J. V. Chakkalakal, J. K. Eibl and B. J. Jasmin, *Appl. Physiol. Nutr. Metab.* 32, 921 (2007).

[148] B. B. Friday, P. O. Mitchell, K. M. Kegley and G. K. Pavlath, *Differentiation* 71, 217 (2003).

[149] N. Al-Shanti and C. E. Stewart, *Biol. Rev. Camb. Philos. Soc.* 84, 637 (2009).

[150] K. Sakuma, R. Nakao, W. Aoi, S. Inashima, T. Fujikawa, M. Hirata, M. Sano and M. Yasuhara, *Acta Neuropathol. (Berl.)* 110, 269 (2005).

[151] K. Sakuma and A. Yamaguchi, *J. Biomed. Biotechnol.* 2010, Article ID721219 (2010).

[152] U. Delling, J. Tureckova, H. W. Lim, L. J. De Windt, P. Rotwein and J. D. Molkentin, *Mol. Cell. Biol.* 20, 6600 (2000).

[153] R. N. Michel, S. E. Dunn and E. R. Chin, *Proc. Nutr. Soc.* 63, 341 (2004).

[154] M. Miyazaki, Y. Hitomi, T. Kizaki, H. Ohno, T. Katsumura, S. Haga and T. Takemasa, *Med. Sci. Sports Exerc.* 38, 1065 (2006).

[155] Y. Oishi, T. Ogata, K. I. Yamamoto, M. Terada, T. Ohira, Y. Ohira, K. Taniguchi and R. R. Roy, *Acta Physiol. (Oxf.)* 192, 381 (2008).

[156] R. J. Talmadge, J. S. Otis, M. R. Rittler, N. D. Garcia, S. R. Spencer, S. J. Lees and F. J. Naya, *BMC Cell Biol.* 5, 28 (2004).

[157] A. Hennebry, C. Berry, V. Siriett, P. O'Callaghan, L. Chau, T. Watson, M. Sharma and R. Kambadur, *Am. J. Physiol. Cell Physiol.* 296, C525 (2009).

[158] K. Sakuma, M. Akiho, H. Nakashima, R. Nakao, M. Hirata, S. Inashima, A. Yamaguchi and M. Yasuhara, *Acta Neuropathol. (Berl.)* 115, 663 (2008).

[159] Y. Hinits and S. M. Hughes, *Development* 134, 2511 (2007).

[160] M. J. Potthoff, M. A. Arnold, J. McAnally, J. A. Richardson, R. Bassel-Duby and E. N. Olson, *Mol. Cell. Biol.* 27, 8143 (2007).

[161] P. M. Garcia-Roves, T. E. Jones, K. Otani, D. H. Han and J. O. Holloszy, *Diabetes* 54, 624 (2005).

[162] C. Grondard, O. Biondi, C. Pariset, P. Lopes, S. Deforges, S. Lécolle, B. D. Gaspera, C. L. Gallien, C. Chanoine and F. Charbonnier, *J. Cell. Physiol.* 214, 126 (2008).

[163] H. Wu, B. Rothermel, S. Kanatous, P. Rosenberg, F. J. Naya, J. M. Shelton, K. A. Hutcheson, J. M. DiMaio, E. N. Olson, R. Bassel-Duby and R. S. Williams, *EMBO J.* 20, 6414 (2001).

[164] L. Lamas, M. S. Aoki, C. Ugrinowitsch, G. E. Campos, M. Regazzini M, A. S. Moriscot and V. Tricoli, *Scand. J. Med. Sci. Sports* 20, 216 (2010).

[165] P. Puigserver, Z. Wu, C. W. Park, R. Graves, M. Wright and B. M. Spiegelman, *Cell* 92, 829 (1998).

[166] R. C. Scarpulla, *Physiol. Rev.* 88, 611 (2008).

[167] A. P. Russell, J. Feilchenfeldt, S. Schreiber, M. Praz, A. Crettenand, C. Gobelet, C. A. Meier, D. R. Bell, A. Kralli, J. P. Giacobino and O. Dériaz, *Diabetes* 52, 2874 (2003).

[168] W. Aoi, Y. Naito, K. Mizushima, Y. Takanami, Y. Kawai, H. Ichikawa and T. Yoshikawa, *Am. J. Physiol. Endocrinol. Metab.* 298, E799 (2010).

[169] M. Sandri, J. Lin, C. Handschin, W. Yang, Z. P. Arany, S. H. Lecker, A. L. Goldberg and B. M. Spiegelman, *Proc. Natl. Acad. Sci. U. S. A.* 103, 16260 (2006).

[170] J. J. Brault, J. G. Jespersen and A. L. Goldberg, *J. Biol. Chem.* 285, 19460 (2010).

[171] D. J. Baker, A. C. Betik, D. J. Krause and R. T. Hepple, *J. Gerontol. A Biol. Sci. Med. Sci.* 61, 675 (2006).

[172] S. Crunkhorn, F. Dearie, C. Mantzoros, H. Gami, W. S. da Silva, D. Espinoza, R. Faucette, K. Barry, A. C. Bianco and M. E. Patti, *J. Biol. Chem.* 282, 15439 (2007).

[173] T. K. Roberts-Wilson, R. N. Reddy, J. L. Bailey, B. Zheng, R. Ordas, J. L. Gooch and S. R. Price, *Biochim. Biophys. Acta Mol. Cell Res.* 1803, 960 (2010).

[174] G. Fuster, S. Busquets, E. Ametller, M. Olivan, V. Almendro, C. C. de Oliveira, M. Figueras, F. J. López-Soriano and J. M. Argilés, *Cancer Res.* 67, 6512 (2007).

[175] C. Handshin, S. Chin, P. Li, F. Liu, E. Maratos-Flier, N. K. Lebrasseur, Z. Yan and B. M. Spiegelman, *J. Biol. Chem.* 282, 30014 (2007).

[176] T. Wenz, S. G. Rossi, R. L. Rotundo, B. M. Spiegelman and C. T. Moraes, *Proc. Natl. Acad. Sci. U. S. A.* 106, 20405 (2009).

[177] J. Hanai, P. Cao, P. Tanksale, S. Imamura, E. Koshimizu, J. Zhao, S. Kishi, M. Yamashita, P. S. Phillips, V. P. Sukhatme and S. H. Lecker, *J. Clin. Invest.* 117, 3940 (2007).

[178] P. O. Hasselgren, *J. Cell. Physiol.* 213, 679 (2007).

[179] S. Miura, Y. Kai, M. Ono and O. Ezaki, *J. Biol. Chem.* 278, 31385 (2003).

[180] S. Miura, E. Tomitsuka, Y. Kamei, T. Yamazaki, Y. Kai, M. Tamura, K. Kita, I. Nishino and O. Ezaki, *Am. J. Pathol.* 169, 1129 (2006).

[181] D. Yin, W. Gao, J. D. Kearbey, H. Xu, K. Chung, Y. He, C. A. Marhefka, K. A. Veverka, D. D. Miller and J. T. Dalton, *J. Pharmacol. Exp. Ther.* 304, 1334 (2003).

[182] S. Bhasin, L. Woodhouse and T. W. Storer, *Growth Horm. IGF Res.* 13, S63 (2003).

[183] S. Bhasin, T. W. Storer, N. Berman, C. Callegari, B. Clevenger, J. Phillips, T. J. Bunnell, R. Tricker, A. Shirazi and R. Casaburi, *N. Engl. J. Med.* 335, 1 (1996).

[184] S. Bhasin, L. Woodhouse, R. Casaburi, A. B. Singh, R. P. Mac, M. Lee, K. E. Yarasheski, I. Sinha-Hikim, C. Dzekov, J. Dzekov, L. Magliano and T. W. Storer, *J. Clin. Endocrinol. Metab.* 90, 678 (2005).

[185] I. Sinha-Hikim, J. Artaza, L. Woodhouse, N. Gonzalez-Cadavid, A. B. Singh, M. I. Lee, T. W. Storer, R. Casaburi, R. Shen and S. Bhasin, *Am. J. Physiol. Endocrinol. Metab.* 283, E154 (2002).

[186] H. E. MacLean and D. J. Handelsman, *Endocrinology* 150, 3437 (2009).

[187] P. M. Cawthon, K. E. Ensrud, G. A. Laughlin, J. A. Cauley, T. T. Dam, E. Barrett-Connor, H. A. Fink, A. R. Hoffman, E. Lau, N. E. Lane, M. L. Stefanick, S. R. Cummings and E. S. Orwoll, *J. Clin. Endocrinol. Metab.* 94, 3806 (2009).

[188] U. Srinivas-Shankar, S. A. Roberts, M. J. Connolly, M. D. O'Connell, J. E. Adams, J. A. Oldham and F. C. Wu, *J. Clin. Endocrinol. Metab.* 95, 639 (2010).

[189] F. Claessens, S. Denayer, N. Van Tilborgh, S. Kerkhofs, C. Helsen and A. Haelens, *Nucl. Recept. Signal* 6, e008 (2008).

[190] A. Wyce, Y. Bai, S. Nagpal and C. C. Thompson, *Mol. Endocrinol.* 24, 1665 (2010).

[191] S. Vlahopoulos, W. E. Zimmer, G. Jenster, N. S. Belaguli, S. P. Balk, A. O. Brinkmann, R. B. Lanz, V. C. Zoumpourlis and R. J. Schwartz, *J. Biol. Chem.* 280, 7786 (2005).

[192] A. L. Amir, M. Barua, N. C. McKnight, S. Cheng, X. Yuan and S. P. Balk, *J. Biol. Chem.* 278, 30828 (2003).

[193] B. S. Keenan, G. E. Richards, S. W. Ponder, J. S. Dallas, M. Nagamani and E. R. Smith, *J. Clin. Endocrinol. Metab.* 76, 996 (1993).

[194] A. Ulloa-Aguirre, R. M. Blizzard, E. Garcia-Rubi, A. D. Rogol, K. Link, C. M. Christie, M. L. Johnson and J. D. Veldhuis, *J. Clin. Endocrinol. Metab.* 71, 846 (1990).

[195] S. Bhasin, L. Woodhouse, R. Casaburi, A. B. Singh, D. Bhasin, N. Berman, X. Chen, K. E. Yarasheski, L. Magliano, C. Dzekov, J. Dzekov, R. Bross, J. Phillips, I. Sinha-Hikim, R. Shen and T. W. Storer, *Am. J. Physiol. Endocrinol. Metab.* 281, E1172 (2001).

[196] K. Raskin, K. De Gendt, A. Duittoz, P. Liere, G. Verhoeven, F. Tronche and S. Mhaouty-Kodja, *J. Neurosci.* 29, 4461 (2009).

[197] C. Serra, S. Bhasin, F. Tangherlini, E. R. Barton, M. Ganno, A. Zhang, J. Shansky, H. H. Vandenburgh, T. G. Travison, R. Jasuja and C. Morris, *Endocrinology* 152, 193 (2011).

[198] T. Xu, Y. Shen, H. Pink, J. Triantafillou, S. A. Stimpson, P. Turnbull and B. Han, *J. Steroid Biochem. Mol. Biol.* 92, 447 (2004).

[199] A. C. McPherron, A. M. Lawler and S. J. Lee, *Nature* 387, 83 (1997).

[200] S. Reisz-Porszasz, S. Bhasin, J. N. Artaza, R. Shen, I. Sinha-Hikim, A. Hogue, T. J. Fielder and N. F. Gonzalez-Cadavid, *Am. J. Physiol. Endocrinol. Metab.* 285, E876 (2003).

[201] R. Singh, S. Bhasin, M. Braga, J. N. Artaza, S. Pervin, W. E. Taylor, V. Krishnan, S. K. Sinha, T. B. Rajavashisth and R. Jasuja, *Endocrinology* 150, 1259 (2009).

[202] J. X. Zhao, J. Hu, M. J. Zhu and M. Du, *Domest. Anim. Endocrinol.* 40, 222 (2011).

[203] H. Amthor, G. Nicholas, I. McKinnell, C. F. Kemp, M. Sharma, E. Kambadur and K. Patel, *Dev. Biol.* 279, 19 (2004).

[204] H. Gilson, O. Schakman, S. Kalista, P. Lause, K. Tsuchida and J. P. Thissen, *Am. J. Physiol. Endocrinol. Metab.* 297, E157 (2009).

[205] I. M. Conboy and T. A. Rando, *Dev. Cell* 3, 397 (2002).

[206] V. Dubois, M. Laurent, S. Boonen, D. Vanderschueren and F. Claessens, *Cell. Mol. Life Sci.* 69, 1651 (2012).

[207] D. Brown, A. P. Hikim, E. L. Kovacheva and I. Sinha-Hikim, *J. Endocrinol.* 201, 129 (2009).

[208] E. L. Kovacheva, A. P. Hikim, R. Shen, I. Sinha and I. Sinha-Hikim, *Endocrinology* 151, 628 (2010).

[209] M. F. Buas and T. Kadesch, *Exp. Cell Res.* 316, 3028 (2010).

[210] S. Tsivitse, *Int. J. Biol. Sci.* 6, 268 (2010).

[211] K. Y. Ho, J. D. Veldhuis, M. L. Johnson, R. Furlanetto, W. S. Evans, K. G. Alberti and M. O. Thorner, *J. Clin. Invest.* 81, 968 (1988).

[212] A. Giustina, G. Mazziotti and E. Canalis, *Endocr. Rev.* 29, 535 (2008).

[213] A. Moran, D. R. J. Jacobs, J. Steinberger, P. Cohen, C. Hong, R. Prineas and A. R. Sinaiko, *J. Clin. Endocrinol. Metab.* 87, 4817 (2002).

[214] J. G. Ryall, J. D. Schertzer and G. S. Lynch, *Biogerontology* 9, 213 (2008).

[215] J. R. Florini, D. Z. Ewton and S. A. Coolican, *Endocr. Rev.* 17, 481 (1996).

[216] D. Le Roith, C. Bondy, S. Yakar, J. L. Liu and A. Butler, *Endocr. Rev.* 22, 53 (2001).

[217] R. A. Frost, G. J. Nystrom and C. H. Lang, *Endocrinology* 143, 492 (2002).

[218] S. Perrini, A. Natalicchio, L. Laviola, A. Cignarelli, M. Melchiorre, E. De Stefano, C. Caccioppoli, A. Leonardini, S. Martemucci, G. Belsanti, S. Miccoli, A. Ciampolillo, A. Corrado, F. P. Cantatore, R. Giorgino and F. Giorgino, *Endocrinology* 149, 1302 (2008).

[219] J. A. Hansen, K. Lindberg, D. J. Hilton, J. H. Nielsen and N. Billestrup, *Mol. Endocrinol.* 13, 1832 (1999).

[220] A. Sotiropoulos, M. Ohanna, C. Kedzia, R. K. Menon, J. J. Kopchick, P. A. Kelly and M. Pende, *Proc. Natl. Acad. Sci. U. S. A.* 103, 7315 (2006).

[221] M. Kadowaki and T. Kanazawa, *J. Nutr.* 133, 2052S (2003).

[222] P. Grumati, L. Coletto, P. Sabatelli, M. Cescon, A. Angelin, E. Bertaggia, B. Blaauw, A. Urciuolo, T. Tiepolo, L. Merlini, N. M. Maraldi, P. Bernardi, M. Sandri and P. Bonald, *Nat. Med.* 16, 1313 (2010).

[223] G. Marino, J. A. Uria, X. S. Puente, V. Quesada, J. Bordallo and C. Lopez-Otin, *J. Biol. Chem.* 278, 3671 (2003).

[224] T. Ogata, Y. Oishi, M. Higuchi and I. Muraoka, *Biochem. Biophys. Res. Commun.* 394, 136 (2010).

[225] M. F. N. O'Leary and D. A. Hood, *Autophagy* 5, 230 (2009).

[226] S. C. Bodine, E. Latres, S. Baumhueter, V. K. Lai, L. Nunez, B. A. Clarke, W. T. Poueymirou, F. J. Panaro, E. Na, K. Dharmarajan, Z. Q. Pan, D. M. Valenzuela, T. M. DeChiara, T. N. Stitt, G. D. Yancopoulos and D. J. Glass, *Science* 294, 1704 (2001).

[227] P. Costelli, P. Reffo, F. Penna, R. Autelli, G. Bonelli and F. M. Baccino, *Int. J. Biochem. Cell Biol.* 37, 2134 (2005).

[228] I. J. Smith and S. L. Dodd, *Exp. Physiol.* 92, 561 (2007).

[229] J. Du, X. Wang, C. Miereles, J. L. Bailey, R. Debigare and B. Zheng, *J. Clin. Invest.* 113, 115 (2004).

[230] V. Moresi, A. Presterá, B. M. Scicchitano, M. Molinaro, L. Teodori, D. Sassoon, S. Adamo and D. Coletti, *Stem Cells*, 26, 997 (2008).

[231] A. Hershko and A. Ciechanover, *Annu. Rev. Biochem.* 51, 335 (1982).

[232] M. H. Glickman and A. Ciechanover, *Physiol. Rev.* 82, 373 (2002).

[233] D. Attaix, E. Aurousseau, L. Combaret, A. Kee, D. Larbaud, C. Ralliére, B. Souweine, D. Taillandier and T. Tilignac, *Reprod. Nutr. Dev.* 38, 153 (1998).

[234] J. Khal, A. V. Hine, K. C. H. Fearon, C. H. C. Dejong and M. J. Tisdale, *Int. J. Biochem. Cell Biol.* 37, 2196 (2005).

[235] J. Khal, S. M. Wyke, S. T. Russell, A. V. Hine and M. J. Tisdale, *Br. J. Cancer* 93, 774 (2005).

[236] M. J. Tisdale, *J. Support. Oncol.* 3, 209 (2005).

[237] K. S. Kwak, X. Zhou, V. Solomon, V. E. Baracos, J. Davis, A. W. Bannon, W. J. Boyle, D. L. Lacey and H. Q. Han, *Cancer Res.* 64, 8193 (2004).

[238] D. Cai, J. D. Frantz, N. E. Tawa Jr, P. A. Melendez, B. C. Oh, H. G. Lidov, P. O. Hasselgren, W. R. Frontera, J. Lee, D. J. Glass and S. E. Shoelson, *Cell* 119, 285, (2004).

[239] M. Sandri, C. Sandri, A. Gilbert, C. Skurk, E. Calabria, A. Picard, K. Walsh, S. Schiaffino, S. H. Lecker and A. L. Goldberg, *Cell* 117, 399 (2004).

[240] R. J. Deshaies, *Annu. Rev. Cell Dev. Biol.* 15, 435 (1999).

[241] O. Coux, K. Tanaka and A. L. Goldberg, *Annu. Rev. Biochem.* 65, 801 (1996).

[242] A. Rivett, *Biochem. J.* 291, 1 (1993).

[243] W. Baumeister, J. Walz, F. Zühl and E. Seemüller, *Cell 92, 367* (1998).

[244] M. D. Gomes, S. H. Lecker, R. T. Jagoe, A. Navon and A. L. Goldberg, *Proc. Natl. Acad. Sci. U. S. A.* 98, 14440 (2001).

[245] A. Csibi, L. A. Tintignac, M. P. Leibovitch and S. A. Leibovitch, *Cell Cycle* 7, 1698 (2008).

[246] C. A. Joazeiro and A. M. Weissman, *Cell* 102, 549 (2000).

[247] J. A. Spencer, S. Eliazer, R. L. Ilaria Jr., J. A. Richardson and E. N. Olson, *J. Cell Biol.* 150, 771 (2000).

[248] T. Centner, J. Yano, E. Kimura, A. S. McElhinny, K. Pelin, C. C. Witt, M. -L. Bang, K. Trombitas, H. Granzier, C. C. Gregorio, H. Sorimachi and S. Labeit, *J. Mol. Biol.* 306, 717 (2001).

[249] S. H. Lecker, R. T. Jagoe, A. Gilbert, M. Gomes, V. Baracos, V. Bailey, S. R. Price, W. E. Mitch and A. L. Goldberg, *FASEB J.* 18, 39 (2004).

[250] E. Zhu, C. S. Sassoon, R. Nelson, H. T. Pham, L. Zhu, M. J. Baker and V. J. Caiozzo, *J. Appl. Physiol.* 99, 747 (2005).

[251] J. Lagirand-Cantaloube, N. Offner, A. Csibi, M. P. Leibovitch, S. Batonnet-Pichon, L. A. Tintignac, C. T. Segura and S. A. Leibovitch, *EMBO J.* 27, 1266 (2008).

[252] H. H. Li, V. Kedar, C. Zhang, H. McDonough, R. Arya, D. Z. Wang and C. Patterson, *J. Clin. Invest.* 114, 1058 (2004).

[253] L. A. Tintignac, J. Lagirand, S. Batonnet, V. Sirri, M. P. Leibovitch and S. A. Leibovitch, *J. Biol. Chem.* 280, 2847 (2005).

[254] J. Lagirand-Cantaloube, K. Cornille, A. Csibi, S. Batonet-Pichon, M. P. Leibovitch and S. A. Leibovitch, *PLoS One* 4, e4973 (2009).

[255] M. Jogo, S. Shiraishi and T. A. Tamura, *FEBS Lett.* 583, 2715 (2009).

[256] S. Cohen, J. J. Brault, S. P. Gygi, D. J. Glass, D. M. Valenzuela, C. Gartner, E. Latres and A. L. Goldberg, *J. Cell Biol.* 185, 1083 (2009).

[257] J. Fielitz, M. S. Kim, J. M. Shelton, S. Latif, J. A. Spencer, D. J. Glass, J. A. Richardson, R. Bassel-Duby and E. N. Olson, *J. Clin. Invest.* 117, 2486 (2007).

[258] T. J. Zhao, Y. B. Yan, Y. Liu and H. M. Zhou, *J. Biol. Chem.* 282, 12022 (2007).

[259] M. Llovera, N. Carbó, C. García-Martínez, P. Costelli, L. Tessitore, F. M. Baccino, N. Agell, G. J. Bagby, F. J. López-Soriano and J. M. Argilés, *Biochem. Biophys. Res. Commun.* 221, 653 (1996).

[260] P. Costelli, M. Muscaritoli, M. Bossola, F. Penna, P. Reffo, A. Bonetto, S. Busquets, G. Bonelli, F. J. Lopez-Soriano, G. B. Doglietto, J. M. Argilés, F. M. Baccino and F. Rossi Fanelli, *Am. J. Physiol. Regul. Integr. Comp. Physiol.* 291, R674 (2006).

[261] P. K. Paul, S. K. Gupta, S. Bhatnagar, S. K. Panguluri, B. G. Darnay, Y. Choi and A. Kumar, *J. Cell Biol.* 191, 1395 (2010).

[262] V. C. Foletta, L. J. White, A. E. Larsen, B. Léger and A. P. Russell, *Pflügers Arch.* 461, 325 (2011).

[263] S. W. Jones, R. J. Hill, P. A. Krasney, B. O'Conner, N. Pierce and P. L. Greenhalf, *FASEB J.* 18, 1025 (2004).

[264] S. A. Whitman, M. J. Wacker, S. R. Richmond and M. P. Godard, *Pflügers Arch.* 450, 437 (2005).

[265] T. Ogawa, H. Furochi, M. Mameoka, K. Hirasaka, Y. Onishi, N. Suzue, M. Oarada, M. Akamatsu, H. Akima, T. Fukunaga, K. Kishi, N. Yasui, K. Ishidoh, H. Fukuoka and T. Nikawa, *Muscle Nerve* 34, 463 (2006).

[266] B. Léger, R. Cartoni, M. Praz, S. Lamon, O. Dériaz, A. Crettenand, C. Gobelet, P. Rohmer, M. Konzelmann, F. Luthi and A. P. Russell, *J. Physiol.* 576, 923 (2006).

[267] B. Léger, L. Vergani, G. Soraru, P. Hespel and A. P. Russell, *FASEB J.* 20, 583 (2006).

[268] M. L. Urso, Y. W. Chen, A. G. Scrimgeour, P. C. Lee, K. F. Lee and P. M. Clarkson, *J. Physiol.* 579, 877 (2007).

[269] M. Salanova, G. Schiffl, B. Püttmann, B. G. Schoser and D. Blottner, *J. Anat.* 212, 306 (2008).

[270] B. Léger, W. Derave, K. De Bock, P. Hespel and A. P. Russell, *Rejuvenat. Res.* 11, 163B (2008).

[271] B. Léger, R. Senese, A. W. Al-Khodairy, O. Deriaz, C. Gobelet, J. P. Giacobino and A. P. Russell, *Muscle Nerve* 40, 69 (2009).

[272] K. Sakuma, K. Watanabe, N. Hotta, T. Koike, K. Ishida, K. Katayama and H. Akima, *Acta Physiol. (Oxf.)* 197, 151 (2009).

[273] T. Gustafsson, T. Osterlund, J. N. Flanagan, F. Von Waldén, T. A. Trappe, R. M. Linnehan and P. A. Tesch, *J. Appl. Physiol.* 109, 721 (2010).

[274] P. Sundaram, Z. Pang, M. Miao, L. Yu, and S. S. Wing, *Am. J. Physiol. Endocrinol. Metab.* 297, E1283 (2009).

[275] B. Levine and D. J. Klionsky, *Dev. Cell* 6, 463 (2004).

[276] A. J. Meijer and P. Codogno, *Int. J. Biochem. Cell Biol.* 36, 2445 (2004).

[277] Z. Xie and D. J. Klionsky, *Nat. Cell Biol.* 9, 1102 (2007).

[278] M. Sandri, *Am. J. Physiol. Cell Physiol.* 298, C1291 (2010).

[279] M. Sandri, *Curr. Opin. Clin. Nutr. Metab. Care,* 14, 223 (2011).

[280] H. K. Lee and L. Marzella, *Int. Rev. Exp. Pathol.* 35, 39 (1994).

[281] A. M. Cuervo, *Mol. Cell Biochem.* 263, 55 (2004).

[282] N. Raben, C. Schreiner, R. Baum, S. Takikita, S. Xu, T. Xie, R. Myerowitz, M. Komatsu, J. H. Van der Meulen, K. Nagaraju, E. Ralston and P. H. Plotz, *Autophagy* 6, 1078 (2010).

[283] C. W. Wang and D. J. Klionsky, *Mol. Med.* 9, 65 (2003).

[284] N. Mizushima, B. Levine, A. M. Cuervo and D. J. Klionsky, *Nature* 451, 1069 (2008).

[285] I. Tanida, T. Ueno and E. Kominami, *J. Biol. Chem.* 279, 47704 (2004).

[286] I. Tanida, T. Ueno, and E. Kominami, *Int. J. Biochem. Cell Biol.* 36, 2503 (2004).

[287] G. Bjørkøy, T. Lamak, A. Brech, H. Outzen, M. Perander, A. Overvatn, H. Stenmark and T. Johansen, *J. Cell Biol.* 171, 603 (2005).

[288] S. Pankiv, T. H. Clausen, T. Lamak, A. Brech, J. A. Bruun, H. Outzen, A. Øvervatn A, G. Bjørkøy and T. Johansen, *J. Biol. Chem.* 282, 24131 (2007).

[289] E. Masiero, L. Agatea, C. Mammucari, B. Blaauw, E. Loro, M. Komatsu, D. Metzger, C. Reggiani, S. Schiaffino and M. Sandri, *Cell Metab.* 10, 507 (2009).

[290] M. Gaugler, A. Brown, E. Merrell, M. Disanto-Rose, J. A. Rathmacher and T. H. Reynolds 4th, *J. Appl. Physiol.* 111, 192 (2011).

[291] C. A. McMullen, A. L. Ferry, J. L. Gamboa, F. H. Andrade and E. E. Dupont-Versteegden, *Exp. Gerontol.* 44, 420 (2009).

[292] S. E. Wohlgemuth, A. Y. Seo, E. Marzetti, H. A. Lees and C. Leeuwenburgh, *Exp. Gerontol.* 45, 138 (2010).

[293] P. J. Plant, D. Brooks, M. Faughnan, T. Bayley, J. Bain, L. Singer, J. Correa, D. Pearce, M. Binnie and J. Batt, *Am. J. Respir. Cell Mol. Biol.* 42, 461 (2010).

[294] D. Yamamoto, T. Maki, E. H. Herningtyas, N. Ikeshita, H. Shibahara, Y. Sugiyama, S, Nakanishi, K. Iida, G. Iguchi, Y. Takahashi, H. Kaji, K. Chihara and Y. Okimura, *Muscle Nerve* 41, 819 (2010).

[295] A. Noglaska, C. Terracciano, C. D'Agostino, W. K. Engel and V. Askanas, *Acta Neuropathol. (Berl.)* 118, 407 (2009).

[296] D. H. Cho and S. J. Tapscott, *Biochim. Biophys. Acta Mol. Basis Dis.* 1772, 195 (2007).

[297] P. S. Harper, In: W. B. Saunders (ed.), *Myotonic Dystrophy,* London (2001).

[298] E. Loro, F. Rinaldi, A, Malena, G. Novelli, C. Angelini, V. Romeo, M. Sandri, A. Botta and L. Vergani, *Cell Death Diff.* 17, 1315 (2010).

[299] F. Barthélémy, N. Wein, M. Krahn, N. Lévy and M. Bartoli, *Mol. Med.* 17, 875 (2011).

[300] V. Askanas and W. K. Engel, *Neurology* 66, S39 (2006).

[301] E. Fujita, Y. Kouroku, A. Isoai, H. Kumagai, A. Misutani, C. Matsuda, Y. K. Hayashi and T. Momoi, *Hum. Mol. Genet.* 16, 618 (2007).

[302] S. J. Lee, *Annu. Rev. Cell Dev. Biol.* 20, 61 (2004).

[303] S. J. Lee and A. M. McPherron *Proc. Natl. Acad. Sci. U. S. A.* 98, 9306 (2001).

[304] N. M. Wolfman, A. C. McPherron, W. N. Pappano, M. V. Davies, K. Song, K. N. Tomkinson, J. F. Wright, L. Zhao, S. M. Sebald, D. S. Greenspan and S. J. Lee, *Proc. Natl. Acad. Sci. U. S. A.* 100, 15842, (2003).

[305] T. A. Zimmers, M. V. Davies, L. G. Koniaris, P. Haynes, A. F. Esquela, K. N. Tomkinson, A. C. McPherron, N. M. Wolfman and S. J. Lee, *Science* 296, 1486 (2002).

[306] H. D. Kollias and J. C. McDermont, *J. Appl. Physiol.* 104, 579 (2008).

[307] D. Joulia-Ekaza and G. Cabello, *Curr. Opin. Pharmacol.* 7, 310 (2007).

[308] J. Yang, T. Ratovitski, J. P. Brady, M. B. Solomon, K. D. Wells and R. J. Wall, *Mol. Reprod. Dev.* 60, 351 (2001).

[309] M. M. Matzuk, N. Lu, H. Vogel, K. Sellheyer, D. R. Roop and A. Bradley, *Nature* 374, 360 (1995).

[310] J. J. Hill, M. V. Davies, A. A. Pearson, J. H. Wang, R. M. Hewick, N. M. Wolfman and Y. Qiu, *J. Biol. Chem.* 277, 40735 (2002).

[311] J. J. Hill, Y. Qiu, R. M. Hewick and N. M. Wolfman, *Mol. Endocrinol.* 17, 1144 (2003).

[312] S. B. Anderson, A. L. Goldberg and M. Whitman, *J. Biol. Chem.* 283, 7027 (2008).

[313] A. C. McPherron and S. J. Lee, *J. Clin. Invest.* 109, 595 (2002).

[314] A. Rebbapragada, H. Benchabane, J. L. Wrana, A. J. Celeste and L. Attisano, *Mol. Cell. Biol.* 23, 7230 (2003).

[315] B. J. Feldman, R. S. Streeper, R. V. Farese Jr. and K. R. Yamamoto, *Proc. Natl. Acad. Sci. U. S. A.* 103, 15675 (2006).

[316] J. Lin, H. B. Arnold, M. A. Della-Fera, M. J. Azain, D. L. Hartzell and C. A. Baile, *Biochem. Biophys. Res. Commun.* 291, 701 (2002).

[317] M. Wehling, B. Cai and J. G. Tidball, *FASEB J.* 14, 103 (2000).

[318] M. E. Carlson, M. Hsu and I. M. Conboy, *Nature* 454, 528 (2008).

[319] F. Haddad and G. R. Adams, *J. Appl. Physiol.* 100, 1188 (2006).

[320] N. F. Gonzalez-Cadavid, W. E. Taylor, K. Yarasheski, I. Sinha-Hikim, K. Ma, S. Ezzat, R. Shen, R. Lalani, S. Asa, M. Mamita, G. Nair, S. Arver and S. Bhasin, *Proc. Natl. Acad. Sci. U. S. A.* 95, 14938 (1998).

[321] M. Sharma, R. Kambadur, K. G. Matthews, W. G. Somers, G. P. Devlin, J. V. Conaglen, P. J. Fowke and J. J. Bass, *J. Cell. Physiol.* 180, 1 (1999).

[322] K. G. Shyu, M. J. Lu, B. W. Wang, H. Y. Sun and H. Chang, *Eur. J. Clin. Invest.* 36, 713 (2006).

[323] J. Heineke, M. Auger-Messier, J. Xu, M. Sargent, A. York, S. Welle and J. D. Molkentin, *Circulation* 121, 419 (2010).

[324] P. J. Plant, D. Brooks, M. Faughnan, T. Bayley, J. Bain, L. Singer, J. Correa, D. Pearce, M. Binnie and J. Batt, *Am. J. Respir. Cell Mol. Biol.* 42, 461 (2010).

[325] D. Testelmans, T. Crul, K. Maes, A. Agten, M. Comach, M. Decramer and G. Gayan-Ramirez, *Eur. Respir. J.* 35, 549 (2010).

[326] M. Hayot, J. Rodriguez, B. Vernus, G. Carnac, E. Jean, D. Allen, L. Goret, P. Obert, R. Candau and A. Bonnieu, *Mol. Cell. Endocrinol.* 332, 38 (2011).

[327] B. Langley, M. Thomas, A. Bishop, M. Sharma, S. Gilmour and R. Kambadur, *J. Biol. Chem.* 277, 49831 (2002).

[328] M. Thomas, B. Langley, C. Berry, M. Sharma, S. Kirk, J. Bass and R. Kambadur, *J. Biol. Chem.* 275, 40235 (2000).

[329] W. Yang, Y. Zhang, Y. Li, Z. Wu and D. Zhu, *J. Biol. Chem.* 282, 3799 (2007).

[330] N. Bakkar, H. Wackerhage and D. Guttridge, *Signal Transduction* 5, 202 (2005).

[331] C. McFarlane, E. Plummer, M. Thomas, A. Hennebry, M. Ashby, N. Ling, H. Smith, M. Sharma and R. Kambadur, *J. Cell. Physiol.* 209, 501 (2006).

[332] S. Tajbakhsh, D. Rocancourt, G. Cossu and M. Buckingham, *Cell* 89, 127 (1997).

[333] K. M. Mulder, *Cytokine Growth Factor Rev.* 11, 23 (2000).

[334] N. Nicolas, G. Marazzi, K. Kelley and D. Sassoon, *Dev. Biol.* 281, 171 (2005).

[335] Z. A. Felton-Edkins, J. A. Fairley, E. L. Graham, I. M. Johnston, R. J. White and P. H. Scott, *EMBO J.* 22, 2422 (2003).

[336] W. Yang, Y. Chen, Y. Zhang, X. Wang, N. Yang and D. Zhu, *Cancer Res.* 66, 1320 (2006).

[337] M. R. Morissette, S. A. Cook, C. Buranasombati, M. A. Rosenberg and A. Rosenzweig, *Am. J. Physiol. Cell Physiol.* 297, C1124 (2009).

[338] D. L. Allen and T. G. Unterman, *Am. J. Physiol. Cell Physiol.* 292, C188 (2007).

[339] P. T. Bhaskar and N. Hay, *Dev. Cell* 12, 487 (2007).

[340] A. U. Trendelenburg, A. Meyer, D. Rohner, J. Boyle, S. Hatakeyama and D. J. Glass, *Am. J. Physiol. Cell Physiol.* 296, C1258 (2009).

[341] L. Bradley, P. J. Yaworsky and F. S. Walsh, *Cell Mol. Life Sci.* 65, 2119 (2008).

[342] S. Bogdanovich, T. O. Krag, E. R. Barton, L. D. Morris, L. A. Whittemore, R. S. Ahima and T. S. Khurana, *Nature* 420, 418 (2002).

[343] E. L. Holzbaur, D. S. Howland, N. Weber, K. Wallace, Y. She, S. Kwak, L. A. Tchistiakova, E. Murphy, J. Hinson, R. Karim, X. Y. Tan, P. Kelley, K. C. McGill, G. Williams, C. Hobbs, P. Doherty, M. M. Zaleska, M. N. Pangalos and F. S. Walsh, *Neurobiol. Dis.* 23, 697 (2006).

[344] K. R. Wagner, J. L. Fleckenstein, A. A. Amato, R. J. Barohn, K. Bushby, D. M. Escolar, K. M. Flanigan, A. Pestronk, R. Tawil, G. I. Wolfe, M. Eagle, J. M. Florence, W. M. King, S. Pandya, V. Straub, P. Juneau, K, Meyers, C. Csimma, T. Araujo, R. Allen, S. A. Parsons, J. M. Wozney, E. R. Lavallie and J. R. Mendell, *Ann. Neurol.* 63, 561 (2008).

[345] V. Siriett, L. Platt, M. S. Salerno, N. Ling, R. Kambadur and M. Sharma, *J. Cell. Physiol.* 209, 866 (2006).

[346] N. K. Lebrasseur, T. M. Schelhorn, B. L. Bernardo, P. G. Cosgrove, P. M. Loria and T. A. Brown, *J. Gerontol. A Biol. Sci. Med. Sci.* 64, 940 (2009).

[347] K. T. Murphy, R. Koopman, T. Naim, B. Léger, J. Trieu, C. Ibebunjo and G. S. Lynch, *FASEB J.* 24, 4433 (2010).

[348] X. Zhou, J. L. Wang, J. Lu, Y. Song, K. S. Kwak, Q. Jiao, R. Rosenfeld, Q. Chen, T. Boone, W. S. Simonet, D. L. Lacey, A. L. Goldberg and H. Q. Han, *Cell* 142, 531 (2010).

[349] S. Acharyya, S. A. Villalta, N. Bakkar, T. Bupha-Intr, P. M. Janssen, M. Carathers, Z. W. Li, A. A. Beg, S. Ghosh, Z. Sahenk, M. Weinstein, K. L. Gardner, J. A. Rafael-Fortney, M. Karin, J. G. Tidball, A. S. Baldwin and D. C. Guttridge, *J. Clin. Invest.* 117, 889 (2007).

[350] M. S. Hayden and S. Ghosh, *Cell* 132, 344 (2008).

[351] U. Spate and P. C. Schulze, *Curr. Opin. Clin. Nutr. Metab. Care* 7, 265 (2004).

[352] T. Phillips and C. Leeuwenburgh, *FASEB J.* 19, 668 (2005).

[353] L. Fernandez-Celemin, N. Pasko, V. Blomart and J. P. Thissen, *Am. J. Physiol. Endocrinol. Metab.* 283, E1279 (2002).

[354] A. Kumar, Y. Takada, A. M. Boriek and B. B. Aggarwal, *J. Mol. Med.* 82, 434 (2004).

[355] H. Li, S. Malhotra and A. Kumar, *J. Mol. Med.* 86, 1113 (2008).

[356] D. Van Gammeren, J. S. Damrauer, R. W. Jackman and S. C. Kandarian, *FASEB J.* 23, 362 (2009).

[357] S. M. Senf, S. L. Dodd, J. M. McClung and A. R. Judge, *FASEB J.* 23, 362 (2009).

[358] S. L. Dodd, B. Hain, S. M. Senf and A. R. Judge, *FASEB J.* 23, 3415 (2009).

[359] S. M. Senf, S. L. Dodd and A. R. Judge, *Am. J. Physiol. Cell Physiol.* 298, C38 (2010).

[360] M. B. Reid and Y. P. Li, *Respir. Res.* 2, 269 (2001).

[361] R. B. Hunter, E. Stevenson, A. Koncarevic, H. Mitchell-Felton, D. A. Essig and S. C. Kandarian, *FASEB J.* 15, 529 (2002).

[362] R. B. Hunter and S. C. Kandarian, *J. Clin. Invest.* 114, 1504 (2004).

[363] A. R. Judge, A. Koncarevic, R. B. Hunter, H. C. Liou, R. W. Jackman and S. C. Kandarian, *Am. J. Physiol. Cell Physiol.* 292, C372 (2007).

[364] F. Mourkioti, P. Kratsios, T. Luedde, Y. H. Song, P. Delafontaine, R. Adami, V. Parente, R. Bottinelli, M. Pasparakis and N. Rosenthal, *J. Clin. Invest.* 116, 2945 (2006).

[365] L. A. Schaap, S. M. F. Pluijim, D. J. H. Deeg, T. B. Harris, S. B. Kritchevsky, A. B. Newman, L. H. Colbert, M. Pahor, S. M. Rubin, F. A. Tylavsky and M. J. Visser, *J. Gerontol. A Biol. Sci. Med. Sci.* 64A, 1183 (2009).

[366] H. Y. Chung, M. Cesari, S. Anton, E. Marzetti, S. Giovannini, A. Y. Seo, C. Carter, B. P. Yu and C. Leeuwenburgh, *Ageing Res. Rev.* 8, 18 (2009).

[367] W. Aoi and K. Sakuma, *Curr. Aging Sci.* 4, 101 (2011).

[368] S. -J. Meng and L. -J. Yu, *Int. J. Mol. Sci.* 11, 1509 (2010).

[369] M. Bar-Shai, E. Carmeli, R. Coleman, N. Rozen, S. Perek, D. Fuchs and A. Z. Reznick, *Mech. Ageing Dev.* 126, 289 (2005).

[370] V. Moresi, A. H. Williams, E. Meadows, J. M. Flynn, M. J. Potthoff, J. McAnally, J. M. Shelton, J. Backs, W. H. Klein, J. A. Richardson, R. Bassel-Duby and E. M. Olson, *Cell* 143, 35 (2010).

[371] E. E. Dupont-Versteegden, *Exp. Gerontol.* 40, 473 (2005).

[372] A. Riva, B. Tandler, E. J. Lesnefsky, G. Conti, F. Loffredo, E. Vazquez and C. L. Hoppel, *Mech. Ageing Dev.* 127, 917 (2006).

[373] P. J. Adhihetty, V. Ljubicic, K. J. Menzies and D. A. Hood, *Am. J. Physiol. Cell Physiol.* 289, C994 (2005).

[374] N. N. Daniel and S. J. Korsmeyer, *Cell* 116, 205 (2004).

[375] M. O. Hengartner, *Nature* 407, 770 (2000).

[376] Z. B. Wang, Y. Q. Liu and Y. F. Cui, *Cell Biol. Int.* 29, 489 (2005).

[377] T. Nakagawa and J. Yuan, *J. Cell Biol.* 150, 887 (2000).

[378] P. Jezek and L. Hlavata, *Int. J. Biochem. Cell Biol.* 37, 2478 (2005).

[379] F. M. Yakes and H. B. Van, *Proc. Natl. Acad. Sci. U. S. A.* 94, 514 (1997).

[380] Y. H. Wei and H. C. Lee, *Exp. Biol. Med. (Maywood)* 227, 671 (2002).

[381] C. Wang and R. J. Youle, *Ann. Rev. Genet.* 43, 95 (2009).

[382] D. S. Tews, W. Behrhof and S. Schindler, *Muscle Nerve* 31, 175 (2005).

[383] D. S. Tews, W. Behrhof and S. Schindler, *Mol. Morphol.* 16, 66 (2008).

[384] D. S. Tews and H. H. Goebel, *Exp. Neurol.* 56, 150 (1997).

[385] D. S. Tews and H. H. Goebel, *Neuromuscul. Disord.* 6, 265 (1996).

[386] P. M. Siu and S. E. Alway, *J. Physiol.* 565, 309 (2006).

[387] E. Marzetti, C. S. Carter, S. E. Wohlgemuth, H. A. Lees, S. Giovannini, B. Anderson, L. S. Quinn and C. Leeuwenburgh, *Mech. Ageing Dev.* 130, 272 (2009).

[388] W. Song, H. B. Kwak and J. M. Lawler, *Antioxid. Redox Signal.* 8, 517 (2006).

[389] E. A. Bua, S. H. McKiernan, J. Eanagat, D. McKenzie and J. M. Aiken, *J. Appl. Physiol.* 92, 2617 (2002).

[390] J. Wanagat, Z. Cao, P. Pathare and J. M. Aiken, *FASEB J.* 15, 322 (2001).

[391] J. M. Argiles, F. J. Lopez-Soriano and S. Busquets, *Int. J. Biochem. Cell Biol.* 40, 1674 (2008).

[392] D. J. Baker and R. T. Hepple, *Exp. Gerontol.* 41, 1149 (2006).

[393] E. Marzetti, C. S. Carter, S. E. Wohlgemuth, H. A. Lees, S. Giovannini, B. Anderson, L. S. Quinn and C. Leeuwenburgh, *Mech. Ageing Dev.* 130, 272 (2009).

[394] S. Y. Park, H. Y. Kim, J. H. Lee, K. H. Yoon, M. S. Change and S. K. Park, *Cell. Mol. Biol. Lett.* 15, 1 (2010).

[395] J. C. Bruusgaard and K. Gundersen, *J. Clin. Invest.* 118, 1450 (2008).

[396] K. Gundersen and J. C. Bruusgaard, *J. Physiol.* 586, 2675 (2008).

[397] P. J. Adhihetty, M. F. O'Leary, B. Chabi, K. L. Wicks and D. A. Hood, *J. Appl. Physiol.* 102, 1143 (2007).

[398] D. L. Allen, J. K. Linderman, R. R. Roy, A. J. Bigbee, R. E. Grindeland, V. Mukku and V. R. Edgerton, *Am. J. Physiol. Cell Physiol.* 273, C579 (1997).

[399] S. E. Always, H. Degens, G. Krishnamurthy and A. Chaudhrai, *J. Gerontol. A Biol. Sci. Med. Sci.* 58, 687 (2003).

[400] S. E. Always, J. K. Martyn, J. Ouyang, A. Chaudhrai and Z. S. Murlasits, *Am. J. Physiol. Regul. Integr. Comp. Physiol.* 284, R540 (2003).

[401] P. M. Siu, E. E. Pistilli and S. E. Always, *Am. J. Physiol. Regul. Integr. Comp. Physiol.* 289, R1015 (2005).

[402] H. Tang, W. M. Cheung, F. C. Ip and N. Y. Ip, *Mol. Cell. Neurosci.* 16, 127 (2000).

[403] D. S. Tews, H. H. Goebel, I. Schneider, A. Gunkel, E. Stennert and W. F. Neiss, *Neuropathol. Appl. Neurobiol.* 23, 141 (1997).

[404] J. P. Hyatt, R. R. Roy, K. M. Baldwin, A. Wernig and V. R. Edgerton, *Muscle Nerve* 33, 49 (2006).

[405] S. H. Lecker, V. Solomon, W. E. Mitch and A. L. Goldberg, *J. Nutr.* 129, 227S (1999).

[406] E. Lofberg, A. Gutierrez, J. Wernerman, B. Anderstam, W. E. Mitch, S. R. Price, J. Bergstrom and A. Alvestrand, *Eur. J. Clin. Invest.* 32, 345 (2002).

[407] J. L. Kostyo and A. F. Redmond, *Endocrinology* 79, 531 (1966).

[408] O. J. Shah, S. R. Kimball and L. S. Jefferson, *Biochem. J.* 347, 389 (2000).

[409] Z. Liu, G. Li, S. R. Kimball, L. A. Jahn and E. J. Barrett, *Am. J. Physiol. Endocrinol. Metab.* 287, E275 (2004).

[410] M. F. Te Pas, P. R. De Jong and F. J. Verburg, *Mol. Biol. Reports* 27, 87 (2000).

[411] H. Yang, M. J. Menconi, W. Wei, V. Petkova and P. O. Hasselgren, *J. Cell. Biochem.* 94, 1058 (2005).

[412] H. Gilson, O. Schakman, L. Combaret, P. Lause, L. Grobet, D. Attaix, J. M. Ketelslegers and J. P. Thissen, *Endocrinology* 148, 452 (2007).

[413] S. Y. Ying, D. C. Chang and S. L. Lin, *Mol. Biotechnol.* 38, 257 (2008).

[414] N. Bushati and S. M. Cohen, *Annu. Rev. Cell Dev. Biol.* 23, 175 (2007).

[415] Y. Ge and J. Chen, *Cell Cycle* 10, 441 (2011).

[416] J. F. Chen, E. M. Mandel, J. M. Thomson, Q. Wu, T. E. Callis, S. M. Hammond, F. L. Conlon and D. Z. Wang, *Nat. Genet.* 38, 228 (2006).

[417] P. K. Rao, R. M. Kumar, M. Farkhondeh, S. Baskerville and H. F. Lodish, *Proc. Natl. Acad. Sci. U. S. A.* 103, 8721 (2006).

[418] N. S. Sokol and V. Ambros, *Genes Dev.* 19, 2343 (2005).

[419] J. F. Chen, Y. Tao, J. Li, Z. Deng, Z. Yan, X. Xiao and D. Z. Wang, *J. Cell Biol.* 190, 867 (2010).

[420] M. I. Rosenberg, S. A. Georges, A. Asawachaicharn, E. Analau and S. J. Tapscott, *J. Cell Biol.* 175, 77 (2006).

[421] B. K. Dey, J. Gagan, and A. Dutta, *Mol. Cell. Biol.* 31, 203 (2011).

[422] B. Cardinali, L. Castellani, P. Fasanaro, A. Basso, S. Alema, F. Martelli and G. Falcone, *PLoS One* 4, e7607 (2009).

[423] J. J. McCarthy and K. A. Esser, *J. Appl. Physiol.* 102, 306 (2007).

[424] L. Elia, R. Contu, M. Quintavalle, F. Varrone, C. Chimenti, M. A. Russo, V. Cimino, L. De Marinis, A. Frustaci, D. Catalucci and G. Condorelli, *Circulation* 120, 2377 (2009).

[425] D. L. Allen, E. R. Bandstra, B. C. Harrison, S. Thorng, L. S. Stodieck, P. J. Kostenuik, S. Morony, D. L. Lacey, T. G. Hammond, L. L. Leinwand, W. S. Argraves, T. A. Bateman and J. L. Barth, *J. Appl. Physiol.* 106, 582 (2009).

[426] L. A. Tintignac, J. Lagirand, S. Batonnet, V. Sirri, M. P. Leibovitch and S. A. Leibovitch, *J. Biol. Chem.* 280, 2847 (2005).

[427] S. K. Panguluri, S. Bhatnagar, A. Kumar, J. J. McCarthy, A. K. Srivastava, N. G. Cooper and R. F. Lundy, *PLoS One* 5, e8760 (2010).

[428] S. F. Jeng, C. S. Rau, P. C. Liliang, C. J. Wu, T. H. Lu, Y. C. Chen, C. J. Lin and C. H. Hsieh, *J. Neurotrauma* 26, 2345 (2009).

[429] C. S. Rau, J. C. Jeng, S. F. Jeng, T. H. Lu, Y. C. Chen, P. C. Liliang, C. J. Wu, C. J. Lin and C. H. Hsieh, *BMC Musculoskelet. Disord.* 11, 181 (2010).

[430] J. J. McCarthy, K. A. Esser, C. A. Peterson and E. E. Dupont-Versteegden, *Physiol. Genomics* 39, 219 (2009).

[431] A. Lewis, J. Riddoch-Contreras, S. A. Natanek, A. Donaldson, W. D. -C. Man, J. Moxham, N. S. Hopkinson, M. I. Polkey and R. R. Kemp, *Thorax* 67, 26 (2012).

[432] K. Yuasa, Y. Hagiwara, M. Ando, A. Nakamura, S. Takeda and T. Hijikata, *Cell Struct. Funct.* 33, 163 (2008).

[433] A. V. Greco, G. Mingrone, A. Giancaterini, M. Manco, M. Morroni, S. Cinti, M. Granzotto, R. Vettor, S. Camastra and E. Ferrannini, *Diabetes* 51, 144 (2002).

[434] S. Gambardella, F. Rinaldi, S. M. Lepore, A. Viola, E. Loro, C. Angelini, L. Vergani, G. Novelli and A. Botta, *J. Transl. Med.* 8, 48 (2010).

[435] R. Perbellini, S. Greco, G. Sarra-Ferraris, R. Cardani, M. C. Capogrossi, G. Meola and F. Martelli, *Neuromuscul. Disord.* 21, 81 (2011).

[436] A. H. Williams, G. Valdez, V. Moresi, X. Qi, J. McAnally, J. L. Elliott, R. Bassel-Duby, J. R. Sanes and E. N. Olson, *Science* 326, 1549 (2009).

[437] M. J. Drummond, J. J. McCarthy, C. S. Fry, K. A. Esser and B. B. Rasmussen, *Am. J. Physiol. Endocrinol. Metab.* 295, E1333 (2008).

[438] K. Sakuma and A. Yamaguchi, *Curr. Aging Sci.* 3, 90 (2010).

[439] K. Sakuma and A. Yamaguchi, In: J. W. Perloft, and A. H. Wong (ed.), *Cell Aging,* pp. 93, Nova Science Publisher, New York (2011).

[440] J. Y. Lee, N. S. Hopkinson and P. R. Kemp, *Biochem. Biophys. Res. Commun.* 415, 632 (2011).

[441] S. Sriram, S. Subramanian, D. Sathiakumar, R. Venkatesh, M. S. Salerno, C. D. McFarlane, R. Kambadur and M. Sharma, *Aging Cell* 10, 231 (2011).

In: Skeletal Muscle
Editor: Mark Willems

ISBN: 978-1-62417-271-7
© 2013 Nova Science Publishers, Inc.

Chapter 10

SKELETAL MUSCLE LOSS AND THE ROLE OF BRANCHED-CHAIN AMINO ACIDS

Stefan M. Pasiakos[*] *and James P. McClung*

Military Nutrition Division, United States Army Research
Institute of Environmental Medicine, Natick, Massachusetts, US

ABSTRACT

Energy restriction, aging, and disease are conditions characterized by a loss of skeletal muscle mass caused by an imbalance between rates of muscle protein synthesis and breakdown. Skeletal muscle loss is associated with a myriad of negative consequences, including increased susceptibility to injury, decrements in physical performance, and ultimately increased morbidity and mortality. Branched-chain amino acids (BCAA) are anti-catabolic nutrients that may attenuate proteolysis and promote muscle anabolism.

It has been reported that the BCAA leucine is particularly important given the unique ability of leucine to stimulate muscle protein synthesis by triggering complex intracellular signaling pathways that regulate messenger ribonucleic acid translational efficiency. Other evidence suggests that leucine (or total BCAA) may reduce proteolysis by mitigating the irreversible degradation of muscle proteins through the ubiquitin-proteasome system.

This chapter will focus on the underlying mechanisms contributing to the loss of skeletal muscle mass in response to physiological stress. Populations that may experience severe muscle loss will also be addressed, to include military personnel, aging adults, as well as individuals suffering from cachectic conditions such as cancer. Furthermore, this review will explore the potential benefit of BCAA administration for the treatment and prevention of muscle loss, and highlight the available literature examining the cellular properties of leucine on skeletal muscle protein metabolism.

---

[*] Email address: stefan.pasiakos@us.army.mil

INTRODUCTION

Maintaining skeletal muscle mass is essential for optimal metabolic function, strength, physical performance, and for the prevention of injuries [1]. Skeletal muscle functions as the principal reservoir of amino acids, which contribute to whole-body protein metabolism and glycemic regulation by serving as precursors for protein synthesis of vital organs, and tissues [2], and substrates for hepatic gluconeogenesis [3]. Maintaining muscle mass may be achieved by consuming adequate energy and dietary protein in combination with regular physical activity. However, many pathophysiological conditions including energy restriction, aging, and chronic cachectic conditions, such as cancer, kidney disease, and infection, are characterized by significant declines in skeletal muscle mass. The loss of muscle mass is often progressive, and the severity is dictated by the metabolic demand for energy and amino acids that causes accelerated proteolysis resulting in imbalanced rates of muscle protein synthesis (MPS) and breakdown (MPB). Clinical manifestations of severe muscle loss include compromised basal metabolic rate, increased susceptibility to injury, decrements in performance, and ultimately increased morbidity and mortality.

Branched-chain amino acids (BCAA, leucine, isoleucine, and valine) are essential amino acids necessary for the maintenance of normal cellular function. BCAA are required for glutamine and alanine metabolism, and can be readily oxidized in skeletal muscle, serving as a key energy source during periods of increased metabolic demand such as exercise, energy restriction, cancer, infection, and other chronic catabolic conditions [4]. More importantly, circulating BCAA levels contribute to the maintenance of muscle mass by influencing the regulation of skeletal muscle protein turnover (SMPTO) [5]. The unique regulatory role of BCAA on SMPTO may be attributed to leucine, which functions as an anabolic cellular signal by modulating messenger ribonucleic acid (mRNA) translation efficiency through the cell hypertrophy pathway that involves the mammalian target of rapamycin complex 1 (mTORC1) [6]. Increasing blood and intramuscular leucine levels stimulates MPS [7, 8]. Other studies have demonstrated that increasing leucine levels reduces MPB by attenuating the irreversible degradation of muscle proteins through the ubiquitin-proteasome system (UP) [9]. BCAA have long been studied for efficacy as therapeutic aids to counteract alterations in SMPTO that contribute to the loss of skeletal muscle mass. As such, this chapter will focus on the underlying cellular mechanisms regulating skeletal muscle mass, address how those mechanisms may account for exaggerated muscle loss in some pathophysiological conditions or diseases, and explore the potential benefit of BCAA for the treatment and prevention of muscle loss.

REGULATION OF SKELETAL MUSCLE MASS

Protein turnover in skeletal muscle is the fundamental biological process that encompasses both the synthesis of new protein, and the breakdown of existing protein. Skeletal muscle mass is preserved when the intricate cycling of amino acids through MPS and MPB is balanced in response to anabolic and catabolic stimuli. SMPTO is regulated in part by nutrition, as dietary energy and protein intake influence MPS and MPB [10, 11]. In the post absorptive state, MPB rises with a concomitant reduction in MPS, in large part to provide hepatic gluconeogenic precursors to maintain blood glucose [12]. However, feeding stimulates MPS and attenuates MPB, and protein gained in response to feeding typically compensates for losses that occur during fasting, thus muscle protein mass remains constant [1]. Combining an anabolic stimulus such as resistive-type exercise with feeding can potentiate the SMPTO response, which may contribute to hypertrophic gains in skeletal muscle [12, 13].

The regulation of SMPTO occurs through a choreographed series of events involving complex intracellular networks (Figure 1). The regulation of MPS occurs primarily through cellular modulation of mRNA translation initiation and elongation [14, 15]. This process involves the insulin/insulin-like growth factor-phosphoinositol-3 kinase (insulin/IFG-PI3K) intracellular signaling cascade that converges at the multiunit protein complex termed mTORC1. The major functions of mTORC1 include the inactivation of the repressor of mRNA translation, eukaryotic initiation factor 4E-binding protein (4E-BP1), and the activation of 70 kDa ribosomal protein S6 kinase (p70^{S6K}). The phosphorylation states of these critical intracellular signaling proteins affect mRNA translation initiation and elongation. Energy status, growth factors (e.g., insulin), nutrition (e.g., amino acids), and exercise all influence mTORC1 and therefore MPS by modulating mRNA translation [8, 11].

The process of MPB is highly regulated, essentially irreversible, and occurs primarily through the ATP dependent UP [16]. The muscle proteolytic process involves the 'tagging', or ubiquitination, of proteins through a discrete series of reactions catalyzed by three distinct enzyme complexes.

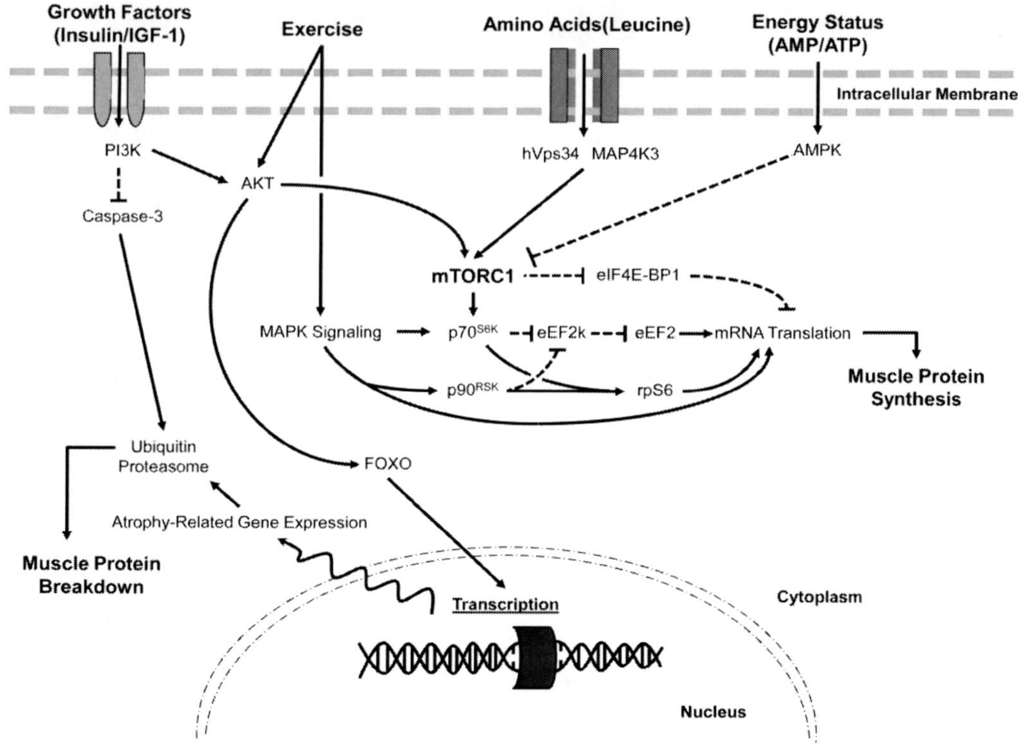

Figure 1. Intracellular regulation of skeletal muscle mass. Stimulatory mechanisms are depicted in solid lines, whereas inhibitory mechanisms are depicted in dashed lines.

This process begins with caspase 3, a cysteine protease that initiates skeletal muscle proteolysis through the UP by cleaving myofibrils into smaller, more accessible actin and myosin monomers [17]. These muscle fragments are then marked for degradation by covalent binding of multiple ubiquitin molecules catalyzed by the enzymes E1 (ATP dependent-ubiquitin-activating enzyme), E2 (ubiquitin-conjugating enzyme), and E3 (ubiquitin-ligating enzyme). The 26S proteasome, a large multi-subunit proteolytic complex consisting of a central catalytic core (20S proteasome) and two terminal regulator complexes (19S complexes), recognizes the polyubiquitin chain. The 19S regulator complex confers a central role in the recognition and unfolding of the proteins and guides them to the catalytic core. The 20S catalytic core is responsible for the hydrolysis of the protein resulting in short oligopeptides. Inflammation and conditions associated with increased metabolic demand regulate the activity and expression of these UP components in part through the insulin/IGF-PI3K pathway (Figure 1) [18]. This likely occurs through the inactivation of protein kinase B (Akt), an insulin sensitive upstream regulator of mTORC1 activity, by increasing the expression of key UP components. Akt phosphorylates the forkhead box-o (FOXO) family of transcription factors in response to an elevation in insulin levels and other growth factors. In the absence of Akt-mediated phosphorylation, these FOXO transcription factors migrate into the nucleus of skeletal muscle cells and increase expression of a number of atrophy-related genes.

CONDITIONS ASSOCIATED WITH THE LOSS OF SKELETAL MUSCLE MASS

Energy restriction alone, or in combination with increased physical activity, are the most common approaches to lose body weight and reduce body fat [19, 20]. Although energy restriction results in beneficial declines in body fat, skeletal muscle mass is also lost in response to prolonged periods of negative energy balance. In general, energy restricted diets elicit an average weight loss of 5-10% initial body mass; however during such diets, more than a quarter of the reduction in body mass typically results from a loss of skeletal muscle [21]. Declines in skeletal muscle mass are accompanied by reductions in nitrogen retention [22, 23], whole-body protein turnover [24], MPS, and mTORC1 intracellular signaling [11]. Despite the down-regulation of protein turnover, the increased metabolic demand for amino acids during energy restriction causes an increase in MPB that generally exceeds MPS [1]. Increased physical activity may also be catabolic in the absence of adequate nutritional intake. Sustained aerobic-type physical activity elicits a reduction in MPS with a simultaneous increase in MPB and BCAA oxidation [12]. Without consuming adequate energy and protein during recovery, post-exercise elevations in MPS will not exceed MPB. As such, the skeletal muscle response to sustained aerobic-type physical activity, which is typically performed in combination with energy restriction to elicit weight loss, is indicative of elevated proteolysis and a loss of muscle proteins [25]. In overweight and obese individuals, reductions in muscle mass may impede additional weight loss and compromise weight management [26]. Healthy, normal weight individuals, such as athletes and military personnel, may also undergo periods of unavoidable negative energy balance and diminished muscle mass, potentially degrading physical performance and increasing susceptibility to injury.

Reductions in skeletal muscle mass coupled with an increase in body fat are among the most debilitating and consistent changes associated with aging. Age-related muscle loss is referred to as sarcopenia, which is clinically defined as presentation with muscle mass at least two standard deviations below the average muscle mass of a younger adult 35 years of age [27, 28]. The rate of skeletal muscle loss can be much as 1% per year after 30 years of age, which continues until the end of life [29]. The etiology of sarcopenia is complex and multifactorial. However, the most evident metabolic explanation for the decline in muscle mass is an imbalance between MPS and MPB [30]. Muscle inactivity (e.g., bed rest) [31], nutritional deficiencies [32], chronic inflammation [33], declines in production and sensitivity to anabolic hormones [34], and blunted MPS responses to feeding [35], are among many factors that may contribute to age-related alterations in SMPTO and subsequent declines in muscle mass. The reduction in muscle mass with aging can degrade strength, metabolic rate, aerobic capacity, and muscular function, thereby reducing quality of life [36], and predisposing elderly individuals to illness, fractures, and other chronic metabolic conditions such as obesity and type 2 diabetes [37], which ultimately increase mortality [38].

Cachexia is a metabolic condition that is associated with an underlying illness and characterized by anorexia, weight loss, with disproportionate losses of muscle mass compared to the loss of fat mass, resulting in weakness, and extreme fatigue [39]. Although most commonly associated with cancer, cachexia is perhaps the most common clinical manifestation of many advanced malignant diseases including congestive heart failure, chronic obstructive pulmonary disease, chronic kidney disease, rheumatoid arthritis, and AIDS [40]. Cachexia is highly predictive of mortality, and in the case of cancer, accounts for more than a quarter of cancer-related deaths [40]. The loss of skeletal muscle can be severe, and has been attributed to elevated UP-mediated proteolysis, and reduced MPS, caused in large part by the chronic inflammatory state that persists during cachectic conditions [41, 42]. Increased gene expression of several components associated with the UP [43,44] and elevated catabolic activity of the 20S proteasome have been observed in skeletal muscle obtained from adults with cancer [45]. Rapid declines in skeletal muscle mass are likely unavoidable, as muscle proteins (e.g., actomyosin, actin, and myosin) are targeted for degradation by the UP under cachectic conditions [46]. Reduced expression and increased degradation of muscle proteins accompany chronic elevations of inflammatory cytokines such as tumor necrosis factor-α (TNF-α), and interferon-γ, indicating that the targeting of muscle proteins for degradation by the UP is highly linked inflammation [40]. Cytokines can also activate nuclear transcription factor κB (NF- κB), which in turn, may result in lower rates of MPS [47]. Reductions in circulating anabolic hormones such as testosterone and IFG-1, coupled with increased glucocorticoid secretion, can exacerbate declines in MPS and elevations in rates of MPB typically observed with cachexia [40].

ATTENUATION OF MUSCLE LOSS
WITH BRANCHED-CHAIN AMINO ACIDS

The effects of BCAA availability on the skeletal muscle protein metabolic responses to pathophysiological stressors have been well described [48, 49, 50]. During conditions associated with increased metabolic demand, providing BCAA increases MPS, anabolic intracellular signaling, and attenuates proteolysis; beneficial effects largely attributed to the anabolic properties of leucine [51, 52]. During extended periods of negative energy balance, nutritional interventions that provide dietary protein in excess of the current recommended dietary allowance (RDA, 0.8 $g \cdot kg^{-1} \cdot d^{-1}$) can confer protection against the loss of skeletal muscle mass [53], probably due to enhanced BCAA availability. Increased BCAA levels during energy restriction can support gluconeogenesis, maintain whole-body protein synthesis and MPS, and attenuate nitrogen excretion, whole-body proteolysis, and MPB. For aging adults, consuming dietary protein in excess of the RDA may counteract sarcopenia. Proteins obtained by consuming dairy, lean meats, soy, and eggs are rich in BCAA compared to other dietary protein sources, and therefore likely able to better overcome the anabolic resistance typically observed in older adults, which contributes in large part to age-related muscle loss [35, 52, 54]. An alternative nutritional approach to overcome the blunted post-prandial MPS that accompanies aging is to consume between 20-30 grams of high-quality protein with each meal, rather than globally increasing the RDA for older adults [55, 56]. Regardless of the nutritional approach, it has become increasingly evident that consuming dietary protein sources rich in BCAA may be an effective strategy to counteract skeletal muscle loss that occurs not only with energy restriction, but also with sarcopenia.

BCAA protection against accelerated declines in muscle mass associated with cachectic conditions has not been clearly defined. Daily administration of BCAA suppressed weight loss, attenuated MPB, increased MPS, and anabolic intracellular signaling in cachectic tumor-bearing mice [57].

The apparent anti-catabolic and anabolic stimulatory effect has been confirmed using similar animal cancer models [58], but results of human trials have not consistently demonstrated a protective effect of BCAA for the nutritional management of cachexia [28, 49]. Early reports from studies that provided post-operative cancer patients total parenteral nutrition (TPN) enriched with BCAA (45-50% of total protein) failed to enhance nitrogen retention and amino acid utilization compared to TPN solutions providing a lower concentration of BCAA (20-25%) [59, 60, 61]. However, other reports demonstrated enhanced whole-body protein turnover, albumin synthesis rates, and leucine balance in metastatic cancer patients receiving TPN formulas enriched with BCAA [62]. Attenuated weight loss, shorter hospital stays, reduced morbidity, and improved quality of life have also been reported in cancer patients supplemented with BCAA [63, 64]. In contrast, BCAA supplementation during cancer treatment had no effect on clinical outcomes including mortality [65].

Examining the efficacy of BCAA supplementation for the nutritional support of other highly catabolic states such as sepsis, burn, and trauma have yielded similar inconclusive results [66]. The most consistent finding is that delivery of BCAA (and protein) to cachectic patients provides substrate for protein synthesis, but nutritional supplementation often fails to conserve skeletal muscle mass [28].

The seemingly inconsequential effects of BCAA have been attributed to poor study design, small sample size, inappropriate outcome measures, and inadequate overall nutritional support that hampered the utilization of the BCAA, particularly leucine. As such, the concept of BCAA supplementation has been abandoned for a more focused nutritional strategy that emphasizes the unique cellular properties of leucine [66].

LEUCINE: A THERAPEUTIC NUTRIENT TO COUNTERACT MUSCLE LOSS

Leucine is a potent independent stimulator of MPS in cell models and in vivo animal models through enhanced cellular regulation of mRNA translation [15]. However, the anabolic properties of leucine on human SMPTO and associated intracellular signaling have not been clearly defined. In recent studies, consuming a 10 gram leucine-enriched essential amino acid (L-EAA; 3.5 grams leucine) supplement failed to enhance MPS and mTORC1 signaling compared to an isonitrogenous EAA (EAA; 1.9 grams leucine) supplement in healthy resting adults [67]. However, consuming the same L-EAA mixture did enhance MPS and reduce whole-body proteolysis in healthy adults during recovery from sustained aerobic-type physical activity [8]. The greater skeletal muscle and whole-body demand for leucine during exercise likely accounted for the discordant data between studies. As such, optimizing the leucine content of the diet by providing a protein-containing supplement, or selecting foods providing high-quality protein may be an effective strategy to conserve skeletal muscle mass in response to prolonged periods of physiological stress. The RDA for leucine is currently 1-3 grams per day, yet the amount of leucine needed to maximize the stimulation of intracellular pathways that promote MPS, and contribute to the preservation of muscle mass during stressors such as energy restriction, may be as high as 7-12 grams per day [53]. Further study is certainly required to define optimal leucine intake for adults predisposed to muscle loss. Nonetheless, based on recent findings, consuming a diet with protein at levels above the RDA, derived from high-quality sources, distributed equally throughout the day, likely confers protection against severe declines in skeletal muscle mass in response to sustained periods of negative energy balance.

The leucine content of a meal, or the diet, may be especially important for aging adults to counteract the debilitating loss of muscle associated with sarcopenia. In one recent study, the SMPTO responses to consuming two isonitrogenous EAA solutions with differing concentrations of leucine were assessed in older and young adults. Ingestion of both leucine concentrations increased rates of MPS in young adults, but in older adults, MPS was increased only when consuming the EAA supplement with the greater concentration of leucine [35]. Others have confirmed this report [68]. The beneficial effects of leucine in older adults may be due to intramuscular expression of the novel class 3 PI3K, hVps34 (vacuolar protein sorting 34), which was recently demonstrated to be elevated in skeletal muscle cells of older compared to younger adults [69]. Although the actions of hVps34 have not yet confirmed in human muscle, hVps34 has been identified as a unique amino acid sensing mechanism that affects mTORC1 activation in cell models [70]. Greater intramuscular hVps34 expression in older adults could provide a mechanism by which aging skeletal muscle may overcome the blunted MPS response to feeding that generally accompanies advanced age.

In recent years, a number of cell culture and animal studies have assessed the efficacy of leucine for the treatment and management of cachexia. In tumor-bearing mice, supplementing with leucine, or the leucine metabolite β-hydroxy-β-methylbutyrate, preserved skeletal muscle mass by attenuating UP activity and increasing MPS [57, 71]. The expression of 26S catalytic core subunits was also reduced in tumor-bearing mice that consumed a leucine-enriched diet [72]. Reduced protein expression corresponded with lower MPB, and higher

MPS, which suggests that leucine may mitigate muscle loss during cachectic conditions by both attenuating UP-mediated proteolysis and stimulating MPS. Consuming a leucine-enriched diet also limits cachexia-induced weight loss, and more importantly, conserves the myosin content of rat skeletal muscle [73]. Another recent animal study demonstrated that combining low levels of aerobic-type physical activity with a leucine-enriched diet potentiates the beneficial effects of leucine by further attenuating muscle wasting and lowering the expression of key components of the UP compared to consuming a leucine-enriched diet without additional physical activity [74]. Positive results from these cell culture and animal studies provide a foundation the implementation of clinical trials designed to assess whether leucine functions as an effective therapeutic aid to counteract the severe loss of muscle associated with cachexia.

CONCLUSION/RECOMMENDATIONS

Skeletal muscle plays a critical role in metabolic homeostasis, strength, functionality, and overall quality of life. However, many pathophysiological conditions are associated with decrements in skeletal muscle mass. Severe declines in skeletal muscle mass increase susceptibility to injury, decrease performance, and ultimately increase morbidity and mortality. However, nutritional interventions that promote increased consumption of BCAA, in particular leucine, may attenuate declines in skeletal mass by attenuating MPB and stimulating MPS through modulation of nutrient-sensitive signaling steps in the intracellular regulation of SMPTO. Nutritional interventions that promote consuming high-quality dietary protein in excess of the RDA have proven to be advantageous for the protection of skeletal muscle in response to aging and prolonged energy restriction. Although the protective effect of leucine on skeletal muscle in response to chronic cachectic conditions has not been elucidated, recent evidence from animal studies suggests that protein-based nutrition countermeasures could likely be proven effective for preventing muscle loss under that pathological state.

DISCLAIMER

The opinions or assertions contained herein are the private views of the authors and are not to be construed as official or as reflecting the views of the Army or Department of Defense. Any citations of commercial organizations and trade names in this report do not constitute an official Department of the Army endorsement of approval of the produces or services of these organizations.

REFERENCES

[1]   R. R. Wolfe, *Am. J. Clin. Nutr.* 84, 475, (2006).
[2]   P. Felig, O. E. Owen, J. Wahren and G. F. Cahill Jr., *J. Clin. Invest.* 48, 584, (1969b).

[3]   P. Felig, E. Marliss, O. E. Owen and G. F. Cahill Jr., *Adv. Enzyme Regul.* 7, 41, (1969a).

[4]   A. E. Harper, R. H. Miller and K. P. Block, *Annu. Rev. Nutr.* 4, 409, (1984).

[5]   R. R. Wolfe, *J. Nutr.* 132, 3219S, (2002).

[6]   M. J. Drummond and B. B. Rasmussen, *Curr. Opin. Clin. Nutr. Metab. Care* 11, 222, (2008).

[7]   M. G. Buse and S. S. Reid, *J. Clin. Invest.* 56, 1250, (1975).

[8]   S. M. Pasiakos, H. L. McClung, J. P. McClung, L. M. Margolis, N. E. Andersen, G. J. Cloutier, M. A. Pikosky, J. C. Rood, R. A. Fielding and A. J. Young, *Am. J. Clin. Nutr.* 94, 809, (2011).

[9]   K. Nakashima, A. Ishida, M. Yamazaki and H. Abe, *Biochem. Biophys. Res. Commun.* 336, 660, (2005).

[10]   D. R. Bolster, M. A. Pikosky, P. C. Gaine, W. Martin, R. R. Wolfe, K. D. Tipton, D. Maclean, C. M. Maresh and N. R. Rodriguez, *Am. J. Physiol. Endocrinol. Metab.* 289(4), E678, (2005).

[11]   S. M. Pasiakos, L. M. Vislocky, J. W. Carbone, N. Altieri, K. Konopelski, H. C. Freake, J. M. Anderson, A. A. Ferrando, R. R. Wolfe and N. R. Rodriguez, *J. Nutr.* 140, 745, (2010b).

[12]   L. E. Norton and D. K. Layman, *J. Nutr.* 136, 533S, (2006).

[13]   S. M. Phillips, J. W. Hartman and S. B. Wilkinson, *J. Am. Coll. Nutr.* 24, 134S, (2005).

[14]   M. J. Drummond, H. C. Dreyer, C. S. Fry, E. L. Glynn and B. B. Rasmussen, *J. Appl. Physiol.* 106, 1374, (2009a).

[15]   S. M. Pasiakos and J. P. McClung, *Nutr. Rev.* 69, 550, (2011).

[16]   S. H. Lecker, A. L. Goldberg and W. E. Mitch, *J. Am. Soc. Nephrol.* 17, 1807, (2006).

[17]   V. R. Rajan and W. E. Mitch, *Pediatr. Nephrol.* 23, 527, (2008).

[18]   S. C. Kandarian and R. W. Jackman, *Muscle Nerve* 33, 155, (2006).

[19]   T. Andreyeva, M. W. Long, K. E. Henderson and G. M. Grode, *J. Am. Diet Assoc.* 110, 535, (2010).

[20]   K. E. Friedl and G. P. Bathalon, *Mil. Med.* 171(6), vi, (2006).

[21]   E. M. Weinheimer, L. P. Sands and W. W. Campbell, *Nutr. Rev.* 68, 375, (2010).

[22]   D. H. Calloway and H. Spector, *Am. J. Clin. Nutr.* 2, 405, (1954).

[23]   D. H. Calloway, *J. Nutr.* 105, 914, (1975).

[24]   T. P. Stein, W. V. Rumpler, M. J. Leskiw, M. D. Schluter, R. Staples and C. E. Bodwell, *Metabolism,* 40, 478, (1991).

[25]   S. M. Pasiakos, H. L. McClung, J. P. McClung, M. L. Urso, M. A. Pikosky, G. J. Cloutier, R. A. Fielding and A. J. Young, *Int. J. Sport Nutr. Exerc. Metab.* 20, 282, (2010a).

[26]   E. Ravussin, S. Lillioja, W. C. Knowler, L. Christin, D. Freymond, W. G. Abbott, V. Boyce, B. V. Howard and C. Bogardus, *N. Engl. J. Med.* 318, 467, (1988).

[27]   R. N. Baumgartner, K. M. Koehler, D. Gallagher, L. Romero, S. B. Heymsfield, R. R. Ross, P. J. Garry and R. D. Lindeman, *Am. J. Epidemiol.* 147, 755, (1998).

[28]   W. J. Evans, *Am. J. Clin. Nutr.* 91, 1123S, (2010).

[29]   J. E. Morley, *Nutrition* 17, 660, (2001).

[30]   Y. Boirie, *J. Nutr. Health Aging* 13, 717, (2009).

[31]   P. Kortebein, A. Ferrando, J. Lombeida, R. Wolfe and W. J. Evans, *JAMA* 297, 1772, (2007).

[32] J. E. Morley, J. M. Argiles, W. J. Evans, S. Bhasin, D. Cella, N. E. Deutz, W. Doehner, K. C. Fearon, L. Ferrucci, M. K. Hellerstein, K. Kalantar-Zadeh, H. Lochs, N. MacDonald, K. Mulligan, M. Muscaritoli, P. Ponikowski, M. E. Posthauer, F. F. Rossi, M. Schambelan, A. M. Schols, M. W. Schuster and S. D. Anker, *J. Am. Med. Dir. Assoc.* 11, 391, (2010).

[33] M. Cesari, S. B. Kritchevsky, R. N. Baumgartner, H. H. Atkinson, B. W. Penninx, L. Lenchik, S. L. Palla, W. T. Ambrosius, R. P. Tracy and M. Pahor, *Am. J. Clin. Nutr.* 82, 428, (2005).

[34] J. E. Morley, *J. Am. Geriatr. Soc.* 51, S333, (2003).

[35] C. S. Katsanos, H. Kobayashi, M. Sheffield-Moore, A. Aarsland and R. R. Wolfe, *Am. J. Physiol. Endocrinol. Metab.* 291(2), E381, (2006).

[36] B. H. Goodpaster, S. W. Park, T. B. Harris, S. B. Kritchevsky, M. Nevitt, A. V. Schwartz, E. M. Simonsick, F. A. Tylavsky, M. Visser and A. B. Newman, *J. Gerontol. A Biol. Sci. Med. Sci.* 61, 1059, (2006).

[37] S. W. Park, B. H. Goodpaster, E. S. Strotmeyer, R. N. de Rekeneire, T. B. Harris, A. V. Schwartz, F. A. Tylavsky and A. B. Newman, *Diabetes,* 55, 1813, (2006).

[38] A. B. Newman, V. Kupelian, M. Visser, E. M. Simonsick, B. H. Goodpaster, S. B. Kritchevsky, F. A. Tylavsky, S. M. Rubin and T. B. Harris, *J. Gerontol. A Biol. Sci. Med. Sci.* 61, 72, (2006).

[39] W. J. Evans, J. E. Morley, J. Argiles, C. Bales, V. Baracos, D. Guttridge, A. Jatoi, K. Kalantar-Zadeh, H. Lochs, G. Mantovani, D. Marks, W. E. Mitch, M. Muscaritoli, A. Najand, P. Ponikowski, F. F. Rossi, M. Schambelan, A. Schols, M. Schuster, D. Thomas, R. Wolfe and S. D. Anker, *Clin. Nutr.* 27, 793, (2008).

[40] J. E. Morley, D. R. Thomas and M. M. Wilson, *Am. J. Clin. Nutr.* 83, 735, (2006).

[41] S. Acharyya and D. C. Guttridge, *Clin. Cancer Res.,* 13, 1356, (2007).

[42] F. Dworzak, P. Ferrari, C. Gavazzi, C. Maiorana and F. Bozzetti, *Cancer,* 82, 42, (1998).

[43] M. Bossola, M. Muscaritoli, P. Costelli, R. Bellantone, F. Pacelli, S. Busquets, J. Argiles, F. J. Lopez-Soriano, I. M. Civello, F. M. Baccino, F. Rossi Fanelli and G. B. Doglietto, *Am. J. Physiol. Regul. Integr. Comp. Physiol.* 280, R1518, (2001).

[44] A. Williams, X. Sun, J. E. Fischer and P. O. Hasselgren, *Surgery,* 126, 744, (1999).

[45] M. Bossola, M. Muscaritoli, P. Costelli, G. Grieco, G. Bonelli, F. Pacelli, F. Rossi Fanelli, G. B. Doglietto and F. M. Baccino, *Ann. Surg.,* 237, 384, (2003).

[46] S. Acharyya, K. J. Ladner, L. L. Nelsen, J. Damrauer, P. J. Reiser, S. Swoap and D. C. Guttridge, *J. Clin. Invest.* 114, 370, (2004).

[47] D. C. Guttridge, M. W. Mayo, L. V. Madrid, C. Y. Wang and A. S. Baldwin Jr., *Science* 289, 2363, (2000).

[48] V. E. Baracos and M. L. Mackenzie, *J. Nutr.,* 136, 237S, (2006).

[49] H. A. Choudry, M. Pan, A. M. Karinch and W. W. Souba, *J. Nutr.,* 136, 314S, (2006).

[50] S. Fujita and E. Volpi, *J. Nutr.* 136, 277S, (2006).

[51] A. A. Ferrando, B. D. Williams, C. A. Stuart, H. W. Lane and R. R. Wolfe, *J. Parenter. Enteral Nutr.,* 19, 47, (1995).

[52] D. K. Walker, J. M. Dickinson, K. L. Timmerman, M. J. Drummond, P. T. Reidy, C. S. Fry, D. M. Gundermann and B. B. Rasmussen, *Med. Sci. Sports Exerc.* 43, 2249, (2011).

[53] D. K. Layman, *J. Am. Coll. Nutr.* 23, 631S, (2004).

[54] D. Paddon-Jones, M. Sheffield-Moore, C. S. Katsanos, X. J. Zhang and R. R. Wolfe, *Exp. Gerontol.*, 41, 215, (2006).

[55] L. Breen and S. M. Phillips, *Nutr. Metab. (Lond).* 8, 68, (2011).

[56] D. Paddon-Jones and B. B. Rasmussen, *Curr. Opin. Clin. Nutr. Metab. Care*, 12, 86, (2009).

[57] H. L. Eley, S. T. Russell and M. J. Tisdale, *Biochem. J.* 407, 113, (2007b).

[58] S. Busquets, B. Alvarez, F. J. Lopez-Soriano and J. M. Argiles, *J. Cell. Physiol.* 191, 283, (2002).

[59] R. A. Bonau, M. Jeevanandam and J. M. Daly, *J. Parenter. Enteral Nutr,* 8, 622, (1984).

[60] R. A. Bonau, M. Jeevanandam, L. Moldawer, G. L. Blackburn and J. M. Daly, *Surgery,* 101, 400, (1987).

[61] J. M. Daly, M. H. Mihranian, J. E. Kehoe and M. F. Brennan, *Surgery,* 94, 151, (1983).

[62] J. A. Tayek, B. R. Bistrian, D. J. Hehir, R. Martin, L. L. Moldawer and G. L. Blackburn, *Cancer,* 58, 147, (1986).

[63] S. T. Fan, C. M. Lo, E. C. Lai, K. M. Chu, C. L. Liu and J. Wong, *N. Engl. J. Med.,* 331, 1547, (1994).

[64] R. T. Poon, W. C. Yu, S. T. Fan and J. Wong, *Aliment. Pharmacol. Ther.* 19, 779, (2004).

[65] W. C. Meng, K. L. Leung, R. L. Ho, T. W. Leung and W. Y. Lau, *Aust. N. Z. J. Surg.* 69, 811, (1999).

[66] J. P. De Bandt and L. Cynober, *J. Nutr.* 136, 308S, (2006).

[67] E. L. Glynn, C. S. Fry, M. J. Drummond, K. L. Timmerman, S. Dhanani, E. Volpi and B. B. Rasmussen, *J. Nutr.* 140, 1970, (2010).

[68] I. Rieu, M. Balage, C. Sornet, C. Giraudet, E. Pujos, J. Grizard, L. Mosoni and D. Dardevet, *J. Physiol.* 575, 305, (2006).

[69] M. J. Drummond, M. Miyazaki, H. C. Dreyer, B. Pennings, S. Dhanani, E. Volpi, K. A. Esser and B. B. Rasmussen, *J. Appl. Physiol.* 106, 1403, (2009b).

[70] M. P. Byfield, J. T. Murray and J. M. Backer, *J. Biol. Chem.* 280, 33076, (2005).

[71] H. L. Eley, S. T. Russell, J. H. Baxter, P. Mukerji and M. J. Tisdale, *Am. J. Physiol. Endocrinol. Metab.* 293, E923, (2007a).

[72] G. Ventrucci, L. G. Ramos Silva, M. A. Roston Mello and M. C. Gomes Marcondes, *Nutrition,* 20, 213, (2004).

[73] M. C. Gomes-Marcondes, G. Ventrucci, M. T. Toledo, L. Cury and J. C. Cooper, *Braz. J. Med. Biol. Res.* 36, 1589, (2003).

[74] E. M. Salomao, A. T. Toneto, G. O. Silva and M. C. Gomes-Marcondes, *Nutr. Cancer* 62, 1095, (2010).

In: Skeletal Muscle
Editor: Mark Willems

ISBN: 978-1-62417-271-7
© 2013 Nova Science Publishers, Inc.

Chapter 11

ROLES OF MYOSTATIN PROPEPTIDE IN PROMOTING SKELETAL MUSCLE GROWTH AND METABOLISM

Jinzeng Yang[*]

Department of Human Nutrition, Food and Animal Sciences,
University of Hawaii at Manoa, Honolulu, Hawaii, US

ABSTRACT

Skeletal muscles provide the physiological foundation for physical activities and fitness. Maintaining proper muscle mass in obesity, insulin resistance/type II diabetes is important for effective treatments of these diseases. One key protein factor named myostatin, primarily produced and secreted from skeletal muscle, is a dominant inhibitor of muscle mass. Our earlier studies indicate that depression of myostatin by its propeptide in transgenic mice produces dramatic muscle mass at the growth stage and less fat at older ages. Muscle tissue utilizes a large portion of metabolic energy for its growth and maintenance. We demonstrated that transgenic over-expression of myostatin propeptide in mice fed a high-fat diet enhanced muscle mass and circulating adiponectin while the wild-type mice developed obesity and insulin resistance. To understand the effects of enhanced muscle growth on adipose tissue metabolism, we analyzed adiponectin, PPAR-alpha, and PPAR-gamma mRNA expressions in several fat tissues. Results indicated that muscled transgenic mice fed a high-fat diet displayed increased epididymal adiponectin mRNA expression by 12 times over wild-type littermates. These transgenic mice fed either a high or normal fat diet also displayed significantly high levels of PPAR-alpha and PPAR-gamma expressions above their wild-type littermates in epididymal fat while their expressions in mesenteric fats were not significantly different between transgenic mice and their littermates. These results demonstrate that enhanced muscle growth has positive effects on fat metabolism through increasing adiponectin expression and its regulations. Enhancing skeletal muscle growth and metabolism provide significant benefit for preventing obesity and diabetes.

---

[*] E-mail address: jinzeng@hawaii.edu

INTRODUCTION

The prevalence of obesity is increasing rapidly worldwide. Approximately one-third of the US adult population are obese according to the 2011 CDC report [1]. Most troubling is the dramatic increase in juvenile obesity. The percentage of young people who are overweight has more than doubled since 1980. Type 2 diabetes and insulin resistance, which are typically associated with adult obesity, are expected to increase dramatically in children and adolescents. What causes this obesity prevalence? Industrializations, smart computers and cellular phones, high-tech entertainments and high-speed transportations, all of them make life so easy and wonderful, but need minimal physical efforts and are diminishing the essential use and need of the skeletal muscle system of the body. Sadly, this is probably one of the most serious consequences that we have to live with. These technologies are popular and widespread in almost every occupation and career. Among other factors, dietary changes and profit-driving food production and industry are also contributing significantly to the physiology and nutrient supplies in our bodies every day. Our genes and genetics have not changed over the generation of our grandparents. However, the interactions between the genome and the functional expressions effected by the nutrients and living environments changed. There is not a simple answer to what causes the prevalence of obesity today. To prevent such problems and live with these technologies, I believe that skeletal muscle and maintaining a healthy skeletal muscle physiology hold an important step for a healthy life in the industrialized societies with more mental work and less physical activities. This chapter first begins with the basic introduction of physiological roles of skeletal muscle in preventing obesity and diabetes, then proceeds to a key regulatory protein for muscle mass, namely myostatin, followed by an updated progress in myostatin propeptide research and its roles in the regulation of skeletal muscle growth and metabolism, and implications for obesity and diabetes preventions.

SKELETAL MUSCLE, PHYSICAL EXERCISE AND OBESITY

Obesity could be simplified as a result of energy intake exceeding energy expenditure. A large portion of the body is composed of skeletal muscle and adipose tissues, both of which are involved in energy metabolism. Skeletal muscle is the major organ responsible for insulin-mediated glucose uptake. Insulin can robustly stimulate muscle glucose uptake from the bloodstream through increased glucose transport and glycogen synthesis. Insulin resistance, a defect in the ability of insulin to drive glucose into its major target tissues-skeletal muscle and liver, is very common in obesity and type 2 diabetes. Adipose tissue serves as an energy storage depot, as well as an endocrine organ to maintain lipid homeostasis. Since energy storage as fat promotes survival when food supplies are scarce, evolution has favored certain populations developing "thrifty" genotypes to store energy in adipose tissue [2]. Metabolic problems such as obesity and insulin resistance happen when adipose tissue is overloaded with abundant high-energy nutrients without subsequent expenditure. Increased triglycerides not only disrupt adipocyte endocrinal function, but also cause dysfunctions of skeletal muscle such as insulin resistance.

Skeletal muscles provide the physiological foundation for physical activities and fitness. Active use and mobilization of stored fat and glycogen by muscle tissue through physical activity can be a very effective means in regulation of metabolic activities and energy balance, a great benefit or relief for insulin resistance and obesity. To prevent juvenile obesity and type II diabetes, health professionals recommend appropriate physical activity and balanced diet. American Diabetes Association suggests regular exercising at least 3 d/wk for at least 150 min/wk, with no more than 2 consecutive days without physical activity. Enhanced physical activity modifies body composition and energy metabolism. A single session of physical exercise such as resistance exercise powerfully regulate fat and glucose metabolism and result in a remarkable improvement of insulin sensitivity by stimulating glucose transport. In addition, physical activities such as resistance exercise promote strength gains with better neuromuscular function, and encompass muscle fiber hypertrophy, which are particularly important for the middle-aged and seniors due to aging-induced loss of muscle mass and endurance. Having or maintaining appropriate muscle mass at any age empowers exercise capacity, resulting in significant beneficial effects and treatment effects on diabetes. Most diabetic medicines are targeting reduction of food digestion and glucose production in the digestion system. Enhanced muscle mass and associated exercise capacity would have minimal side effects, but add significant improvement of the drug treatment on diabetes since skeletal muscle is always in dynamic and active metabolism, and muscle mass is directly related to the energy expenditure. In the modern societies with less physical efforts and activities, sufficient skeletal muscle for most adults actually is "the biological weapon", which can be used very effectively to prevent obesity and diabetes.

MYOSTATIN AND SKELETAL MUSCLE MASS

Skeletal muscle is made up of myofibers, which are developed from fusion of many myoblast cells. Studies demonstrated that myoblasts are rapidly dividing cells in culture, but cease the proliferation process or DNA synthesis once they fuse into a myotube [3]. Therefore, postnatal muscle growth primarily depends upon an increase in muscle fiber size or myofiber hypertrophy. The growth of skeletal muscles growth occurs dramatically during adolescence, along with physical skeletal development for the full functional capacity of musculoskeletal system. The rate of growth rapidly decelerates in late stages of adulthood as adipose tissue gradually accumulates more fat when energy intake exceeds expenditure in the body. Myostatin is member of transforming growth factor β (TGF-β) superfamily, it acts as a strong inhibitor of myogenesis and skeletal muscle mass as myostatin-null or knock-out mice showed a two-fold increase in individual muscle mass over the wild-type mice, as well as suppression of body fat accumulation [4, 5]. Myostatin mutations have been associated with dramatic muscle mass in cattle [6], human [7], dogs [8] and sheep [9].

It is believed that myostatin is primarily synthesized in skeletal muscle, and secreted to the intramuscular spaces. After posttranslational processing through latent associated complex mechanism, myostatin binds to activin receptor type IIB on the plasma membranes of the neighboring muscle cells, then activating Smad2/3 signaling pathway to inhibit myoblast cell proliferation and differentiation.

Cyclin-dependent kinases (Cdks) regulate G1 phase transitions to S phase in a cell cycle. Myostatin is able to increase Cdk inhibitor p21 activity, therefore decreasing the Cdk levels, concurrently resulting in myoblast cell cycle arrest in the G1 phase [10, 11]. Myostatin also inhibits MyoD expression and activity via Smad3, which blocks myoblasts from differentiating into myotubes [12, 13]. Therefore, myostatin inhibits both myoblast cell proliferation and differentiation. Myostatin is also expressed in satellite cells and adult myoblasts. It negatively regulates the G1 to S progression of satellite cells to maintain their quiescent status [14]. Muscles from myostatin-null mice have an increased number of satellite cells as well as a higher proportion of activated satellite cells than muscles of wild-type mice [14, 15].

These results clearly suggest that myostatin maintains satellite cells in a quiescent state in adult muscle. However, recent studies indicate that there is not any increase in satellite cells and the rate of satellite cell proliferation in the myostatin-knockout mice compared to wild-type mice [16]. Muscle hypertrophy through myostatin inhibition is mainly caused by its effects on myofibers rather than satellite cells. Age-induced loss in muscle mass and satellite cells also occurs in myostatin-null mice [17]. Therefore, the researchers believed that enhanced muscle mass in myostatin-null mice might not result from the activation of satellite cells in skeletal muscle. As satellite activation occurs mostly when muscle tissue is injured, the degree of satellite cell activation in normal growth and aging is not known, it is worth further investigation on satellite cell activation in myostatin-genetically manipulated animals.

In the mean time, there are several studies that have demonstrated that enhanced muscle mass in myostatin-manipulated mice primarily resulted from myofiber protein metabolism. The effects of myostatin on myofiber protein metabolism was demonstrated by the observation that myostatin induces cachexia by activating the ubiquitin proteolytic system through a FOXO1-dependent mechanism. A positive feedback mechanism between myostatin and FOXO1 pathways may amplify the atrophic response [18]. In our recent study, we have observed that levels of phosphorylated 4E-BP1 (Thr37/46) and p70S6k (Thr389), two key downstream effectors of the mTOR pathway in regulation of protein synthesis, were greater in the transgenic mice with depressed myostatin function compared with wild-type mice. This suggests that myostatin regulates muscle mass probably through the mTOR pathway [19].

MUSCLE MASS LOSS AND MYOSTATIN EXPRESSION IN DIABETES

A loss of skeletal muscle mass is frequently observed in diabetes. In type I diabetes, insulin deprivation increase catabolism in skeletal muscle, which can result in serious weight loss in the form of both muscle and adipose tissue mass without insulin treatment. Similarly, an excessive loss of skeletal muscle mass is also widely observed in older adults with type 2 diabetes while they may not have significant body weight loss as insulin resistance reduced overall whole-body protein synthesis [20]. Endocrine glands specialize in secretion of hormones into blood circulation. Apparently, both skeletal muscle and adipose tissue serve as endocrine or paracrine functions by constitutively and intermittently secreting proteins that regulate metabolic and cellular events. Recent evidence further define that myostatin is one of the secreted bioactive proteins. In Sprague Dawley rats, the levels of myostatin protein increased progressively at 6, 12, and 27 months of age compared with young animals [21]. In streptozotocin-induced diabetes, myostatin expressions in muscle showed significant increases from initial administration, which corresponded well with body weight loss. While insulin treatment reversed the hyperglycemia and body weight loss, myostatin expression in these mice was also attenuated [22]. A recent study in human subjects also demonstrated the increased levels of myostatin in skeletal muscle and plasma in obese women compared with healthy women. The levels of increased myostatin are significantly correlated with the severity of insulin resistance [23]. It is concluded from several studies in healthy, diabetic humans and injection of myostatin in mice that increased myostatin expression is inversely correlated with insulin sensitivity. It was believed that myostatin might directly regulate skeletal muscle glucose uptake [24]. It was hypothesized that myostatin may act as a chalone to limit muscle size [25]. As the size of muscle tissue is dynamic with great plasticity in response to various physiological and biochemical stimuli, myostatin expression may represent a significant contribution to the deterioration or loss of muscle mass in diabetes. As the effects of muscle mass on insulin sensitivity are predominantly significant, the direct effects of myostatin on muscle glucose uptake hardly distinguish from its indirect effects simply due to its primary genetic regulations on muscle mass.

MYOSTATIN PROPEPTIDE AND ENHANCED MUSCLE GROWTH

Like most members of transforming growth factor β (TGF-β) superfamily, the mature form of myostatin is generated by cleavage of the precursor protein at the tetrapeptide (RSRR) site. The remaining N-terminal peptide is named propeptide (previously called prodomain). We and others have demonstrated that transgenic expression of the propeptide cDNA sequence successfully depresses myostatin function, and promoted muscularity phenotype [26-28]. In our transgenic mice, generated by muscle-specific expression of the propeptide (the 5'-region 886 nucleotides) of myostatin we observed significant muscling phenotypes: 20% faster growth rate and 44% more muscle mass than wild-type mice [27]. In comparison with myostatin-knockout mice, the propeptide transgenic mice still produce myostatin in the muscle and maintain normal adipose tissues. To determine whether enhanced muscle growth minimizes the incidence of diet-induced obesity, we studied the response of the propeptide transgenic mice to a high-fat diet. We showed that the propeptide transgene not only enhanced muscle growth, but also prevented dietary fat-induced obesity and insulin resistance [29, 30]. While wild-type mice on a high-fat diet for two months developed serious adiposity, impaired glucose tolerance and insulin resistance, the transgenic mice are normal and healthy, accommodating a metabolic regulatory system that utilizes dietary fat for muscle growth and maintenance. Individual major muscles of transgenic mice were 45–115% heavier than those of wild-type mice, maintained normal blood glucose, insulin sensitivity and fat mass after a two-month regimen with a high-fat diet (45% kcal fat).

In contrast, high-fat diet induced wild-type mice with 170-214% more fat mass than transgenic mice (Figure 1), and developed impaired glucose tolerance and insulin resistance. Insulin signaling, analyzed by Akt phosphorylation, was elevated by 144% in transgenic mice over wild-type mice fed a high-fat diet.

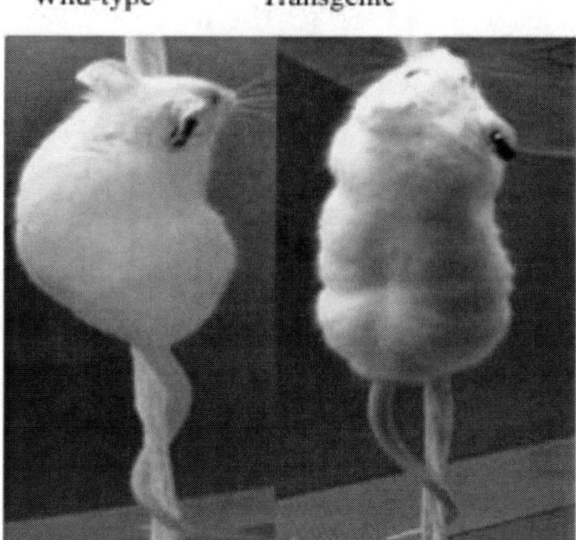

Figure 1. In a wild-type mouse with normal levels of the protein myostatin (left), a high-fat diet leads to obesity and loss of insulin sensitivity. On the same diet, a transgenic mouse that produces less myostatin (right) stays lean and retains its insulin sensitivity.

These findings from animal model clearly suggest that muscle development and buildup plays a fundamental role in balancing the utilization and expenditure of body energy. The results support our hypothesis that well-developed skeletal muscles in early stage of life increase the flexibility of the body in utilizing dietary fat or other energy resource, so that the body would be better equipped to prevent obesity and its associated metabolic disorders in adulthood or later stages of life [29].

Interestingly, the mice have increased adiponectin secretion [31]. Adiponectin is a 30-kDa adipocytokine hormone secreted primarily by the adipose tissue. It is a relatively abundant serum protein and accounts for between 0.01-0.03% of total serum protein in humans [32]. Increased serum levels of adiponectin are associated with increased insulin sensitivity, increased fatty acid oxidation, and decreased hepatic glucose production [32]. Obesity tends to increase the plasma concentrations of most proteins produced by adipose tissue because of the increase in total fat mass, but plasma levels of adiponectin were found to be much lower in obesity [32]. Significantly lower levels of adiponectin were also observed in fat tissues from obese mice and obese humans [33, 34]. Adiponectin can ameliorate insulin resistance by decreasing hepatic glucose production through lowering activity of the gluconeogenic enzyme phosphoenolpyruvate carboxykinase (PEPCK) [35]. Along with its ability to decrease the activity of hepatic gluconeogenic enzymes, adiponectin improves insulin sensitivity by decreasing the concentration of circulating triglycerides and free fatty acids [36]. To investigate the hormonal changes caused by high-fat diet and the propeptide transgene, we measured insulin and several adipocyte hormones [29]. The concentrations of insulin, leptin and resistin were similar in transgenic mice fed either a normal or a high-fat diet; these levels were slightly higher but not significantly different from those seen in wild-type mice fed a normal diet. However, high-fat diet increased serum insulin, leptin and resistin by 97%, 377% and 32%, respectively, in wild-type mice compared to transgenic mice

[29]. High-fat diet induced much higher levels of serum adiponectin concentration in the transgenic mice than that was found in the wild-type mice (9.44 ± 0.79 vs. 7.90 ± 0.68 µg/ml, P=0.008) while transgenic mice fed a normal diet had significantly lower serum adiponectin levels (4.35 ± 0.60 µg/ml).

When we studied the gene expression of adiponectin in adipose tissue, myostatin propeptide transgenic mice fed either normal-fat or high-fat diet showed significant increased levels of adiponectin expression in epididymal fat pad. The relative expression level of adiponectin was the highest in the epididymal fat in the trangenic/high fat group with a value of 3.10±0.67, which is a 12-time increase over its control group-wild type /high fat diet. The subcutaneous fat of the transgenic/normal fat group also showed an increased level over its control group-transgenic/normal fat. Within the epididymal fat pad, the wild-type/high fat group was significantly lower in adiponectin expression than wild-type/normal fat group. These results clearly suggest that an increased secretion of adiponectin may promote energy partition toward skeletal muscles, therefore promoting a beneficial interaction between muscle and adipose tissue [31]. In addition, ex-vivo delivery of a mutant form of myostatin propeptide also functionally improved dystrophic muscle in mice. By AAV-mediated expression of a mutated propeptide, researchers showed a boost in muscle mass and an increase in absolute force in calpain 3-deficient mice. Myostatin inhibition by its propeptide over expression could be effective for therapeutic treatment of atrophic disorder [37]. A similar experiment with mdx mice, a murine model of Duchenne muscular dystrophy, was conducted by adeno-associated virus serotype 8 delivery of myostatin propeptide. There was a significant increase in skeletal muscle mass in the form of fiber hypertrophy after the propeptide virus injection. A grip force test and an in vitro tetanic contractile force test showed improved muscle strength [38].

Myostatin sequence is highly conserved among mammalian species. Mouse, pig, human and chicken myostatin are 100% identical in the amino acid sequence of the mature peptide [4]. Pig is becoming a valuable model animal for biomedical research. We have created a mutated form of pocine myostatin propeptide at the cleavage site of metalloproteinase BMP-1/TLD family, which strongly blocked myostatin activity in A204 cell assay. Administration of this mutated propeptide to mice significantly increased skeletal muscle growth. Enhanced muscle growth by two times of injection of the propetide in neonatal stages can last 4 weeks after injection and primarily resulted from fiber hypertrophy. These results suggest that depression of myostatin activity by BMP-1/TLD proteinase-resistant propeptide can be an effective way to promote muscle growth [39]. The result is consistent with early experimental data with murine myostatin propeptide obtained by Wolfman et al [40], where the concentration of murine propeptide at 500 pM were used for suppressing 1 pM myostatin. However, the result of porcine myostatin propeptide from this study may appear more effective as the concentration of the propeptide at 8 pM (100 ng/ml) depressed the activity of 2pM (20 ng/ml) myostatin. Mice injected with the mutated propeptide showed a significant increase (P<0.05) in body weight from littermate controls as early as one week after the first injection (at the age of 11 days) in females, and one week after the second injection in males (18 days of age). After the second injection, male mice injected with mutated propeptide were 12-15% heavier than their controls from the age of 25 days to 57 days (P<0.05). Whereas, female mice injected with the propeptide were 11-15% heavier than their littermate controls (P<0.05) from the age of 25 to 57 days. The effect of the injections of mutated propeptide on muscle growth was maintained for at least 4 weeks from the age of 25 days to 53 days [39].

Taken together, these findings indicate that suppressing myostatin activity by its propeptide in native or mutated form can be an effective strategy to enhance muscle growth. The size and amount of skeletal muscle certainly are important factors for muscle strength. A therapeutic strategy based on myostatin propetide for muscle dystrophy or any other muscle wasting appears effective in small animals. Further testing and ex-vivo delivery to a large animal model such as pigs would provide initial evidence for application to human therapy.

MOLECULAR MECHANISM OF MYOSTATIN PROPETIDE ON MUSCLE GROWTH

The mechanism by which myostatin function is disrupted by its propeptide has been proposed. Results from both *in vitro* and *in vivo* support the mechanism of myostatin latency during posttranslational processing. The myostatin precursor protein can be cleaved by furin proteases *in vitro*, producing N-terminal propeptide and C-terminal mature peptide or myostatin. The propeptide remains non-covalently bound to the mature C-terminal dimer in a latent state [40- 42], which also appears to circulate in the blood [43]. In transgenic mice with mutated BMP-1/TLD family of metalloproteases, which can block the cleavage pathway, myostatin function is partially inhibited as the muscle mass is increased, along with increased circulating myostatin levels [44]. Therefore, myostatin propeptide is through the latency-associated complex to regulate the release of myostatin peptide.

In the myostatin propeptide transgenic mice in my laboratory, a large, latent form of myostatin complex was detected in skeletal muscle [unpublished]. The molar ratio between propeptide and mature myostatin in serum is approximately 1:1, and no data is available about the molar ratio in muscle. As the propeptide and mature myostatin are generated from the same precursor peptide, we assumed their molar ratio is close to 1:1 in muscle tissue. A transgenic over-expression of the propeptide may promote forming of more myostatin-latent complex, therefore resulting in less biologically active mature myostatin in muscle or even in circulation. In the propeptide transgenic mice, transgene mRNA was not detected in the adipose tissue, excluding its possible direct effects on adipose tissue. We believe that enhanced muscling induced by the transgene during the early-growing stage significantly changed nutrient partitions in the body. Energy demand for muscle growth and maintenance in the transgenic mice was so prominent that less energy was available for fat accumulation, consequently, they maintained normal adipose tissue and a high insulin sensitivity [29, 30]. Interestingly, when we did muscle histology in the old transgenic animals, the observation of the myofiber histology indicated more nuclei were localized in the central and basal lamina of the myofibers of the transgenic mice at one year old (Figure 2). Long and stretched nuclei were also noted in some fibers, suggesting active fiber fusion in these muscles at one year old. In contrast, muscle histology from wild-type littermates did not show such changes in nuclei distributions. The number of nuclei per fiber in both basal and central lamina of the myofiber were significantly higher in transgenic mice than in wild-type littermates by 58.3% and 458%, respectively (P < 0.01). These results provide evidences that transgenic expression of myostatin propeptide supported continuous muscle build-up in adult skeletal muscle tissue [45].

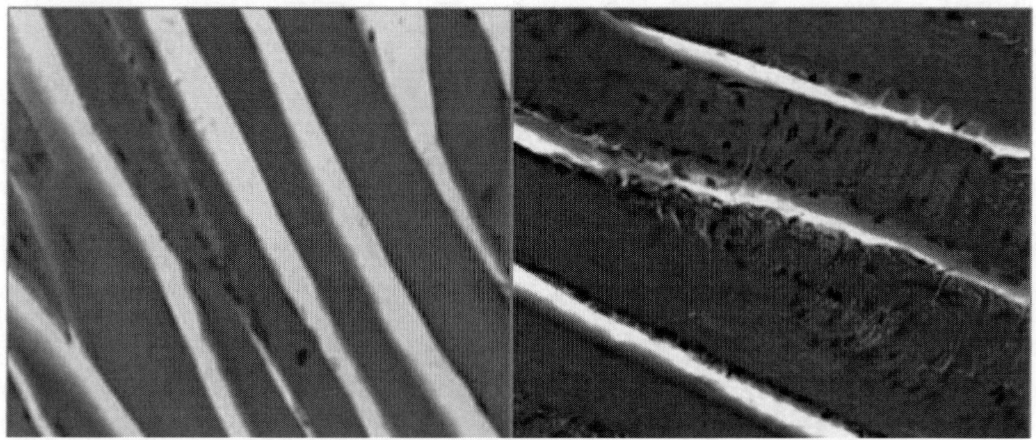

Figure 2. Comparisons of muscle fiber staining of biceps femoris between wild-type and transgenic mice. Mice were sacrificed at 12 months of age. Biceps femoral muscle samples were used for this study, and samples were frozen-sectioned. Hematoxylin and Eosin Staining of both wild-type (left) and transgenic muscle sample (right) shows the myofiber nuclei. Muscle tissue was longitudinally sectioned to show myofiber nuclei.

To understand how the enhanced muscle growth by myostatin propetide are regulated at the whole genome level, we employed microarray analysis. The results from microarray analysis of global gene expression profile, supported by qRT-PCR assays and biochemical analysis, provide distinct muscle growth pathways that integrate low protein degradation and ATP synthesis for efficient muscle build-up and energy utilization. Skeletal muscle build-up is maintained by high-level expressions of myogenin, Cdk inhibitor P21, follistatin-like factor (Fstl), and Rho-associated kinase (Rock1), and ECM components such as procollagen, fibronectin and biglycan. In the meantime, there are decreased levels of protein degradation and mitochondrial ATP synthesis in the transgenic mice, which suggests efficient energy utilization for muscle build-up. Although the profile change and patterns of muscle gene expressions were caused and maintained by the manipulation of a single gene namely myostatin, clearly the genes associated with myofiber fusions and energy utilizations were dynamically changed when expressed at the mRNA levels. In the current myostatin propeptide transgenic mouse model, decreased protein degradation and mitochondrial ATP synthesis may represent an important metabolic type that significantly enhances muscle growth in adult stages. Since adult muscle build-up is complicated by age-induced muscle atrophy, any regulatory and metabolic pathway or mechanism can maintain skeletal muscle mass and growth would provide important directions for future studies. We have begun to define more specific mechanisms of myogenic initiation, maintenance of myogenic states and mitochondrial energy production in adult stages [45]. In addition, further studies with this model may shed light on its potential application to the treatment of muscle dystrophy and cachexia by depressing myostatin activity.

CONCLUDING REMARKS

Obesity and diabetes are becoming a national health problem in the US. Skeletal muscles provide the physiological foundation of physical activity and fitness. Active skeletal muscle effectively regulates metabolic activities and energy balance. Preservation of muscle mass in obesity, insulin resistance/type II diabetes is important for effective treatment and prevention of these diseases. One critical protein, named myostatin, is a dominant inhibitor of muscle mass. Our study in transgenic mice indicates that depression of myostatin function by its propeptide produces dramatic muscle mass at the growth stage and less fat at older ages. A high level of myostatin has been found in skeletal muscle and plasma in severely obese patients. Plasma myostatin is correlated with the severity of insulin resistance. Although myostatin primarily inhibits myoblast cell function during embryo development, it also regulates postnatal muscle buildup and metabolism. Myostatin depression by its propeptide appears very effective in promoting muscle growth and maintaining muscle mass in adult stages. Experiments from the myostatin propeptide transgenic mice and high-fat feeding trial further support the flexibility of skeletal muscle growth and build-up while the dominant muscle gene-myostatin was modified. Consequently, the metabolic pathway for muscle growth are also changed accordingly, with low protein degradation and mitochondrial ATP synthesis in the transgenic mice, skeletal muscle build-up and maintenance appears sustained in adult stages. Muscle mass or size of the skeletal muscle in the body determines the exercise capacity or degree of voluntary physical activities. There have not been strong experimental data to directly link myostatin function and insulin sensitivity in skeletal muscle. However, the indirect effects of the dominant actions of myostatin on muscle mass play a significant role in improving insulin sensitivity in diabetes. Therefore, myostatin inhibition by its propeptide or molecular modulators of its signaling pathways is important for boosting/maintaining skeletal muscle mass, therefore improving insulin sensitivity for diabetes treatment and prevention. Further experiments with myostatin-genetically modified animals in moderate and controlled physical activities can provide much useful information for understanding the dynamics of skeletal muscle for human health.

REFERENCES

[1]   M. Shields, M. D. Carroll and C. L. Ogden, *Phys. Rev.* 117, 1235 (2011).
[2]   J. V. Neel, *World Rev Nutr Diet.* 84, 1 (1999).
[3]   R. Konigsberg, *Science* 140, 1273 (1963).
[4]   C. McPherron, A. M. Lawler and S. J. Lee, *Nature* 387, 83 (1997).
[5]   C. McPherron and S. J. Lee, *J. Clin. Invest.* 109, 595 (2002).
[6]   C. McPherron and S. J. Lee, *Proc Natl Acad Sci USA* 94, 12457 (1997).
[7]   M. Schuelke, K. R. Wagner, L. E. Stolz, C. Hubner, T. Riebel, W. Komen, T. Braun, J. F. Tobin and S. J. Lee, *N. Engl. J. Med.* 350, 2682 (2004).
[8]   D. S. Mosher, P. Quignon, C. D. Bustamante, N. B. Sutter, C. S. Mellersh, H. G. Parker and E. A. Ostrander, *PLoS Genet.* 3, e79, (2007).

[9]   Clop, F. Marcq, H. Takeda, D. Pirottin, X. Tordoir, B. Bibé, J. Bouix, F. Caiment, J. M. Elsen, F. Eychenne, C. Larzul, E. Laville, F. Meish, D. Milenkovic, J. Tobin, C. Charlier and M. Georges, *Nat. Genet.* 38, 813 (2006).

[10] M. Thomas, B. Langley, C. Berry, M. Sharma, S. Kirk, J. Bass and R. Kambadur, *J. Biol. Chem.* 275, 40235 (2000).

[11] W. E. Taylor, S. Bhasin, J. Artaza, F. Byhower, M. Azam, D. H. Willard Jr, F. C. Kull Jr and N. Gonzalez-Cadavid, *Am. J. Physiol. Endocrinol. Metab.* 280, E221 (2001).

[12] Langley, M. Thomas, A. Bishop, M. Sharma, S. Gilmour and R. Kambadur, *J. Biol. Chem.* 277, 49831 (2002).

[13] Joulia, H. Bernardi, V. Garandel, F. Rabenoelina, B. Vernus and G. Cabello, *Exp. Cell Res.* 286, 263 (2003).

[14] S. McCroskery, M. Thomas, L. Maxwell, M. Sharma and R. Kambadur, *J. Cell. Biol.* 162, 1135 (2003).

[15] S. McCroskery, M. Thomas, L. Platt, A. Hennebry, T. Nishimura, L. McLeay, M. Sharma and R. Kambadur, *J. Cell Sci.* 118, 3531 (2005).

[16] H. Amthor, A. Otto, A. Vulin, A. Rochat, J. Dumonceaux, L. Garcia, E. Mouisel, C. Hourdé, R. Macharia, M. Friedrichs, F. Relaix, P. S. Zammit, A. Matsakas, K. Patel and T. Partridge, Proc. Natl. Acad. Sci. U.S.A. 106, 7479 (2009).

[17] Q. Wang and A. C. McPherron, *J. Physiol.* 590(Pt 9), 2151 (2012).

[18] McFarlane, E. Plummer, M. Thomas, A. Hennebry, M. Ashby, N. Ling, H. Smith, M. Sharma and R. Kambadur, *J. Cell Physiol.* 209, 501 (2006).

[19] K. H. Kim, Y. S. Kim and J. Yang, *Muscle Nerve* 43, 700 (2011).

[20] S. W. Park, B. H. Goodpaster, J. S. Lee, L. H. Kuller, R. Boudreau, N. de Rekeneire, T. B. Harris, S. Kritchevsky, F. A. Tylavsky, M. Nevitt, Y. W. Cho and A. B. Newman, *Diabetes Care* 32, 1993 (2009).

[21] P. Baumann, C. Ibebunjo, W. A. Grasser and V. M. Paralkar, *J. Musculoskelet. Neuronal Interact.* 3, 8 (2003).

[22] Y. Chen, L. Cao, J. Ye and D. Zhu, *Biochem. Biophys. Res. Commun.* 388, 112 (2009).

[23] S. Hittel, J. R. Berggren, J. Shearer, K. Boyle and J. A. Houmard, *Diabetes* 58, 30 (2009).

[24] D. L. Allen, D. S. Hittel and A. C. McPherron, *Med. Sci. Sports Exerc.* 43, 1828 (2011).

[25] S. J. Lee, Annu. Rev. Cell Dev. Biol. 20, 61 (2004).

[26] S. J. Lee and A. C. McPherron, *Proc. Natl. Acad. Sci. U.S.A.* 98, 9306 (2001).

[27] J. Yang, T. Ratovitski, J. P. Brady, M. B. Solomon, K. D. Wells and R. J. Wall, *Mol. Reprod. Dev.* 60, 351 (2001).

[28] X. Zhu, M. Hadhazy, M. Wehling, J. G. Tidball and E. M. McNally, *FEBS Lett.* 474, 71 (2000).

[29] Zhao, R. J. Wall and J. Yang, *Biochem. Biophys. Res. Commun.* 337, 248 (2005).

[30] J. Yang and B. Zhao, *Mol. Reprod. Dev.* 73, 462 (2006).

[31] S. T. Suzuki, B. Zhao and J. Yang, *Biochem. Biophys. Res. Commun.* 369, 767 (2008).

[32] H. Berg, T. P. Combs, X. Du, M. Brownlee and P. E. Scherer, *Nat. Med.* 7, 947 (2002).

[33] Y. Arita, S. Kihara, N. Ouchi, M. Takahashi, K. Maeda, J. Miyagawa, K. Hotta, I. Shimomura, T. Nakamura, K. Miyaoka, H. Kuriyama, M. Nishida, S. Yamashita, K. Okubo, K. Matsubara, M. Muraguchi, Y. Ohmoto, T. Funahashi and Y. Matsuzawa, *Biochem. Biophys. Res. Commun.* 257, 79 (1999).

[34] Hu, P. Liang and B. M. Spiegelman, *J. Biol. Chem.* 271, 10697 (1996).

[35] T. P. Combs, A. H. Berg, S. Obici, P.E. Scherer and L. Rossetti, *J. Clin. Invest.*, 108, 1875 (2001).

[36] M. Muoio, J. M. Way, C. J. Tanner, D. A. Winegar, S. A. Kliewer, J. A. Houmard, W. E. Kraus and G. L. Dohm, *Diabetes* 51, 901 (2002).

[37] M. Bartoli, J. Poupiot, A. Vulin, F. Fougerousse, L. Arandel, N. Daniele, C. Roudaut, F. Noulet, L. Garcia, O. Danos and I. Richard, *Gene Ther.* 14, 733 (2007).

[38] Qiao, J. Li, J. Jiang, X. Zhu, B. Wang, J. Li and X. Xiao, *Hum Gene Ther.* 19, 241 (2008).

[39] Z. Li, B. Zhao, Y. S. Kim, C. Y. Hu and J. Yang, *Molecul. Reprod. Dev.* 77, 76 (2009).

[40] N. M. Wolfman, A. C. McPherron, W. N. Pappano, M. V. Davies, K. Song, K. N. Tomkinson, J. F. Wright, L. Zhao, S. M. Sebald, D. S. Greenspan and S. J. Lee, *Proc. Natl. Acad. Sci. U.S.A.* 100, 15842 (2003).

[41] C. A. Harrison, S. L. Al-Musawi and K. L. Walton, *Growth Factors* 29, 174 (2011).

[42] R. S. Thies, T. Chen, M. V. Davies, K. N Tomkinson, A. A. Pearson, Q. A. Shakey and N. M. Wolfman, *Growth Factors* 18, 251 (2001).

[43] J. J. Hill, M. V. Davies, A. A. Pearson, J. H. Wang, R. M. Hewick, N. M. Wolfman and Y. Qiu, *J. Biol. Chem.* 277, 40735 (2002).

[44] S. J. Lee, *PLoS One* 3, e1628, (2008).

[45] B. Zhao, E. J. Li, R. J. Wall and J. Yang, *BMC Genomics* 10, 305 (2009).

INDEX

#

2,6-diisopropylphenol, 51
70 kDa ribosomal protein S6 kinase (p70^{S6K}), 197
70kDa ribosomal protein S6 kinase (p70S6K), 149
8-hydroxy-20-deoxyguanosine, 74

A

A-band, 5, 7, 9
abatacept, 56
acetyl-CoA, 43
actin, 11, 15, 17, 108, 116, 130, 136, 143, 150, 152, 153, 155, 198, 199
activator protein 1 (AP-1), 79
activin receptor type IIB, 209
activin receptor-like kinase (ALK) 4, 165
ActRIIA/B receptors, 165
acute-phase proteins, 73
adenine nucleotide translocator1 (ANT1), 60
adenine nucleotides, 81
adenosine, 42, 92, 156
adhesion molecules, 57, 73, 74, 168
adipocyte, 208, 212
adiponectin, 207, 212, 213
adipose tissue, 30, 42, 45, 47, 48, 59, 154, 165, 207, 208, 209, 210, 211, 212, 213, 214
ADP, 41, 42, 43, 46, 48, 95
aging, vii, 25, 26, 28, 29, 30, 31, 32, 33, 34, 35, 44, 48, 91, 92, 141, 155, 158, 163, 168, 170, 172, 195, 196, 199, 200, 201, 202, 209, 210
Akt, 97, 98, 100, 142, 144, 148, 149, 150, 151, 153, 157, 163, 166, 167, 170, 174, 176, 198, 211
Akt/protein kinase b (PKB), 97
Akt1, 149, 151
Akt2, 149

alanine, 196
albumin, 74, 200
ALK5, 165
alpha1-syntrophin, 146
alveolar-arterial oxygen gradient, 81
amino acids, viii, 150, 157, 158, 159, 163, 173, 176, 195, 196, 197, 198
AMP, 92, 93, 95
amyotrophic lateral sclerosis (ALS), 144
androgens, 155, 156
aneurysm, 72, 82
angioedema, 84
annexin A1, 113
annexin V, 75
ANO5 gene, 105, 119
anoctamin, 118
anterior pituitary gland, 157
antioxidant, 44, 56, 62, 65, 66, 74, 83, 84
anti-oxidant therapies, 76, 77
AP-1, 79
Apaf-1, 171
apoptosis, 27, 29, 31, 34, 35, 41, 43, 74, 111, 116, 134, 141, 142, 143, 148, 149, 158, 171, 172, 174
apoptosis-inducing factor (AIF), 172
apoptosis-programmed cell death, 43
asymptomatic hyperCKemia, 115
Atg12, 162, 163, 164
ATP, vii, 1, 2, 3, 4, 5, 16, 41, 42, 43, 44, 45, 46, 47, 48, 51, 63, 64, 65, 76, 80, 82, 93, 95, 155, 158, 197, 198, 215, 216
atractyloside, 83
atrophy, viii, 25, 26, 29, 31, 33, 34, 35, 60, 113, 143, 144, 145, 149, 152, 154, 155, 158, 159, 160, 161, 162, 163, 165, 166, 167, 168, 169, 171, 172, 173, 175, 176, 198, 215
autocorrelation, 6, 8, 9, 10, 11, 12, 15, 22
autoimmune pancreatitis, 63

autophagins, 158
autophagy, 65, 110, 142, 144, 149, 155, 158, 162,
    163, 164, 176
axonal Charcot-Marie-Tooth disease type 2B1, 109

B

B cells, 56, 79
Bardet-Biedl Syndrome, 116
basal lamina, 130, 143, 173, 214
Bax/Bcl2, 81
Bcl-3, 169
Bcl-XL, 61, 62
Becker muscular dystrophy, 111, 117, 167
Bethlem myopathy, 107
biglycan, 30, 215
biomarker, 73, 75
blood vessels, 30, 59
bupivacaine, 50

C

C2C12, 28, 93, 97, 99, 153, 157, 162, 166, 167, 174,
    175
C3a, 77
C5a, 77
cachectic patients, 200
cachexia, 144, 160, 164, 166, 167, 168, 173, 176,
    199, 200, 201, 210, 215
calcineurin (Cn), 34
calcium (Ca^{2+}), 93, 130
calorie restriction, 34
calpain-3, 106, 111, 112, 118, 119
calpains, 111, 133, 158
calpastatin, 158
CaMK, 92, 93, 96
CaMK signalling pathway, 93
cancer, 154, 155, 158, 159, 160, 163, 165, 167, 168,
    195, 196, 199, 200
cancer cachexia, 154, 155, 158, 160, 165, 167, 168
CAPN3, 106, 111, 112, 113, 115, 119
cardiomyocytes, 50, 80
cardiomyopathy, 106, 109, 110, 115, 116, 117, 119
cardioprotection, 80
CArG boxes, 151
caspase-12, 171
caspase-3, 171, 172
caspase-8, 171
caspase-9, 171, 172
caspases, 158, 171
catalase (CAT), 62
cathepsin L, 155

CAV3 gene, 109, 113
caveolae, 109
caveolin-3, 108, 110, 113, 136
CD11b, 82
CD11c, 77
cell differentiation, 43, 148, 153
chaperone-mediated autophagy, 162
chemokine, 35
chromosomes, 28
chronic diseases, 92, 100
chronic inflammation, 199
chronic obstructive pulmonary disease (COPD), 155
citric acid cycle (Krebs Cycle), 43
c-Jun NH2-terminal kinase, 156
c-Met, 146
coagulation, 74, 75, 77, 79
Cockayne syndrome, 29
collagen, 30, 31, 63
collagen-induced arthritis, 63
comparative genome hybridization (CGH), 107
compartment syndrome, 72, 75
complement, 64, 72, 75, 77, 84, 114
complex I, 50, 57, 60, 62, 76, 79
complex II, 60, 62, 76
complex III, 60, 76
concentric, 131
congenital muscular dystrophy, 107
congenital myastenic syndrome, 107
connective tissue, 27, 30, 32, 114
contractile proteins, 27
COS-7 cells, 94
c-Rel, 167, 169
Cre-LoxP, 152
cross-bridges, vii, 1, 2, 5, 10, 11, 14, 15, 16, 24
cross-linking, 5, 11, 12, 13, 17, 30
cyclin D, 33
cyclin-dependent kinase (CDK), 32
cyclosporine A (CsA), 153
cytochrome b, 60
cytochrome c, 159, 171, 172
cytokine IL-18, 57
cytokines, 35, 55, 57, 58, 61, 62, 65, 73, 74, 75, 77,
    78, 79, 84, 141, 145, 168, 199
cytosol, 42, 43, 93, 94, 96, 158

D

DAG1, 106, 117
damage associated molecular patterns (DAMPs), 56,
    78
decorin, 30
dehydroepiandrosterone (DHEA), 31
dendritic cells (DC), 57

denervation, 34, 141, 154, 155, 158, 159, 163, 169, 170, 172, 173
dermatomyositis (DM), 56
desflurane, 50
desmin, 110
D-fragment, 77
diabetes, 42, 44, 48, 92, 158, 159, 165, 199, 207, 208, 209, 210, 216
DNA, 3, 28, 29, 60, 63, 93, 99, 107, 108, 112, 117, 141, 143, 144, 151, 154, 156, 157, 168, 169, 170, 171, 172, 209
DNAJB6, 108, 110
Duchenne muscular dystrophy, 115, 129, 134, 136, 137, 167, 213
dye, 4, 5, 6, 7, 9, 11, 16, 21, 24, 131, 132, 133, 136
dysferlin, 113, 114, 119, 164
dysferlin mRNA, 113, 114
dysferlinopathy, 113, 114
dystrophic mice model (mdx mice), 31
dystrophin, 31, 105, 113, 114, 116, 117, 129, 130, 131, 135, 136, 137, 146, 151, 172
dystrophin gene, 31, 105, 129
dystrophin-associated glycoprotein (DAG), 136
dystrophin-associated glycoprotein complex, 130

E

eccentric contractions, 131, 132, 151
ECSIT, 61, 62
ED1-positive macrophages, 153
E-fragment, 77
eIF 3 subunit 5 (eIF3-f), 159
eIF-4^E, 149, 173
eIF-4E-BP1, 173
electron transport chain (ETC), 43, 93
elongation initiation factor, 142, 149
elongation initiation factor (eIF)-2, 149
Emery-Dreifuss muscular dystrophy, 109
endoplasmic reticulum (ER), 164
endothelial cells, 57, 65, 72, 77, 78
endothelial nitric oxide synthase (eNOS), 81
endothelin, 73
energy restriction, 196, 198, 200, 201, 202
enzyme, 28, 59, 60, 61, 77, 92, 95, 131, 159, 197, 198, 212
enzymes, 43, 48, 55, 74, 92, 94, 105, 107, 117, 159, 161, 171, 198, 212
epidermal growth factor, 146
E-selectin, 74
etanercept, 57
ethane, 74
eukaryotic initiation factor 4E-binding protein (4E-BP1), 197

exercise, vii, 27, 30, 31, 34, 42, 45, 51, 60, 61, 66, 91, 92, 94, 95, 96, 97, 98, 99, 100, 113, 117, 135, 146, 147, 154, 157, 176, 196, 197, 198, 201, 209, 216
exercise training, vii, 92, 97, 98
exome, 108
exons, 107, 112, 114, 116
exosomes, 75
extracellular matrix (ECM), 30
extracellular signal-regulated kinase (ERK) 1/2, 166

F

factor Xa, 74
familial hypertrophic cardiomyopathy, 109
familial partial lipodystrophy, 109
Fas-associate death domain protein, 62
fast-twitch muscle fibers, 34
fat metabolism, 207
FGF2, 31
fibrillin, 30
fibrin, 73, 77
fibrinogen, 73
fibroblast growth factor 2 (FGF2), 30
fibronectin, 30, 215
filamin C, 108
FKRP gene, 106, 117, 119
flow cytometry, 75
fluorescence, vii, 1, 2, 3, 5, 6, 7, 8, 9, 10, 11, 12, 16, 23, 24, 107, 132, 133
fluorescence lifetime, 6, 7, 9, 16, 23
focal adhesion kinase (FAK), 149, 151
follistatin, 156, 165, 170, 215
forkhead box-o (FOXO), 198
formyl-peptide receptors (FPRs), 64
Fos Jun, 79
FOXO1, 166, 210
free radicals, 29, 44, 72
FSH muscular dystrophy, 105
Fukuyama congenital muscular dystrophy, 117

G

GABARAP, 162
GASP-1, 142, 165
GATE16, 162
Gd^{3+}, 133, 134
gene transcription, 94, 95, 97, 99, 141, 153
genetic mutation, 26, 33
genomic stability, 28
GH/IGF-I axis, 31
glucocorticoids, 173

glutamine, 196
glutathione, 44, 81, 83
glycogen synthase 3-β (GSK-3β), 149
glycogen synthase kinase, 97
Golgi apparatus, 117
Gomoritrichrome, 60, 61
Gowers sign, 109
growth and differentiation factor 8 (GDF8), 33
growth factors, 30, 31, 35, 141, 145, 150, 197, 198
growth hormone (GH), 155
GSK-3β, 142, 149, 151, 173

H

hand-heart syndrome of Slovenian type, 109
hemostatic cofactor von Willebrand factor, 73
heparin sulfate proteoglycan, 30
hepatocyte growth factor (HGF), 30, 145
Hes, 32, 52, 54
Hey, 32
HGF, 31, 142, 145, 146, 148, 174
high-energy phosphates, 80
high-intensity interval training, 95
HIV, 65
Homer 1, 136, 137
HSP-40, 108, 110
Hutchinson-Gilford progeria syndrome, 109
hydrocarbons, 74
hypertrophy, viii, 34, 109, 117, 143, 144, 145, 146,
    147, 148, 149, 150, 151, 153, 154, 155, 156, 159,
    164, 174, 176, 196, 209, 210, 213
hypochlorous acid, 79
hypoxanthine, 76

I

ICAM-1, 65, 73, 74
Id family, 153
IFN-β, 58, 63
IGF-I, 31, 142, 145, 147, 148, 150, 151, 152, 153,
    155, 156, 157, 163, 166, 168, 170, 173, 174
IGF-IEa, 31
IKKα, 168
IKKβ, 168, 170
IL-1, 33, 35, 57, 58, 62, 63, 64, 65, 72, 74, 77, 78,
    79, 168, 169
IL-1β, 72, 78
IL-1Ra, 57
IL-1ß, 77
IL-6, 33, 35, 57, 58, 62, 64, 72, 74, 78, 79, 168, 169
IL-6 messenger RNA, 57
IL-8, 64, 77, 79

immunohistochemistry, 105, 107, 135
inflammation, 48, 55, 56, 57, 59, 60, 61, 62, 63, 64,
    65, 66, 72, 73, 74, 75, 77, 78, 79, 81, 83, 84, 114,
    168, 169, 171, 199
Inflammatory myopathies (IM), 55
INF-γ, 58, 61, 62
innervation, 26, 34, 35
insulin, 48, 97, 98, 142, 145, 148, 150, 163, 166,
    170, 173, 197, 198, 207, 208, 209, 210, 211, 212,
    214, 216
insulin resistance, 207, 208, 209, 210, 211, 212, 216
insulin/insulin-like growth factor-phosphoinositol-3
    kinase (insulin/IFG-PI3K), 197
Insulin-like growth factor-I (IGF-I), 141
interferon, 57, 142, 168, 199
interferon-γ, 168, 199
interleukin 1 (IL-1), 73
intracellular domain of Notch (NICD), 32
intron, 107
ischemia, 48, 49, 71, 72, 73, 76, 77, 80, 81, 82, 83
ischemia-reperfusion, vii, 51, 72, 73, 75, 76, 77, 78,
    79, 81, 83, 84
ischemic conditioning, 80
ischemic preconditioning (IPreC), 80
isoflurane, 42, 50
isoleucine, 196
isometric, 1, 2, 11, 15, 16, 131, 132
isoprostanes, 74
IκB, 79, 142, 168, 169
IκB kinases (IKK), 168

J

Janus kinase (JAK) 1, 147

K

kinesin-1, 2

L

lactate, 80
lamin A/C, 108, 109
laminin, 30, 117, 172
laminin α-2, 117
L-arginine, 146
laser Doppler flowmetry, 83
LC3, 142, 162, 163, 164
LC3-I, 162, 163, 164
LC3-II, 162, 163, 164
leucine, 151, 159, 173, 195, 196, 200, 201, 202
leukemia inhibitory factor (LIF), 145

leukocytes, 35, 57, 72, 74, 77, 78, 79, 84
lever arm, vii, 1, 2, 3, 6, 7, 10, 11, 12, 14, 15, 21
LGMD1, 106, 107, 108, 110, 111
LGMD2, 106, 107, 111, 119
LGMD2P, 106, 117
LIF mRNA, 147
like tumor necrosis factor alpha (TNF-α), 73
limb-girdle muscular dystrophies (LGMD), 105
lipid peroxidation, 74
L-selectin, 73

M

macroautophagy, 162
macrophage, 35, 63
magnetic resonance imaging (MRI), 26
mammalian target of rapamycin complex 1
    (mTORC1), 196
mandibuloacral dysplasia, 109
mannose binding lectin (MBL) pathways, 77
Marfan syndrome, 30
matrix proteins, 29, 35
M-cadherin–β-catenin complex, 111
MDA-5, 57, 58, 63
mdx fibers, 131, 133, 134, 137
mdx mice, 31, 130, 132, 133, 135, 213
mdx myotubes, 134, 136
mechano-growth factor (MGF), 31
MEF2, 99, 148, 154, 156, 174
membrane depolarization, 131
membrane potential, 41, 43, 45, 46, 47, 49, 51, 61,
    62
membrane-anchored ligand Delta, 32
messenger ribonucleic acid (mRNA), 196
microautophagy, 162
microparticles, 74
microRNAs, 108, 156, 174
miR-1, 148, 174, 175
miR-133, 148, 174, 175
miR-133a, 148, 174, 175
miR-133b, 175
miR-206, 148, 174, 175
miR-208, 174
miR-221, 174
miR-222, 174
miR-23, 175
mitochondria, vii, 29, 42, 43, 44, 46, 47, 50, 51, 55,
    59, 60, 61, 62, 63, 64, 65, 66, 92, 155, 162, 171,
    172
mitochondrial ATP-sensitive potassium channel
    (mK_{ATP}), 82
mitochondrial biogenesis, vii, 91, 92, 95, 96, 97, 98,
    99, 100, 155
mitochondrial coupling, 41, 46, 62
mitochondrial density, 92
mitochondrial DNA (mtDNA), 55, 60
mitochondrial DNA polymerase gamma (POLG1),
    60
mitochondrial gene transcription, 92, 93, 94, 96, 99,
    100
mitochondrial membrane potential preservation, 43
mitochondrial permeability transition pore (mPTP),
    83
mitochondrial respiratory chain, 41, 61, 62, 72, 76,
    77, 81, 92
mitochondrial transcription factor A (Tfam), 93
mitochondrial uncoupling, vii, 41, 45, 46, 48, 49, 50,
    51, 52, 61
mitochondrion, 80
mitogen-activated protein kinases (MAPKs), 79
Mito-TEMPO, 65
mitotic activity, 27, 173
Miyoshi myopathy, 106, 113, 118, 164
motor neurons, 30, 34
motor unit types, 34
MRTF-B, 152, 153
mTOR, 143, 144, 148, 149, 150, 151, 153, 157, 163,
    167, 170, 176, 210
mTORC1, 143, 148, 149, 150, 151, 197, 198, 201
mTORC2, 149
multiple sclerosis, 63
MuRF-1, 152, 155, 159, 160, 162, 164, 167, 169,
    170
MuRF-2, 152
MuRFs, 152
muscle biopsy, 56, 57, 92, 98, 105, 106, 108, 109,
    110, 111, 112, 114, 116, 117, 119
muscle carnitine insufficiency, 60
muscle contraction, 16, 34, 93, 99, 114, 129, 130,
    132, 156
muscle creatine kinase, 151, 155
muscle fatigue, 34
muscle fibers, 15, 26, 27, 28, 30, 31, 33, 34, 58, 59,
    109, 119, 130, 131, 133, 134, 135, 136, 143, 144,
    145, 146, 147, 148, 152, 154, 155, 164, 171, 173
muscle glycogen, 95
muscle graft, 29
muscle growth, 27, 147, 148, 157, 160, 164, 207,
    208, 211, 213, 214, 215, 216
muscle mass, viii, 25, 26, 31, 33, 34, 35, 141, 143,
    149, 150, 155, 156, 165, 166, 167, 169, 172, 173,
    176, 196, 198, 199, 200, 201, 202, 207, 208, 209,
    210, 211, 213, 214, 216
muscle metabolic function, 92
muscle necrosis, 83
muscle oxidative capacity, 29

muscle protein synthesis (MPS), 196
muscle regeneration, 26, 27, 28, 30, 32, 33, 34, 35
muscle ring finger (MuRF)-2, 152
muscle rippling movements, 106
muscle stem cells, 25, 26, 27, 29, 33, 35
muscle weakness, vii, 30, 105, 106, 107, 108, 109,
    110, 111, 116, 118, 119, 129, 164
Muscle-eye brain syndrome, 117
Mutation, 29, 95, 112, 118
myalgia, 106, 109, 117, 119
myeloperoxidase (MPO), 79
myoblast, 31, 58, 147, 148, 152, 153, 157, 174, 209,
    210, 216
myoblast proliferation, 147, 148, 174
myoblasts, 32, 148, 153, 157, 174, 209, 210
myocardial infarction, 80
myocardin-related transcription factor-A (MRTF-A),
    152
myocardium, 50, 80, 82
myocytes, 51, 83, 147, 149
MyoD, 33, 145, 148, 152, 153, 159, 166, 170, 174,
    175, 210
myofiber hypertrophy, 209
myofibril, 1, 5, 6, 7, 8, 9, 11, 12, 13, 16, 17, 23, 24,
    93
myofibrillar disarray, 131
myofibrillar proteins, 27, 158
myofibrillogenesis,, 118
myofilaments, 13, 16
myogenicity, 31
myogenin, 148, 152, 153, 170, 173, 174, 215
myoglobinuria, 106, 119
myonuclear domain, 26, 144, 145
myonuclear domain theory, 26
myonuclei, 26, 29, 111, 145, 171, 172
myosin, 1, 2, 3, 5, 6, 7, 11, 14, 15, 16, 17, 21, 57,
    116, 142, 143, 151, 152, 160, 198, 199, 202
myosin essential light chain 1 (LC1), 1
myosin light chain, 2, 151
myostatin, viii, 33, 34, 144, 145, 150, 153, 154, 156,
    160, 164, 165, 166, 167, 170, 173, 176, 207, 208,
    209, 210, 211, 212, 213, 214, 215, 216
myostatin mRNA, 154, 165, 166
myostatin propeptide, 165, 207, 208, 213, 214, 215,
    216
myotilin, 108, 119
myotonic dystrophy type 2, 107

neuromuscular junction (NMJ), 34
neuropathy, 106, 108, 172
neutrophil, 81, 82, 83
next-generation sequencing (NGS), 107
NF-κB, 33, 62, 63, 64, 65, 78, 79, 142, 143, 167,
    168, 169, 170, 176
NF-κB signaling, 33, 168, 169, 170, 176
nicotinamide-adenine-dinucleotide phosphate
    (NADPH), 77
nitric oxide (NO), 146
nitric oxide synthase (NOSs), 146
nitrogen retention, 198, 200
nNOS mRNA, 146
NO˙, 74
NOD-like Receptors (NLRs), 57
Notch, 32, 33, 35, 156
Notch antagonists, 32
Notch ligands, 32
Notch pathway, 32
Notch receptor, 32
Notch1, 156
Notch2, 156
NRF-2, 94
nuclear respiratory factor (NRF)-1, 93
Numb, 32
nutrition, 176, 196, 197, 200, 202

O

obesity, 42, 48, 52, 199, 207, 208, 209, 211, 212, 216
occlusion, 80, 81, 82
oedema, 77
overloading, 144, 145, 147, 148, 152, 174
oxidative capacity, 81, 83, 92, 155
oxidative cellular damage, 74
oxidative damage, 29, 51, 55, 74, 171, 172
oxidative defense, 29
oxidative phosphorylation, vii, 42, 43, 45, 47, 49, 50,
    60, 61
oxidative stress, 29, 41, 44, 45, 48, 49, 51, 52, 55,
    60, 61, 62, 63, 73, 74, 75, 83, 97
oxidative stress biomarkers, 74
oxygen, vii, 29, 42, 43, 44, 46, 48, 55, 72, 75, 76, 77,
    92, 133, 143, 155

P

p160 myb binding protein, 96, 99
p21, 148, 157, 210
p300-C/EBPβ, 173
p38 mitogen-activated protein kinase (MAPK), 156
p50, 168, 169

N

NADH tetrazoliumreductase (NADH-TR), 59
NBR1, 152
neural drive, 27

p62/SQSTM-1, 152
pathogen-associated molecular patterns (PAMPs), 56
Pax3, 148, 166, 174
Pax7, 174
pentane, 74
perlecan, 30
peroxidation, 74
peroxisome proliferator-activated receptor γ
    (PPARγ), 92
PGC1α mRNA, 154
phenotype, 26, 28, 33, 51, 62, 106, 109, 110, 111,
    113, 115, 116, 117, 118, 135, 155, 211
phospholipase A$_2$, 133
phospholipase D (PLD), 151
phosphorus magnetic resonance spectroscopy (^{31}P-
    MRS), 60
photobleaching, 17
physical activity, 154, 196, 198, 201, 202, 209, 216
piroxicam, 50
plasma membrane, 31, 135, 136, 137, 209
plasma metalloproteases, 73
plasmalemma, 109
plasminogen activator inhibitor PAI-1, 73
platelet-derived growth factor, 73, 146
platelet-derived growth factor BB, 146
polymyositis (PM), 56
polyunsaturated fatty acids, 74
Pompe disease, 107
POMT1, 117
POMT2, 117
porins, 42
postnatal muscle growth, 145, 154, 209
post-power stroke state, 14
power-stroke cycles, 11, 14
PPAR-alpha, 207
PPAR-gamma, 207
pre-power stroke state, 14
pro-caspase-8, 62
pro-inflammatory cytokines, 56, 57, 62, 63, 64, 72,
    73, 74, 77, 78, 79, 168
propofol, 51
proteasome subunit, 33, 158
protein arginine methyl transferase (PRMT) 1, 98
protein C, 74, 98, 160
protein kinase C, 81, 83, 158
proteolysis, 131, 158, 160, 173, 195, 196, 198, 201,
    202
protons pumps, 43
proximal muscle weakness, 105, 109
P-selectin, 82
pyruvate, 43

R

rapamycin, 142, 143, 144, 149, 150, 163, 170
reactive nitrogen, 74
reactive oxygen species (ROS), 41, 60, 75, 76, 93,
    97, 137, 169
regulatory light chain (RLC), 15
reperfusion injury, 71, 72, 76, 80, 82, 83
restrictive dermopathy, 109
rhabdomyolysis, 51, 75
Rheb, 150, 151, 170
RhoA protein, 153
rhodamine, 2, 5, 6
ribosomal protein S6 kinase 1 (S6K1), 173
ribosome biogenesis, 149
RIP140, 99
rippling muscle disease, 109
rituximab, 56
RNA, 29, 135, 174, 207
ropivacaine, 50
ROS modulator 1 (Romo1), 61
Rotenone, 63

S

SAC blockers, 131, 133, 134
sarcoglycan, 115, 116
sarcolemma, 16, 105, 113, 114, 115, 116, 119, 130,
    131, 132, 135, 136, 143, 146, 164
sarcomere, 1, 3, 5, 13, 15, 16, 17, 154, 156
sarcomeres, 2, 12, 13, 111, 131
sarcopenia, 25, 31, 33, 144, 155, 158, 162, 167, 171,
    172, 173, 176, 199, 200, 201
sarcoplasm, 143, 171
sarcoplasmic reticulum (SR), 131
sarcotubular myopathy, 117
satellite cells, 27, 28, 29, 30, 31, 32, 33, 34, 35, 141,
    144, 145, 146, 147, 148, 149, 153, 173, 210
scaffolding proteins, 137
sepsis, 48, 75, 158, 173, 200
serum creatine kinase (sCK), 106
serum response factor (SRF), 142, 144
SeTau, 3, 4, 6, 7, 9, 16, 20, 21
SeTau-LC1, 4, 7, 9, 20, 21
sevoflurane, 50
signal transducer and activator of transcription 3
    (STAT3), 79
SIRT1, 93, 96, 97, 100
skeletal muscle, vii, viii, 2, 25, 26, 27, 28, 29, 30, 31,
    32, 33, 34, 35, 42, 45, 48, 51, 55, 56, 57, 59, 72,
    73, 75, 76, 77, 78, 79, 80, 81, 82, 83, 84, 91, 92,
    94, 95, 96, 97, 98, 99, 100, 106, 111, 114, 116,

117, 129, 130, 134, 135, 136, 141, 143, 144, 145, 146, 147, 149, 150, 151, 152, 154, 155, 157, 158, 159, 160, 162, 163, 164, 165, 166, 168, 169, 170, 171, 172, 173, 174, 175, 176, 195, 196, 197, 198, 199, 200, 201, 202, 207, 208, 209, 210, 212, 213, 214, 215, 216
Smac/DIABLO, 172
Smad, 165, 166, 170
Smad3, 166, 167, 170, 210
SOD, 79
soluble VEGF-receptor-1, 73
spider venom toxin GsMTx-4, 133
spinal muscular atrophy type III, 107
spontaneous nephritis, 63
sporadic inclusion body myositis (sIBM), 56
STARS, 143, 150, 152, 153
state 2, 46
state 3, 46
state 4, 46
stem cells, 25, 26, 27, 28, 29, 30, 143
streptomycin, 133, 134
stretch-activated channels (SACs), 130, 131
stretch-activated ion channels, vii
stretch-induced muscle damage, 131, 133, 134, 136
succinate dehydrogenase (SDH), 60
SUMOylation, 98
superoxide, 44, 61, 62, 77, 79, 83
superoxide dismutases (SOD), 62
systemic inflammation, 73

T

T cells, 56, 57, 58
TCAP (titin cap) gene, 116
telethonin, 116
telomerase, 28, 31
telomere, 26, 28, 35
telomere shortening, 26, 28, 35
tenascin, 30
testosterone, 31, 155, 156, 199
TF/FVII complex, 74
TGF-β, 30, 32, 33, 35, 146, 165, 166, 209, 211
TGF-β pathway, 30
TGF-β1, 32, 33
TGF-β1/pSmad3 levels, 33
thermogenesis, 42, 45, 48
thrombin, 74
thymidine phosphorylase (ECGF1), 60
tibial muscular dystrophy, 106, 113, 118
titin, 111, 116, 118, 152, 158, 160
TNF receptor associated factor 6 (TRAF6), 79
TNF receptor-associated factor 2, 62

TNF-like weak promoter of apoptosis (TWEAK), 168
TNF-α,, 33, 62, 72, 78, 79, 168, 169, 172
TNF-α/TNF receptor (TNFR), 79
Toll like receptors (TLRs), 56
total internal reflection fluorescence (TIRF), 2
tourniquet, 72, 80, 81
toxicity, 72, 74
TRAF2, 79
TRAF6, 61, 62, 79, 162
training, 91, 92, 94, 141, 143, 146, 147, 152, 153, 154, 176
transcriptional intermediary factor 2 (TIF2), 42
transforming growth factor β, 209, 211
transient receptor potential (TRP), 135
triads, 111
TRIM32 gene, 106, 116
tropomyosin, 151
troponin I, 77
TRPC, 135
TRPC1, 135, 136
TRPC3, 135
TRPM, 135
TRPV, 135
TRPV2, 136
tuberin (TSC-2), 151
tuberous sclerosis complex (TSC), 149
twinkle (C10orf2), 60
Type 2 diabetes, 48, 52, 208
type I fibers, 29, 170
type II fibers, 29

U

ubiquitin, 33, 97, 116, 142, 143, 144, 155, 158, 159, 160, 161, 162, 163, 167, 195, 196, 198, 210
ubiquitin ligase (E3), 159
ubiquitin-activating enzyme (E1), 159
ubiquitination, 97, 159, 160, 161, 197
ubiquitin-conjugated enzyme, 33
ubiquitin-conjugating enzyme (E2), 159
ubiquitin-proteasome pathway, 116
UCP1, 42, 45, 51
UCP2, 42, 45, 48, 49, 51, 52, 61, 62, 63
UCP3, 42, 45, 48, 51, 52, 62
uncoupling protein, 45
utrophin A, 153

V

valine, 196
vascular endothelial growth factor (VEGF), 73

vasodilatation, 72, 82
vasospasm, 82
VCAM-1, 73, 74
VE-cadherin, 78
VEGFA, 31
vitronectin, 30
Vps34, 150, 151

W

Walker-Warburg syndrome, 117
Werner syndrome, 28, 29
Western blot, 94, 105, 107, 111, 112, 113, 114, 116, 117, 135, 163
Wnt, 32, 33, 35

X

xanthine dehydrogenase, 76
xanthine oxidase, 76, 77
xanthine oxidase (XO), 76
X-ray absorptiometry (DEXA), 26

Z

Z-line, 108, 110, 131
Z-line streaming, 131
Z-lines, 108
zymogens, 171

A

α-actin, 108, 151, 152, 153
α-actinin, 108
α-dystroglycan, 106, 107, 117
α-dystroglycan glycosylation, 106, 107
α-sarcoglycanopathy, 115
α-syntrophin, 136, 137

B

β-catenin, 156
β-hydroxy-β-methylbutyrate, 201